Deep-water Coral Reefs
Unique Biodiversity Hot-Spots

Martin Hovland

Deep-water Coral Reefs

Unique Biodiversity Hot-Spots

Published in association with
Praxis Publishing
Chichester, UK

Dr Martin Hovland
Marine Geology Expert
Statoil
Stavanger
Norway

SPRINGER–PRAXIS BOOKS IN LIFE SCIENCES INCORPORATING AQUATIC AND MARINE SCIENCES

ISBN 978-1-4020-8461-4 Springer Dordrecht Berlin Heidelberg New York

Springer is part of Springer-Science + Business Media (springer.com)

Library of Congress Control Number: 2008926733

Apart from any fair dealing for the purposes of research or private study, or criticism or review, as permitted under the Copyright, Designs and Patents Act 1988, this publication may only be reproduced, stored or transmitted, in any form or by any means, with the prior permission in writing of the publishers, or in the case of reprographic reproduction in accordance with the terms of licences issued by the Copyright Licensing Agency. Enquiries concerning reproduction outside those terms should be sent to the publishers.

© Praxis Publishing Ltd, Chichester, UK, 2008
Printed in Germany

The use of general descriptive names, registered names, trademarks, etc. in this publication does not imply, even in the absence of a specific statement, that such names are exempt from the relevant protective laws and regulations and therefore free for general use.

Cover design: Jim Wilkie
Project management: Originator Publishing Services Ltd, Gt Yarmouth, Norfolk, UK

Printed on acid-free paper

Contents

Preface ix
Acknowledgements xiii
List of figures xvii
List of tables xxiii
List of abbreviations and acronyms xxv

1 Introduction 1
   1.1 Introduction 1

2 Coral reefs 7
   2.1 Introduction 7
   2.2 Skeleton formation 10
   2.3 Coral reef ecology 11

3 A modern re-discovery 13
   3.1 Introduction 13
   3.2 Historical background 14
   3.3 The Fugløy reef, 1982 16
   3.4 Pockmark craters 18
   3.5 Why north of the Polar Circle? 22
   3.6 Suspected origin of the Fugløy reef 23
   3.7 Deep-water bioherms at Nyegga 24
        3.7.1 Significance of the tubeworm 29

3.8	Håkon Mosby Mud Volcano (HMMV)		31
	3.8.1 Bacterial mats		33
3.9	Summary		37

4 Scandinavian coral reefs ... 39
- 4.1 Introduction ... 39
- 4.2 Reefs in Norwegian fjords ... 40
  - 4.2.1 Agdenes and Tautra reefs ... 40
  - 4.2.2 Stjernsund and other Norwegian fjord reefs ... 43
- 4.3 Reefs on the continental shelf ... 44
  - 4.3.1 The Haltenpipe reefs ... 44
  - 4.3.2 HRC ... 46
  - 4.3.3 Bivalve association near HRC ... 48
  - 4.3.4 Suspected origin of the HRC ... 55
  - 4.3.5 The Sula Reef Complex ... 57
  - 4.3.6 The Husmus reefs, Draugen field ... 62
  - 4.3.7 The Træna Deep Reef Complex ... 64
- 4.4 Reefs associated with hydrocarbon fields ... 68
  - 4.4.1 Pockmark reefs at the Kristin field ... 69
  - 4.4.2 The "bio-expectations" for DWCRs ... 69
  - 4.4.3 Some relevant emerging results ... 71
  - 4.4.4 Sponge-associated bacteria ... 76
  - 4.4.5 Silted reefs at the Morvin field ... 76
  - 4.4.6 Origin of Kristin and Morvin reefs ... 76
  - 4.4.7 Seep-associated corals and sponges over a leaky field ... 79
- 4.5 Other reef occurrences off mid-Norway and northern Norway ... 80
  - 4.5.1 Reefs in Breisunddjupet, near Ålesund ... 80
  - 4.5.2 The Floholman reefs, with associated bacteria and gas! ... 81
  - 4.5.3 The Malangsreef ... 88
- 4.6 Reefs on the continental edge ... 89
  - 4.6.1 The Røst Reef Complex ... 90
- 4.7 Corals and reefs in the North Sea and Kattegat ... 91
  - 4.7.1 The Fedje reefs ... 91
  - 4.7.2 Corals inside a pockmark, Troll field ... 92
- 4.8 *Lophelia* colonies on man-made structures ... 93
  - 4.8.1 The Brent Spar Buoy ... 93
  - 4.8.2 Statfjord ... 94
  - 4.8.3 UK Block 22/4, blowout site ... 95
- 4.9 Swedish/Norwegian reefs ... 97
  - 4.9.1 Kosterfjord reefs ... 97
- 4.10 Danish "bubbling reefs" ... 98
- 4.11 Reefs off the Faroe Islands ... 100
- 4.12 Summary ... 101

5	**North Atlantic coral reefs and giant carbonate mounds**		103
	5.1	Introduction	103
	5.2	The Atlantic margin	104
		5.2.1 Porcupine Seabight Mounds	106
		5.2.2 Seep or non-seep, could drilling resolve the issue?	114
		5.2.3 The Rockall Trough margin mounds	120
		5.2.4 *Lophelia* within the deep Rockall Trough, the Darwin Mounds	123
		5.2.5 Observed acoustic plumes associated with mounds	127
		5.2.6 Possible sources of seeping fluids off Ireland	127
		5.2.7 Serpentinisation, the most important geobiological process?	128
	5.3	Gulf of Cadiz, west of Gibraltar	129
		5.3.1 The lower slope, mud volcanoes, and gas hydrates	130
		5.3.2 The Pen Duick occurrence	132
		5.3.3 Carbonate mounds off Mauritania	137
	5.4	Summary	139
6	**Other deep-water coral reefs, worldwide**		141
	6.1	Introduction	141
		6.1.1 The Campos and Santos Basin reefs, off Brazil	142
		6.1.2 The Gulf of Mexico (GoM)	144
		6.1.3 Deep-water corals off Belize	147
		6.1.4 The Florida Straits	150
		6.1.5 The Blake Ridge occurrences	153
		6.1.6 Off Nova Scotia	153
		6.1.7 Orphan Knoll, off Newfoundland	154
		6.1.8 The Mid-Atlantic Ridge	155
	6.2	The eastern Atlantic	156
		6.2.1 Small mounds off Congo	156
	6.3	The Mediterranean Sea	157
	6.4	The Pacific Ocean	158
		6.4.1 Coral gardens off the Aleutian Islands	158
		6.4.2 Davison Seamount off California	159
		6.4.3 Australian carbonate knolls	160
	6.5	Summary	161
7	**Ancient and modern analogues**		163
	7.1	Introduction	163
	7.2	Waulsortian-like mounds from the Carboniferous	164
		7.2.1 A key to the present?	165
		7.2.2 Petroleum-filled mounds	167
	7.3	Fossil seep carbonates	168
		7.3.1 Ancient deep-water fossil reefs, Algerian Sahara	169
		7.3.2 Kess-Kess formations, Morocco	170

viii Contents

7.4	Seamounts and carbonates	170
7.5	"Living fossil" structures?	172
	7.5.1 Stromatolites and microbialites	172
	7.5.2 Are they seep bioherms?	173
7.6	Summary	177

8 Competing theories 179
- 8.1 Introduction 179
- 8.2 Reefs that didn't drown 180
- 8.3 A null hypothesis 181
  - 8.3.1 Water-related factors 181
- 8.4 A hydrodynamical case study 184
- 8.5 External or internal control? 186
  - 8.5.1 A case of life and death 189
- 8.6 A question of geophysical interpretation, linear reefs 192
  - 8.6.1 Case 1: the Sula Ridge reefs 193
  - 8.6.2 Case 2: the Træna Deep reefs 195
  - 8.6.3 Case 3: other linear reefs 198
- 8.7 Reefs associated with local depressions 198
  - 8.7.1 Moats and other depressions 198
- 8.8 New support for the hydraulic model 199
- 8.9 Geodiversity and biodiversity, a link? 202
- 8.10 Summary and a re-iteration of the hydraulic theory 204

9 An unintended extinction? 207
- 9.1 Threats to deep-water coral reefs 207
  - 9.1.1 Mechanical disturbance and burial 209
  - 9.1.2 Climate change and acidification 210
- 9.2 A glimmer of hope? 212
  - 9.2.1 Biological response 212
  - 9.2.2 Climate change? 213

10 Conclusions 215
- 10.1 Summary of main conclusions 215

APPENDIXES

- A Some additional photographs and images 217
- B Epilogue 233

References 235

Index 255

A contribution to the resources theme of
the International Year of Planet Earth

Preface

Welcome to a journey beneath the oceans a journey Captain Nemo would have envied us ...

Hovland and Mortensen (1999)

Maybe one day we will see the mountains ahead of us. Maybe one day we will see the seven mountains of our mysterious destiny. Maybe one day we will see that beyond our chaos there could always be a new sunlight, and serenity.

Okri (1993).

When Alice in Wonderland went through the looking glass, she entered a world with different rules. Few of us have had the experience of discovering that the rules that had been guiding our research for the past years or decades were all wrong, that the predictions of our theories were wrong, that the assumptions were wrong, that our professors were wrong, that our textbooks were wrong. This has happened famously in astronomy, cosmology, physics, and chemistry.

Anderson (2007)

Lophelia *is the coral that breaks all the rules. It is a true hard coral produced by a colony of individual polyps but it is found in deep cold water. Unlike its tropical counterparts,* Lophelia *does not contain zooxanthellae and so does not rely on sunlight.*

Roberts (1997)

Seafloor landscapes and features are often fascinating and awe-inspiring. Few of the processes known to occur on land are active on the oceanic seafloor. Therefore, underwater landscapes at depths beyond light penetration need our full imagination for their interpretation. Ever since I got responsibility for explaining seafloor phenomena important for human-made seafloor installations and constructions

x Preface

Figure 0.1. The "crown" of a *Lophelia* reef off mid-Norway, at a water depth of 300 m. This mature *Lophelia pertusa* colony is about 2 m in diameter.

(pipelines, platforms, and well-templates, at the engineering department of Statoil, now StatoilHydro), I have tried to be honest and true to the geophysical, geochemical, and other data we obtained. This has led to numerous discussions with other scientists and marine geologists. These constructive "battles", where facts and "truth" are sought, have often been arduous and long-lasting. In some cases, lacking to understand a process may not be important for engineering and foundation considerations. In other instances it is crucial for the safety of installations. For example, when gas and liquids migrating through deep-sea sediments cause the formation of gas hydrates in near-surface sediments, then any man-made disturbance can lead to seabed instability.

Biology has proven to be an inescapable and intimate part of natural sub-seafloor processes, often being the sole visible manifestation of fluid migration. Wherever gases and liquids emerge through or onto the seafloor, there may be a dramatic effect, especially on the composition of the local fauna. Although this book deals with the deep-water (or cold-water) coral reefs of our planet, my favourite subject over nearly 30 years, that of natural underground fluid migration processes, also permeates the text of this book. In this respect, I may be criticised for having a too personalised description and account of deep-water coral reefs. But, although I have attempted to leave out these discussions, I failed. Fluid migration is indeed intimately linked to deep-water coral reefs, and one cannot present one without the other.

However, it has been a pleasure to put all of our (StatoilHydro's) and some of the other old and newly gained knowledge by many others (mainly academians) on *deep-water coral reefs* and the much larger *carbonate mound structures* together in this way. Therefore, the text is not only descriptive, but also one of discussion and possible ways of understanding (i.e., interpreting) what we observe. In this book, I and my fellow researchers report on biological structures that are not only interesting for an integrated oil (resource) company in several ways, but also for the public at large, and for marine scientists, in general. In this way, the book is also a contribution to the resources theme of the UN International Year of Planet Earth with its emphasis on resource baselines in relation to sustainable use (see *www.yearofplanetearth.org*).

I hope the book will provide a useful overview of various aspects of deep-water coral reefs and carbonate mounds, providing insights into the physical processes, and inspire and stimulate further research into these fascinating natural systems. Although I would have liked to, it has not been possible to include each and every documented deep-water coral reef. Mapping of the seafloor is becoming very efficient and new discoveries are being made every week. Keeping abreast with this development is futile. This book, therefore, presents and discusses only a selection of the world's discovered deep-water coral reefs.

I have chosen to include those reefs and mounds that in my view are best described in the literature I am most familiar with (i.e., literature dealing with seafloor topography, geology, geophysics, and physical aspects). I do not pretend to understand everything about biology, and biologists may therefore find there is too little biological focus in this book. Furthermore, because discoveries made on the virgin seafloor are often unique, first-hand accounts, I have often found it right to quote the actual text used for description by the original authors. In this way, nothing should be lost in translation or by my own second-hand interpretation of their text.

Acknowledgements

The one who tells the stories rules the world.

Hopi, Zona (1994)

Statoil (as of 2007 StatoilHydro), and the partners of the Kristin field development project, the Morvin development project, and the Haltenpipe project are thanked for releasing data for this book. Also I wish to thank Paulo Sumida (Institute of Oceanography, São Paulo University, Brazil), and also Maria Patricia Cubelo Fernandez and her team from the Campos Basin Deep-Sea Coral Assessment Group of Petrobras, Brazil, for sharing information from dredging, coring, and geophysical imaging from offshore Brazil. In StatoilHydro, I wish to thank some of my colleagues who have helped both directly and indirectly. These are Arne Indreeide, Leslie Austdal, Mette Vatland, Glenn Angell, Tore Indreiten, Tone Gimre, Jørgen Leknes, Vigdis Lerang, Bjarne Jensen, Bjørn Bakkevig, Irene Liland, Björn Lindberg, Lars Trodal, Christine Fichler, Hans Konrad Johnsen, Erik Nygaard, Tor Inge Tjelta, Olav Kårstad, Mari Dotterud, Radka Fuxova, Roy Wollvik, Tom Glancy, Arne Erik Bjåstad, and others. Outside of StatoilHydro, I would like to thank Haakon Rueslåtten, Philippe Blondel, Jan Helge Fosså, Pål Buhl-Mortensen, Lene Buhl-Mortensen, Ben de Mol, Jean-Pierre Henriet, Tjeerd van Weering, Dag Ottesen, Reidulv Bøe, Richard Sinding-Larsen, and Harald Nesheim. Furthermore, I want to thank all the personnel involved on the following survey vessels: *Master Surveyor*, *North Sea Surveyor*, *Edda Fonn*, *Seaway Commander*, *Edda Freya*, *Normand Tonjer*, *Acergy Petrel*, *Polar Queen*, *Geosund*, and *Acergy Viking*. Without their professionalism, few of the many reefs would have been documented visually.

"So—all of this consists of coral animals. But then, the whole animal community within a coral reef is very characteristic and special—it is a communty of its own, an exclusive union fully dependent on the *Lophelia*—a child of the coral jungle" (Dons, 1927).

Many of the maps used in this book were produced through the Marine Geoscience Data System, available at *www.geomapapp.com* and *www.marine-geo.org* The producers and maintainers of this system are thanked for their accurate relief base maps.

Reference

Dons, Carl (1927). *Sjøen (The Sea)*. J.W. Cappelens Forlag, Oslo, 108 pp.

To my father, Arvid,
a seasoned mariner,
a lover of the wild

To my beloved wife, Målfrid, and our grandchildren:
Magnus, Tore, Mina Cornelia, Henrik, Mikkel,
Linn Mei, Allis, and Marthe,
who provide perspective

Figures

Cover This photo is by the underwater photographer Erling Svensen on one of the shallowest deep-water (this is rather a cold-water) coral reefs in the world, the Tautra reef, located on a morainic ridge, at 45 m water depth, in the Trondheimsfjord, about 15 km north of the city of Trondheim, Norway. It is a good example of a lush Norwegian *Lophelia* reef.

0.1	The "crown" of a *Lophelia* reef off mid-Norway, at a water depth of 300 m .	x
1.1	The author Martin Hovland and biologist Pål B. Mortensen onboard the *Seaway Commander*.	2
1.2	The general seafloor off mid-Norway looks like this	4
1.3	This image is from the approach to the Malangen deep-water coral reef	5
1.4	A close-up photograph from the Breisunddjupet reefs.	5
2.1	A coral polyp (schematic)	8
2.2	*Lophelia pertusa* with extended tentacles	9
2.3	The *Lophelia* worm (*Eunice norvegica*), from the aquarium in Bergen.	11
3.1	A sketch made in 1990, after the first reef had been documented by ROVs off mid-Norway	14
3.2	The survey vessel *Master Surveyor* used by Statoil for pipeline route-mapping off northern Norway, imaged here, just as it records the Fugløy reef in 1982	17
3.3	Probably the first-ever colour underwater photograph taken of a deep-water coral reef	18
3.4	The impressive crown of the Fugløy reef, photographed in 1982	19
3.5	The *Scorpio* ROV used to photograph and document the Fugløy reef in 1982	19
3.6	A sidescan sonar record from the sedimentary basin, located upstream relative to the Fugløy reef.	20
3.7	Methane gas bubbles emitting from the seafloor at Tommeliten, at 70 m water depth in the North Sea	21
3.8	A large boulder located inside a pockmark near the Fugløy reef	21
3.9	A sketch illustrating how I suggest light hydrocarbons dissolved in porewater seep up along dipping sedimentary rock layers through the seafloor in the Haltenbanken	22

xviii **Figures**

3.10	A bathymetric map, showing the locations of oil-fields, condensate-fields, gas-fields, the Nyegga location, and other locations	25
3.11	A perspective view of a 6 km long and 1 km wide section of the Nyegga seafloor area	26
3.12	Gray and white bacterial mats on the sediment seafloor in the deepest, central part of complex pockmark G11 at Nyegga	27
3.13	Pingo No. 1, located inside complex pockmark G11 at Nyegga	28
3.14	The location of a pingo (No. 2) beneath a carbonate rock overhang	29
3.15	Arrow points at a pingo located between two large carbonate rocks	30
3.16	A shaded relief map showing the detailed topography of HMMV	32
3.17	Close-up of bacterial mat at Gullfaks	34
3.18	Bacterial mats disturbed by an ROV manipulator, Gullfaks	35
4.0	A general map of parts of Scandinavia where deep-water (and cold-water) coral reefs are found	40
4.1	This collage of images has been assembled from many different Norwegian coral reefs	41
4.2	An artist's impression of the *Seaway Commander* located above the prominent *Lophelia* reefs on the Agdenes morainic ridge	42
4.3	Photograph of resting *Sebastes* sp.	43
4.4	Bright-spots (black, acoustically highly reflective horizontal bands in the lower image) caused by gas accumulations	45
4.5	This is how the haystacks appeared on sidescan sonar	46
4.6	More haystacks	47
4.7	A simultaneous sidescan sonar and sub-bottom profiler record of a *Lophelia* reef	48
4.8	A sidescan sonar and sub-bottom profiler record over one of the HRC reefs	49
4.9	Same as in the previous figure, but across two other reefs of the HRC	50
4.10	A photograph showing the top of one of the reefs in HRC	51
4.11	A video-grabbed (low-resolution) image from the lower steep slope of one of the largest HRC reefs	52
4.12	A high-resolution still photograph of the framed portion of the previous image	53
4.13	Clusters of *Acesta excavata* bivalves hanging down from dead and live *Lophelia pertusa* colonies	54
4.14	Drawing illustrating where dead *Lophelia* skeleton pieces were found inside a 10 cm wide sediment core	55
4.15	This digital terrain model, shown in artificially shaded perspective view, has been constructed using two datasets of different resolution	58
4.16	A conceptual diagram explaining how locally produced nutrients feed into one of the HRC reefs and into the water column	59
4.17	An artistic photo-montage showing a fossil deep-water Devonian reef	59
4.18	The Sula Reef and the HRC reefs shown on a shaded digital terrain model	60
4.19	Interpreted image of a 2-D seismic line running across the Sula Ridge flat-iron formation	61
4.20	An interpreted seismic image shows the varying acoustic reflection strength in the up-dipping layers, beneath the Sula Reef	62
4.21	A digital terrain model showing the location of ridges, with coral reefs in the Husmus area	63
4.22	Same as Figure 4.21	64
4.23	Seismic line through the Træna Deep reefs	65

4.24	Dipping strata underneath the Træna Bank reefs	66
4.25	A perspective view of the Træna Deep Reef Complex, based on an integrated sidescan sonar, bathymetric, and high-resolution shallow seismic interpretation	67
4.26	The semi-floating platform at the Kristin field off mid-Norway	70
4.27	A small portion of the coral reef KA2 inside the pockmark KA2 at the Kristin field	71
4.28	An example, from HRC, of various types of sponges at coral reefs	72
4.29	This 3-D perspective image shows how the Kristin semi-floating platform is located above the seafloor	73
4.30	An artificially shaded relief map of parts of the Kristin field, off mid-Norway	74
4.31	Oblique view of parts of the image shown in Figure 4.29	75
4.32	A section of the seafloor mapped using ROV-based MBE	77
4.33	Morvin *Lophelia* reefs	78
4.34	Acoustic evidence of prolific hydrocarbon seepage out of the seafloor	80
4.35	Accumulation of sponges found at the site seen in the previous figure	81
4.36	This photograph was acquired in 1991 from the Block 6407/4 location, off mid-Norway	82
4.37	A beautiful *Lophelia* colony in the Breisunddjupet area	83
4.38	*Lophelia* colonies and a sponge in the Breisunddjupet area	84
4.39	A pink version of the *Lophelia* sp. at Breisunddjupet	84
4.40	A perspective view of the Floholman reefs, based on swath bathymetry data	85
4.41	An underwater photograph of bacterial mats and suspected methane-derived authigenic carbonate crusts	85
4.42	Terrain map from Hola	86
4.43	Echosounder data from Hola	87
4.44	The location of Malangsreef	88
4.45	Photo from the lower portion of the Malangsreef	89
4.46	A beautiful photo of a *Chimaera* sp.	90
4.47	A geological and topographical section through the Fedje reefs	91
4.48	Map of the Fedje reefs	92
4.49	Map of a complex pockmark at Troll	93
4.50	These two large, white and red *Paragorgia arborea* (gum-corals) were found inside the Troll pockmark	94
4.51	A large, apparently layered carbonate rock, a methane-derived authigenic carbonate	95
4.52	Arrows point at living (articulating) *Acesta* individuals	96
4.53	The location of the Troll pockmark corals	97
4.54	A *Lophelia* colony (lower right) on a man-made structure at Statfjord	98
4.55	Same man-made structure as above, but a different colony	99
4.56	Map over parts of the northwest Atlantic margin showing the Faroe Islands, the Rockall bank, and adjacent areas	100
4.57	A rare photograph of a great fork-beard (*Phycis blennoides*)	101
5.1	One of the "Kess-Kess" fossil reefs, Moroccan Sahara	104
5.2	A general shaded relief map of the northeast Atlantic Ocean, Ireland, the British Isles, the North Sea, and parts of Norway and Denmark	105
5.3	The original image that proved the existence of giant carbonate mounds off Ireland	108
5.4	The remarkably similar 2-D seismic image of a live, giant carbonate mound from the Vulcan Sub-Basin, off northwest Australia	108

Figures

5.5	A high-resolution seismic image across the carbonate banks in the Porcupine Seabight	109
5.6	Results of a 2-D hydrocarbon basin modelling (migration pathway) study	109
5.7	This is a photo-montage that figured on the front cover of the *Journal of Petroleum Geology*	111
5.8	Map from the Faroe Islands, in the north, to the Porcupine Basin, in the south, showing various types of coral reefs and carbonate mounds	112
5.9	The resulting geo-profile over the Challenger Mound constructed after the preliminary data from IODP Drilling Expedition 307	116
5.10	This figure shows clearly how unique the location of Belgica Mounds is in relation to the underlying geology	117
5.11	Preliminary data showing the variation in the number of live bacterial cells per cubic square centimetre at the three sites	118
5.12	From the Theresa Reef, Porcupine Slope, Belgica Mounds	119
5.13	One of the giant carbonate mounds on the Rockall Bank	121
5.14	Detailed interpretation of the area of Darwin Mounds	124
5.15	A sidescan sonar record of a typical Darwin Mound with parts of its "tail", seen as the light (low-reflective) sediment on the left	125
5.16	An early drawing showing the W-TR (Wyville–Thomson Ridge) acting as a dam	126
5.17	An interpreted regional geological transect across the Gulf of Cadiz	131
5.18	El Arriche mud volcano geological setting	133
5.19	These two images show perspective views of the topography of the Pen Duick mud volcano and carbonate mound province, off Morocco	134
5.20	A 2-D high-resolution seismic section through the Pen Duick mud volcanoes and carbonate mounds	135
5.21	A 2-D seismic record through the Gemini mud volcano	135
5.22	Mekenes mud volcano	136
5.23	A regional shaded topography map from off Mauritania	137
5.24	A detailed shaded map showing the strange Mauritanian giant carbonate mounds	138
6.0	Deep-water coral reefs, or lithoherms, in the Florida Straits	142
6.1	Echosounder records showing pockmark-associated deep-water coral reefs	143
6.2	A shaded relief map of the Campos and Santos Basins off Brazil	143
6.3	A sidescan sonar record showing a surfacing fault scarp with coral reef overgrowth in the northern part of the West Flower Garden Bank	144
6.4	Map of the northern Gulf of Mexico	146
6.5	A detailed drawing of the southeastern part of East Flower Garden Bank	148
6.6	An interpreted high-resolution shallow seismic section shot across the East Flower Garden Bank	149
6.7	"Fields of Floating Oil", GoM, a map constructed by Captain Soley, 1910	149
6.8	A 25 m high mound located inside a depression on a seismic image from off Belize	150
6.9	Interpretation of a seismic record across reefs at the bottom of the Florida Straits	151
6.10	A shaded relief map of the northwestern part of the Atlantic Ocean and the southeastern USA	152
6.11	This 3-D perspective sketch has been made on the basis of a reflection profile crossing from Florida's Hatteras Slope onto the Blake Plateau	154

6.12	A shaded relief map showing Newfoundland, the fishing ground Flemish Cap, and Orphan Knoll.	155
6.13	A drawing from Gay et al. (2007).	157
6.14	A relief map showing California and the ocean floor west of California.	159
7.1	Illustration showing the typical "canopy" appearance of deep-water coral reefs	164
7.2	Schematic sketch of a (fossil) Waulsortian mound in the Normandville field and sketch of a Waulsortian mound	166
7.3	A photograph of the remarkable fossil deep-water carbonate reefs or mounds located in the Ahnet sedimentary basin, Algerian Sahara	169
7.4	A conceptual sketch illustrating the formation of the Kess-Kess reefs, Morocco	171
7.5	Sketch of submarine spring-controlled growth of bioherms	174
7.6	Sketch showing the suggested growth of bioherms	175
8.1	A sketch that illustrates the hydraulic theory for deep-water coral reefs and giant carbonate mounds.	180
8.2	A fine perspective digital terrain model of the Belgica Mounds	182
8.3	Map from the Porcupine Seabight showing the locations of the three groups of mounds: Magellan, Hovland, and Belgica	185
8.4	A scanned image though a soil core of the Challenger Mound	189
8.5	One of the Magellan Mounds: Perseverance Mound.	190
8.6	An artist's impression of the *Seaway Commander* performing ROV-based documentation of the HRC in 1993	191
8.7	A vertical profile and map representation of the entire 200 km long Haltenpipe route	193
8.8	A relief map of parts of the Sula Ridge.	194
8.9	Same as previous figure, but now only showing the lineations discussed in Figure 8.8.	195
8.10	Relief map of the area between HR and SR	196
8.11	On this map over the general Træna Bank area, there are no obvious iceberg ploughmarks.	197
8.12	Four detailed topographical maps from the Haltenpipe Reefs	199
8.13	A 3-D perspective bathymetric map of the shallow hydrothermal system, located on a coral reef off Taketomi Island, Japan.	200
9.1	The Late Devonian Kess-Kess fossil carbonate mud mounds, which resemble the currently live Porcupine Basin giant mud mounds.	208
A.1	A photograph showing the base of Reef B on the HRC	217
A.2	Some of the strange animals encountered on the Sula Ridge	218
A.3	A large group of Norwegian redfish (*Sebastes viviparus*)	219
A.4	Large blocks of dead *Lophelia* colonies are favoured locations for many types of sponges	220
A.5	A very rare image of a dark-violet, unknown coral species	221
A.6	A video-grabbed image from one of the pockmark reefs at Kristin	222
A.7	One of the early photographs taken on the HRC.	223
A.8	Here is some detail from the HRC in 1997	224
A.9	A beautiful *Paragorgia arborea*.	225
A.10	At the base of *Lophelia* reefs, large sea urchins of the species *Cidaris cidaris* can be found.	226
A.11	A fine photo of the common saithe (*Pollachius* sp.)	227
A.12	On these reefs, which are up to 24 m high, coral colonies grow up and outwards and often form overhangs	228

A.13	A coloured version of the original monochrome sub-bottom profiler and sidescan sonar recordings of the Fugløy reef of 1982	229
A.14	Two video-grabbed images from the centre of complex pockmark G11, at Nyegga.	230
A.15	Two high-resolution underwater photographs taken at the Kristin A2 reef in 2004.	231

Tables

4.1 Geochemical results (occluded, free hydrocarbons, in ppb by volume) from Statoil/BP geochemical sampling. 56

4.2 Geochemical results (adsorbed, sediment-bound hydrocarbons in ppb by volume) from Statoil/BP geochemical sampling 57

4.3 Characteristic size and depth spans of coral reefs and their associated pockmarks, at three locations in the Kristin field 74

Abbreviations and acronyms

AB	Acoustic Basement
ACES	Atlantic Corals Ecosystems Study
AOM	Anoxic Oxidation of Methane
AWI	Alfred Wegener Institute
BIOFAR	Marine Benthic Fauna of the Faroe Islands (programme)
BSR	Bottom Simulating Reflector
CCD	Carbonate Compensation Depth
COMET	COntrols on METhane fluxes
DTM	Digital Terrain Model
DWC	Deep Water Coral
DWCr	Deep Water Coral reef
ECOMOUND	Environmental COntrols on MOUND formation along the European margin
EF	East Flower Garden Bank
ENAW	Eastern North Atlantic Water
EPR	East Pacific Rise
EU	European Union
FFI	Forsvarets Forsknings Institutt (Norwegian Defence Research Institute)
GEOMOUND	GEOlogical controls of carbonate MOUNDs on passive continental margins
GERG	Geochemical Exploration and Research Group
GMV	Gemini Mud Volcano
GoM	Gulf of Mexico
HERMES	Hotspot Ecosystem Research on the Margins of European Seas
HMMV	Håkon Mosby Mud Volcano
HRC	Haltenpipe Reef Cluster

Abbreviations and acronyms

HRC	Haltenpipe Reef Cluster
IFREMER	French Research Institute for Exploration of the Sea
IKU	Continental Shelf Institute
IMR	Institute for Marine Research
IOC	Intergovernmental Oceanographic Commission
IODP	Integrated Ocean Drilling Program
IPCC	Intergovernmental Panel for Climate Change
IS	Inferred Seep
LGM	Last Glacial Maximum
MAR	Mid Atlantic Ridge
MBE	Multi Beam Echosounder
MC	Magma Chamber
MDAC	Methane Derived Authigenic Carbonate
MMV	Mekenes Mud Volcano
MOW	Mediterranean Outflow Water
MYBP	Million Years Before Present
NADW	North Atlantic Deep Water
NGU	Norges Geologiske Undersøkelser (Norwegian Geological Survey)
NIOZ	National Institute for Oceanic Research (the Netherlands)
NOAA	National Oceanographic and Atmospheric Administration
OHCB	Obligate HydrocarbonoClastic Bacteria
PDB	Pedee Belemnite (standard)
ppbv	parts per billion by volume
R&D	Research and Design
RCMG	Renard Centre of Marine Geology
ROV	Remotely Operated Vehicle
SEPM	Society of Sedimentary Geologists
SIP	Stable Isotope Partitioning
SR	Sula Reef
TOBI	Towed Ocean Bottom Instrument
TRIM	TOBI Rockall Irish Margin
TTR	Training Through Research (programme)
UiB	University of Bergen
UNESCO	United Nations Educational, Scientific, and Cultural Organization
UPI	United Press International
VISTA	A project between Vitenskapsakademiet (Norwegian Academy of Science, Oslo) and Statoil
W-TR	Wyville–Thomson Ridge
WF	West Flower Garden Bank
YBP	Years Before Present

1

Introduction

Even if the oceans cover 71 percent of the earth's surface, only 0.2 percent of this area contains one quarter of all marine species. The corals, to say it melodramatically, represent the soul of the ocean, and the ocean is the mother of all life.

Le Page, 1998

1.1 INTRODUCTION

Norwegian and North Atlantic coral reefs are not only *spectacular*, but also very *paradoxical*, the former, because of their size (house-sized, up to hundreds of metres long, tens of metres tall), their numbers (thousands), and their prolific associated fauna (up to several hundred macro-faunal species per reef). They are paradoxical because they thrive in deep, cool, dark (apparently hostile) waters, even north of the Polar Circle, despite any apparent high-grade nutrient source, or apparent help from endosymbionts.[1]

Deep-water coral reefs represent important living, three-dimensional structures in the North Atlantic Ocean and elsewhere. However, one should not think so, when reading some of the modern authoritative scientific books and articles dealing with marine life on the ocean floor; for example, *The Northern North Atlantic: A Changing Environment* (Springer, 2001, 500 pp.). Neither deep-water coral reefs, nor *Lophelia pertusa*, nor *Lophelia* reefs are found in the index, or at all mentioned in the book! Why is it that thousands of large coral reefs, some known to science for at least 150 years, manage to avoid mention (recognition) in such literature? Could it be that researchers sometimes only rely upon "classical marine biological" results? Or, do

---

[1] Endosymbionts are symbiotic organisms living inside their hosts.

Figure 1.1. The author, Martin Hovland (left) and biologist Pål B. Mortensen onboard the *Seaway Commander*, in front of the ROV (remotely operated vehicle) *Solo*, located over the HRC reef off mid-Norway, 1993.

they refuse to write on aspects outside their own special themes, even though they claim: "the Greenland–Iceland–Norwegian Seas can now be considered one of the best studied sub-basins of the world's oceans." But, even so, their information is important, as it provides us with the necessary background knowledge about life on the general seafloor, the seafloor *desert*, outside the coral reefs.

The intention with this book, based on the book *Norske korallrev og prosesser I havbunnen*: "Norwegian Coral Reefs and Processes in the Seafloor" (Hovland and Mortensen, 1999, in Norwegian) is to provide general and specific first- and second-hand information on the occurrence and habitats of *Lophelia* reefs in the North Atlantic and elsewhere. My account is especially detailed and holistic for some of the Norwegian coral reefs, which I have studied first-hand for the last 25 years, and of the fascinating giant carbonate mounds off Ireland. A brief review of the latest research on some other worldwide occurrences of deep-water coral reefs is also included.

A brief statement on *definitions*, the main *facts*, my *inferences*, and the main *objectives*, is appropriate. A "deep-water coral reef" (as used in this book) is a local seafloor mound consisting of accumulations of coral debris, fine-grained and sometimes coarse-grained sediments, and live coral colonies. They occur at water depths between 39 m and 450 m off Norway (including inland fjord waters), but have been reported in up to 3,380 m water depth elsewhere (Mortensen *et al.*, 2001). Off Norway, they measure from 10 m across and 2 m high, to several hundred metres across (length) and up to 40 m high.

"Giant carbonate mounds" (as used in this book) are an order-of-magnitude-larger structures, mainly formed by a local accumulation of sediments and coral debris; for example, such as those occurring in the Porcupine Basin Slope, off Ireland. They are up to two kilometres in diameter and 300 m in height (Henriet *et al.*, 1998).

At 100% certainty, the *facts* about these living structures are

— The reef-forming species are the stony corals (scleractinians) *Lophelia pertusa* (Linnaeus, 1758), *Madrepora oculata* (Linnaeus, 1758), and *Desmophyllum cristagalli* (Milne-Edwards and Haime, 1848).
— At least one large reef on the continental shelf, off Norway, has remained exactly in the same location for at least 8,150 calendar years (i.e., since the end of the Younger Dryas cold spell, Hovland *et al.*, 1998).
— The live coral polyps and fan-shaped octocorals associated with the reefs generally face into the local prevailing current direction. This, generally, means that the reef long-axis grows parallel with the local prevailing current direction.

Over the last 20 years, many inferences (and speculations) about deep-water coral reefs have been published. These mainly refer to their whereabouts on the seafloor. Thus, their locations have been inferred to be dependent on

— Special, local, strong current conditions, such as vortices, breaking internal waves, etc.
— Special water masses, such as mixed, non-turbid, and salinity-defined masses, etc.

— Special sub-stratum relationships, such as firm rock, boulders, and iceberg ploughmarks, in addition to my own preference: that there may be a link with local underground hydrology (ground-water and dissolved gases/chemical species; i.e., "seabed fluid flow"), the so-called "hydraulic theory".

The main objectives of this book are

— to describe deep-water coral reefs experienced first-hand;
— to review second-hand information amassed over the years by many researchers; and
— to discuss why deep-water coral reefs and giant carbonate mounds form.

Whereas many researchers on deep-water coral reefs and carbonate mounds are concerned with the currents and water masses surrounding the structures, external factors, I am particularly concerned with their sub-stratum, internal factors (i.e., what they are perched or anchored on and the underground fluid processes). The chapters in this book deal with description, Scandinavian reefs, reefs and carbonate mounds elsewhere, the substratum, fossil reefs, and theories presented in modern scientific literature. The final chapter deals with the ultimate fate of the reefs and carbonate mounds.

Generally, the seafloor is barren, cool, dark, and, to the untrained observer, looks lifeless and hostile (Figure 1.2). The seafloor itself is typically drab, grey

Figure 1.2. The general seafloor off mid-Norway looks like this. In the background is a tusk (*Brosme brosme*). The stones (cobbles and gravel) have been washed out of the underlying clay-dominated till.

Sec. 1.1] Introduction 5

Figure 1.3. This image is from the approach to the Malangen deep-water coral reef where a school of half-metre-sized *Pollacchius* sp. (saithe) follows the ROV (photograph from *www.mareano.no*, 2007, Lene Buhl-Mortensen, IMR, Bergen).

Figure 1.4. A close-up photograph from the Breisunddjupet reefs. Here there are pink *Lophelia* sp. as well as sea-anemones, shrimp, etc. Note the *Sebastes* sp. (Norway redfish) in the background.

(normally consisting of clay), and in northern waters, with stones dropped by the ice during the last glaciation, about 13,000 years ago. Only a few white sponges, some burying organisms, and some tubeworms can typically be seen. Shoals of fish are, however, often seen when we work on the seafloor with remotely operated vehicles (ROVs). In Norwegian waters, the lights from the ROV attract shoals of saithe (pollock) and sometimes cod (Figure 1.3).

In spectacular contrast to this normal seafloor world, when we move off the general, drab seafloor and onto a *Lophelia* reef it is like coming into an oasis in a sandy desert, on land. The environment suddenly turns from one of apparent desolation to one of teeming life (Figure 1.4). Organisms occur everywhere. They live in abundance, one over the other, and with colonies of the white or pink *Lophelia pertusa* coral standing out in stark contrast to the dark waters. Over the last 25 years, I have conducted and participated in numerous coral reef studies off Norway. This book is mainly a result of that work.

2
Coral reefs

Now the deep ocean's best-kept secrets are being revealed—ancient coral ecosystems hidden from view at great depths.
www.lophelia.org

Paradigm shifts in science seldom involve logic, rational discourse, higher-resolution data, or more accurate calculations.
Anderson (2007)

2.1 INTRODUCTION

Tropical coral reefs are reckoned to be the largest biological structures on Earth. They represent unique habitats, with a very high faunal diversity of corals and other invertebrates, fish, and algae. The term "coral reef" is traditionally used on large tropical shallow-water formations, which represent dangers to navigation. Even though the deep-water coral accumulations do not represent dangerous "reefs", as such, it has become a habit calling them "reefs". A viable, and perhaps more appropriate alternative would be to call them deep-water "bioherms", but, even so, the more common reef concept will be used in this book.

> "The architects and designers, the construction engineers and builders of a coral reef are zoologically classified in the phylum *Coelenterata* or *Cnidaria*" (Bennett, 1971).

This is the way Isobel Bennett introduces us to the formation of coral reefs. Even though the deep-water coral reefs live in total darkness and in cool water, they construct reefs in the same manner as shallow-water corals in the tropics (i.e., by

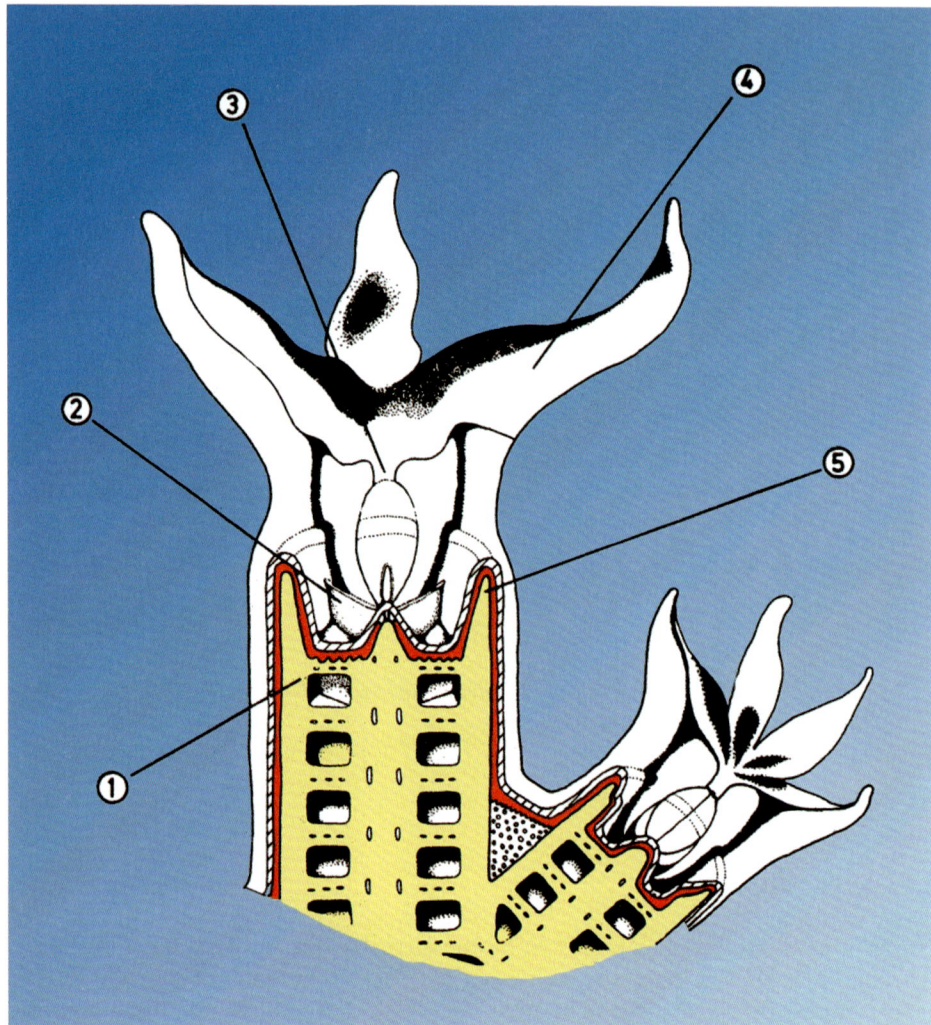

Figure 2.1. A coral polyp (schematic). 1 = skeleton, 2 = septa, 3 = mouth, 4 = tentacle, 5 = coral cup. Red is the endodermis. Scale: The polyp of *Lophelia pertusa* is between 0.5 cm and 1.5 cm across (based on Morton and Morton, 1983).

excreting an exterior skeleton, an exoskeleton). The main reef-building organism on North Atlantic deep-water coral reefs is the stony coral *Lophelia pertusa*, by Norwegian fishermen also called the "glass coral". There are also numerous reefs containing other very similar reef builders, the stony corals *Madrepora oculata* and *Desmophyllum cristagalli*. Even though there may be numerous other types of corals on deep-water reefs, such as the large and beautiful *Paragorgia arborea*, it is only the stony corals that produce robust enough skeletal material to form large mounds (reefs) on the seafloor.

Sec. 2.1] Introduction 9

The corals belong to the phylum *Cnidaria*, the same phylum as the common jellyfish (medusa). Anybody swimming in the ocean will, sooner or later, come into contact with the *Cnidaria* group of animals. The reason for this, is the poisonous nettle cells or nematocysts that paralyse prey and that sting unprotected human flesh. Although the common jellyfish physically does not resemble a coral at all, it does in its larval stage. Sea anemones and hydroids also belong to the same group. Because they have no backbone they are invertebrates. The main difference between the jellyfish and the coral is: the latter excretes a calcium skeleton and builds a foundation, whereas the jellyfish drifts and swims along with the currents. In other words, the coral, being a sessile animal, settles and grows at a suitable location with respect to food and other environmental life conditions.

A coral polyp is a soft, almost transparent animal which has short tentacles (Figure 2.2). Corals that produce a hard skeleton, like the *Lophelia* and *Madrepora*, are called stony corals. Others, with soft skeletons (i.e., the *Arborea*), are called octocorals. Their skeletons are formed within their bodies and occur as small pieces of calcium carbonate embedded in a horny substance that can be bent. When these corals die, their remains soon dissolve, in contrast to the skeletons of stony corals.

Figure 2.2. *Lophelia pertusa* with extended tentacles (clusters of nematocysts can be seen). The polyp to the right has retracted its tentacles, probably because it has managed to get hold of prey. This fragment of a colony was sampled from a coral reef on the Sula Ridge in 1998 and)was kept alive in the aquarium, Bergen, Norway for about 3 years (photo courtesy of P. B. Mortensen). Scale: the polyp on the left is about 1 cm across.

In the tropics, corals have a special way of acquiring a stable food supply: the polyps there contain small plant cells. These are endosymbionts living within the coral polyps, and they consist of algae. Thus, the tropical corals nourish their own internal vegetable gardens. The symbionts (zooxanthellae) are thought to ensure sufficient food at all times. Tropical corals are termed "hermatypic", because of the symbionts they harness:

> "When sea surface temperatures exceed their normal summer high by 1°C or more for a few weeks running, coral polyps, for reasons not entirely understood, expel their zooxanthellae, the symbiotic algae that lend corals color and provide nutrients. The polyps turn pale and starve. If they don't get their zooxanthellae back in a month or so, they die" (Stone, 2007).

The most remarkable feature of deep-water corals is the absence of the algal endosymbionts, the zooxanthellae. Deep-water corals living where no sunlight penetrates do not host the algae and are therefore termed "ahermatypic" corals. But, the question on how they can subsist and survive in apparently hostile environments is a true mystery to science. Unlike most tropical corals, which digest sugars produced by their symbiotic algae, for food, the deep-water corals can snatch theirs as it drifts past. With their tentacles they can even snatch and digest small live animals, such as copepods (Freiwald et al., 2002).

2.2 SKELETON FORMATION

Stony corals have special cells on the outside of their bodies which produce a very hard bicarbonate substance consisting of 90% aragonite ($CaCO_3$), which becomes hard as rock (Figure 2.1). The juvenile coral polyp is called a "planula", and is free-swimming. It attaches itself either to another coral skeleton or to a hard surface, where a sticky mucoid layer is secreted by its basal plate during attachment. This mucoid layer serves as an adhesive substrate for skeletogenesis (skeleton formation). There are two main parts of the primary skeleton of corals: a horizontal basal plate, with organic and inorganic layers, and a vertical set of structures, the septae, theca, and columnella. The basal epiderma (outer) membrane accommodates vertical skeletal growths and circumscribes the horizontal directive of the corallite polyp (Goreau and Hayes, 1977). In this way, the newly formed skeleton and the skeletogenetic zone are isolated from the seawater. The significance of which is that seawater will corrode and dissolve the newly formed $CaCO_3$ (aragonite).

But, in the past, during the Cretaceous period, some corals similar to those living today had calcite skeletons, and not aragonite skeletons. The reason may have been a different climate:

> "Modern coral reefs are built primarily by scleractinian corals, which arose in the Triassic after the Permian extinction. Today, all of these corals form skeletons of aragonite, and this composition has been thought to be typical of fossil scleractinians as well" (Hurtley and Szuromi, 2007).

Stolarski *et al.* (2007) have now identified a Cretaceous scleractinian coral with a primary calcite skeleton. The fine preservation of internal structures and the Mg and Sr chemistry show that the calcite is primary, not diagenetic. This result tightens the evolutionary connection between these corals and rugose corals, which formed calcite skeletons but were eliminated in the Permian extinction. These results suggest that corals may be able to alter their biochemistry in response to changes in seawater chemistry.

2.3 CORAL REEF ECOLOGY

Coral reefs provide a home, food, and shelter for a wide range of animals (see Figure 2.3 and figures in the Appendix): for marine worms, sea-urchins, sponges, sea-fans, molluscs, crabs, shrimps, and other crustaceans, sea-anemones, and not least, many

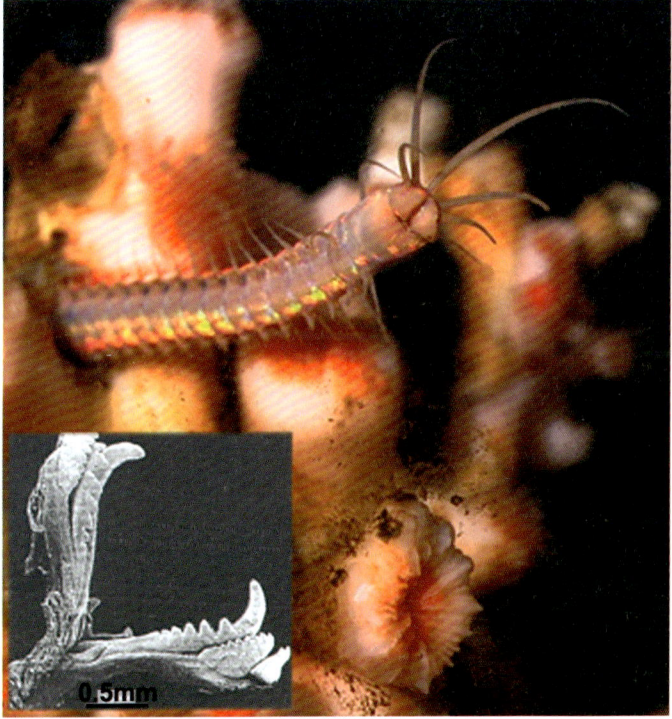

Figure 2.3. The *Lophelia* worm (*Eunice norvegica*), from the aquarium in Bergen, at IMR (by and courtesy of Pål B. Mortensen). The worm is about 0.4 cm wide. The inset is a microscopic view of its jaw apparatus (by and courtesy of André Freiwald). The *Lophelia* worm is observed to feed on food spills from coral polyps. According to Mortensen *et al.* (2002), the worm also aids the *Lophelia* colony by strengthening the physical structure of the skeleton by enhancing the coral's calcification around its dwelling tubes (photo courtesy of P. B. Mortensen, IMR, Norway).

types of fish. In the tropics, where all their energy comes from sunlight, they also house plants. The energy which cycles through the reef is what keeps all these animals and plants alive. In the deep, however, we still lack good models explaining where the necessary energy comes from, although much of it is believed to come from particles originating in the photic zone of the ocean (i.e., near the surface). Being sessile organisms, the corals either rely on currents to bring the nutrients (food) to them, or they rely on catching food with their tentacles. Probably, they rely on all methods of obtaining available food. Sponges, in various colours and shapes, grow on the coral skeleton when it is dead (see Figures A.1 and A.4).

In the tropical hermatypic reefs, there is a vigorous exchange of gases between plants and animals. Thus, the zooxanthellae expel oxygen while performing photosynthesis, during daytime, and use the CO_2 expelled by the polyps. The polyps use that oxygen, which is a win–win situation, typical for an effective symbiosis.

The natural enemies of coral reefs (besides man) are fish that nibble at the coral tentacles and epithelia. Some snails, urchins, worms, and other animals also eat polyps. In the tropics, the coral predators often do this to get the algae inside the polyps. Other species, worms, sponges, mussels, and sea-urchins, bore their way into the coral skeleton, acquiring carbon and calcium, and thus weaken the coral structure.

3

A modern re-discovery

Given the high carbon content available for biomass production, and the high energy content of such highly reduced compounds, it is hardly surprising that many microbes have evolved or acquired the ability to utilize hydrocarbons as sources of carbon and energy.

Yakimov *et al.* (2007)

Discovery commences with the awareness of anomaly, i.e. with the recognition that nature has somehow violated the paradigm-induced expectations that govern normal science.

Kuhn (1970)

3.1 INTRODUCTION

Deep-water coral reefs were originally discovered by fishermen. Scientifically, they were recorded as early as 1752 (Pontoppidan). However, their true structure, prolific associated marine life, and beauty were first visually documented from early in the 1970s (Wilson) up to the turn of the century. One of the first questions raised when confronted with these apparently out-of-place bioherms is "What on Earth makes the structures buzz?". After having seen and studied the consequences of methane seepage at North Sea fluid flow locations, such as Tommeliten and Gullfaks (Hovland and Judd, 1988), a fluid flow-based explanation model was published in 1990 (Hovland, 1990b). Some of the history of my own discovery and realizations on how and why deep-water coral reefs may build up off Norway and elsewhere is provided in this chapter.

14 A modern re-discovery   [Ch. 3

Figure 3.1. A sketch made in 1990, after the first reef had been documented by ROVs off mid-Norway (M. Hovland). Subsequent studies have shown that the ratio between live corals and dead is normally much less than illustrated here. For scale, a typical ROV is a couple of metres long.

3.2 HISTORICAL BACKGROUND

During the last glacial maximum (LGM), around 18,000 years before present (YBP), most of Norway and parts of the rest of Scandinavia were covered by up to 3,000 metres of ice and snow. The volume of water frozen in the two great ice caps, one over northeast Europe, the other over Laurentia (parts of North America/Canada), was so immense that the mean sea level had sunk by about 120 metres, making much of the North Sea and many other shallow seas worldwide dry land! Parts of Norway, Sweden, and Finland are still rebounding, after having been weighted down by the ice cap. This rebound has been especially large near Trondheim (mid-Norway) and near Oslo (eastern Norway). Curiously, some of the oldest Norwegian coral reefs were, therefore, found on dry land. They were then, of course, dead, and had followed the emerging seafloor as it rebounded and rose well above sea level in several places. The following is a translation of O. Nordgaard's (1912) account (originally in Norwegian) of discovering such a fossil coral reef:

> "At Sandvaagen, which is a place just inside of Strømmen in the Borgenfjord (between Sparbuen and Inderøen), a lot of broken *Lophelia prolifera* [he originally called them *Lophohelia*] could be seen in July 1909 along the shore. Some of the pieces were very worn by the waves, as they had laid in the tidal zone over a long

period, but others were apparently newly washed out of the clay. It dawned on me that here lay the remains of an old *Lophelia* reef, similar to that described by M. Sars from near Drøbak. This occurrence is also treated by prof. Brøgger [Brøgger, 1901] in his great work about the late glacial and postglacial elevation changes in the Kristiania area.

The *Lophohelias* near Sandvaagen were not unknown to the local people either. A boy told me that he had collected pieces, which he called "krona" [crowns]. In September the same year [1909] the occurrence was studied more closely. It proved that *Lophelia* occurred in blue clay, beneath a layer of sand and stones in the tidal zone of slight inclination. The *Lophelia*-carrying clay layer could be followed several metres along the tidal zone, and it stretched at least 7–8 metres up onto the kelp-clad beach and about 1 metre above this. However, it was not possible to determine the full size of the patch."

He goes on to describe the remnant animals they found (hydroids, echinoderms, etc.), and by comparing the known live *Lophelia* reefs at that time, concluded that the Sandvaagen reef was a relict and fossil occurrence of *Lophelia* reefs similar to the ones living on the bottom of the Skarnsundet, between Beitstadfjord and Trondheimsfjord (Nordgaard, 1912; Strømgren, 1971).

The early Norwegian descriptions of stranded coral reefs by W. C. Brøgger (1901) are also interesting:

"By own observations, I can fully confirm the accuracy of *Lophelia prolifera* descriptions made by Sars. They are common around Drøbak over the whole seafloor, over an area of about 100 km^2, from a depth of 70–80 fathoms [ca. 150 metres] to sea level. They are found above sea level on Barholmen, Kaholmen [islands], to an elevation of over 30 m. Likewise, they are found further out in the fjord, at Bævø (near Jeløs north point), still near Svelvik, and in large occurrences adjacent to Dramselven [a river] at Ryg, near Mjøndalen [0–10 m above sea level]. Several places, especially near Drøbak and Holmestrand, they occur together with the large [bivalve] *Lima excavata*."

This latter fact is of interest as modern research has found a close association between the *Acesta excavata* bivalve and *Lophelia pertusa*, as will be discussed later.

Nowhere else was so much known about deep-water coral reefs as in Norway during the early 20th century. The first scientific description of *Lophelia pertusa*, besides that catalogued by Carl von Linné (1758), came from observations made by the Norwegian bishops Erik Pontoppidan (1752) and Johan Ernst Gunnerus (1768) in the Trondheimsfjorden, Norway. The name *Lophelia* derives from Greek *lophos* (tuft of) and *helioi* (sun), referring to the individual Sun-like coral polyps.

The oceanographer Michael Sars (1865) was the first to recognise that *Lophelia pertusa* produced mound-shaped reefs on the seafloor. In 1868 live *Lophelia* were found at 969 m depth on the Wyville–Thomson Ridge northwest of Scotland (Thomson, 1873). And, contemporaneously with the work of Carl Dons (1944), off Norway, the French biologist, E. Le Danois studied the fauna associated with *Lophelia* occurrences off France (Le Danois, 1948).

Then, about 25 years later, John Wilson did his pioneering studies on the Rockall Bank:

> "In June 1973, Dr. John Wilson and pilot George Colquohoun dived in the *Pisces* III submersible in search of *Lophelia* on Rockall Bank. These pioneering dives took video and still images of *Lophelia* reefs on the Rockall Bank 350 miles offshore and were used by John Wilson in his classic descriptions of Lophelia coral growth and patch development published in 1979" (*www.lophelia.org*, SAMS, 2007).

Corals and coral localities occurring off Norway were reported by several workers, in addition to the early pioneers mentioned above: Brøgger (1901), Nordgaard (1912), Dons (1927, 1944), Strømgren (1971), and Mikkelsen *et al.* (1982). It is only the latter two of these researchers and groups that reported findings in the English language (i.e., Strømgren, and Mikkelsen *et al.*), the others were in Norwegian.

3.3 THE FUGLØY REEF, 1982

"Coral reefs north of the Polar Circle!?—You must be joking!" the journalist from United Press International (UPI) burst out, when he heard about the deep-water coral reefs, in 1984, off Norway. As with others, we also believed that coral reefs only belonged in tropical and sub-tropical waters, far away from our cold coasts. However, just by chance, we came across one of the reefs a July morning, in 1982, near Fugløya island, north of Tromsø. We were busy mapping the seafloor with the survey vessel *Master Surveyor* of Ålesund, a converted trawler, for a potential pipeline from the Askeladden field in the Barents Sea to Lyngenfjord, in Finnmark, Norway. With the sidescan sonar towing over the stern, we suddenly recorded a mysterious looking cone-shaped feature on the seafloor (Figure 3.2). It was 280 metres below us on a sill, perched on a slope leading down to a sediment basin at 320 m water depth. But, it did not look natural. Was this, perhaps, a human-made submarine listening device, of the Cold War? Rising 15 metres up over the surrounding seafloor, and being 50 m wide at its base, we concluded that this feature, although being unnaturally regular, must be a trick of nature. The vessel was halted and the sidescan sonar was reeled in. The 500 kg heavy, and 3 m long "gravity corer" was readied for action. On the wire immediately above the corer, we attached a small acoustic transponder, by which its exact underwater position could be determined.

Soon, the corer hovered at 260 metres depth, right above the top of the unidentified strange cone. Suddenly the corer was dropped until the wire went slack. As it was winched in, excitement rose. Most of us expected just another sticky clay core. But we had luck and a sigh of astonishment went round as several pieces of white coral bits fell tinkling onto the steel deck, it sounded like bits of glass falling.

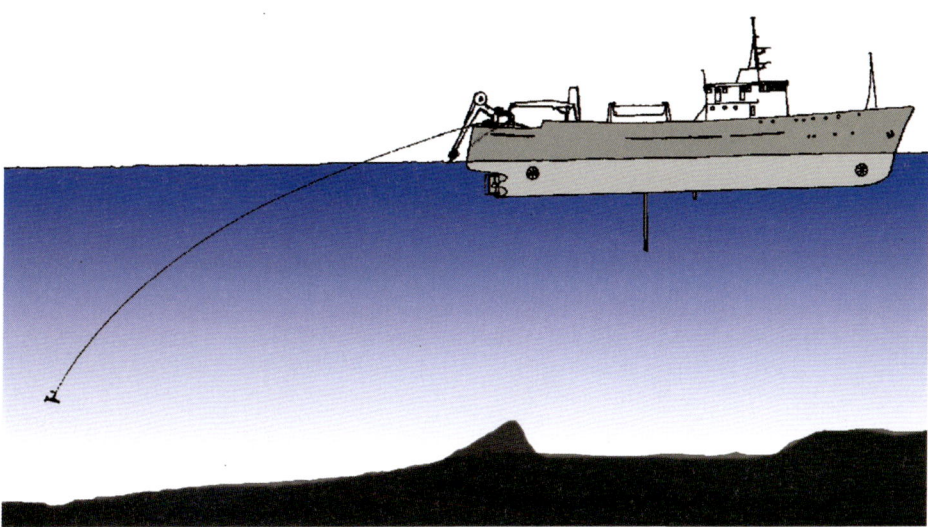

Figure 3.2. The survey vessel *Master Surveyor* used by Statoil for pipeline route-mapping off northern Norway, imaged here, just as it records the Fugløy reef in 1982. Note the towed sidescan sonar fish (it has not detected the reef yet, although the hull-mounted echosounder has).

Some of the bits were pure white and others were slightly pink. Their slimy, transparent coating told us that the corals down there were fully alive! Here we had stumbled across a new object. The task was to map a potential route, but also to perform a sound interpretation of the data we collected. The ROV *Scorpio* was therefore prepared for a dive. At that time, this rather ungainly-looking vehicle was box-shaped, measuring 3 m by 2 m (Figure 3.5). ROVs are any child's dream toy, as it enables one to find out exactly what is down there on the seafloor. For us, it is just another piece of necessary equipment. However, as this was mainly a reconnaissance survey, and because of its slowness compared with towed and hull-mounted equipment, we only used it for special spot checks. After about half an hour, *Scorpio* had been launched and reached the seafloor, about 100 metres from the cone. The vehicle moved slowly across the stone-strewn, drab seafloor. In 1982, there were only black/white video cameras on ROVs (Figure 3.5). However, there was a still camera with colour film mounted on the side, which we could use if something interesting was found.

    Even with the black/white images on the video screens, the live images were stunning and unexpected. The flat, boring seafloor suddenly gave way to a steeply inclined structure consisting of large, white, bulbous forms, the living *Lophelia* colonies. The living wall stretched ahead and upwards, out of the restricted field of view. The British *Scorpio* operators nearly refused to believe their eyes. They had long experience under water, but had never seen anything like this. *Scorpio* manoeuvred up the steep side of the reef, where new bulbous colonies appeared,

18 A modern re-discovery   [Ch. 3

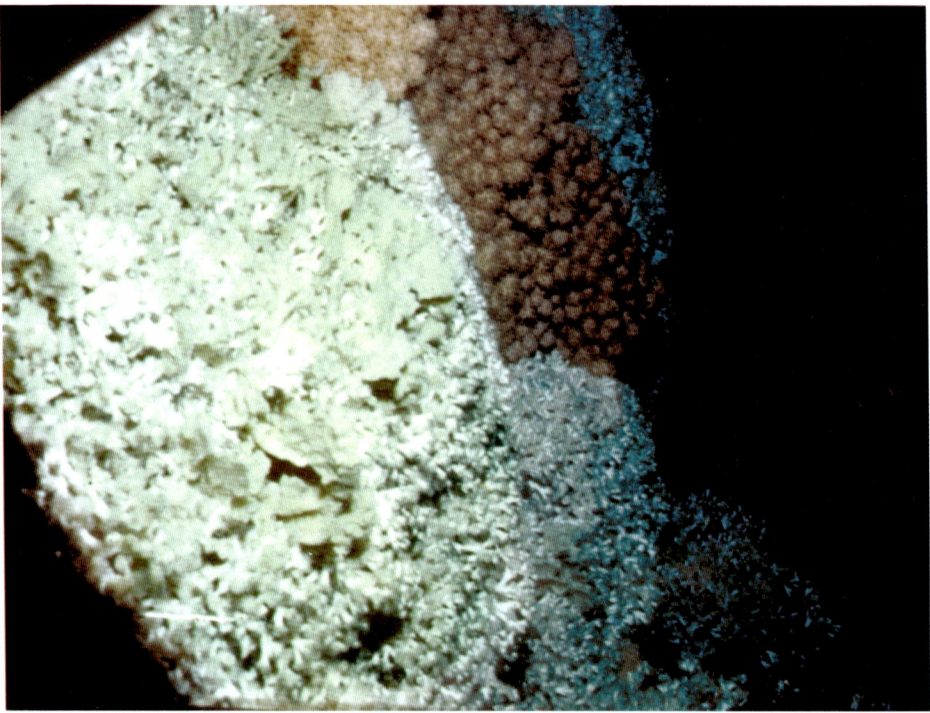

Figure 3.3. Probably the first-ever colour underwater photograph taken of a deep-water coral reef. It is from the Fugløy reef, taken by a *Scorpio* in July 1982.

one stacked on top of the other (Figure 3.3). Colour photographs were taken at the base of the reef, mid-reef, and also from the impressive summit (Figure 3.4), 15 m above the surrounding seafloor. From these images, it was clear that the reef consisted of many more than a single coral species. Furthermore, numerous fish were swimming in and out of the coral clumps, peering at the strange "animal" with bright "eyes". I telephoned professor Tore Vorren, at Tromsø University, who said that coral reefs were known in those areas, and that the name of the white and pink coral was *Lophelia pertusa* (Linné). Although the university had arranged many sampling cruises to these reefs, they had never photographed or filmed them. This, then, was the first time they had been documented by modern imagery (Figures 3.3 and 3.4).

3.4 POCKMARK CRATERS

After having studied the reef with *Scorpio*, the vehicle was manoeuvred down the slope towards the sedimentary basin at 320 metres depth. On the flats of this basin, we had noticed a series of large circular craters on the sidescan sonar records (Figure 3.6). These so-called "pockmarks", which are normally found on the continental

Sec. 3.4]
Pockmark craters 19

Figure 3.4. The impressive crown of the Fugløy reef, photographed in 1982. It came as a surprise that there were more than one large deep-water coral species living here. Besides the stoney coral *Lophelia pertusa*, we here also see variants of paragorgians (gum corals).

Figure 3.5. The *Scorpio* ROV used to photograph and document the Fugløy reef in 1982. The aluminium cylinder was specially made to sample the brittle *Lophelia pertusa* stony coral.

20 A modern re-discovery   [Ch. 3

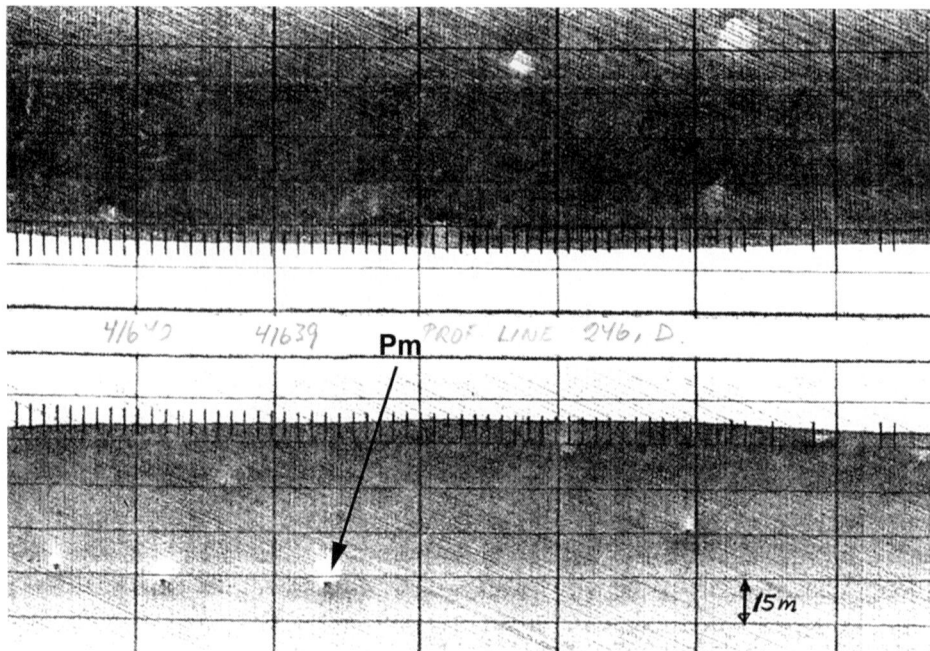

Figure 3.6. This is a sidescan sonar record from the sedimentary basin, located upstream relative to the Fugløy reef. There is a regular pattern of pockmarks here. The pockmarks are formed by focused fluid flow through the seafloor sediments. The empty space in the middle of the record is the water column to each side and beneath the towed sidescan sonar fish. Note that some of the pockmarks have strong acoustic reflections (probably from stones) in their centres.

shelf, are caused when water and gas seeps through the seafloor sediments from below (Figure 3.7). They are the surface manifestation of dewatering, as if groundwater on land came to the surface in puddles (King and MacLean, 1970; Judd and Hovland, 2007). Moving up to one of the craters, the ROV had to back up violently, to avoid falling into the bottom of the pockmark. When the water cleared, Scorpio was perched on the edge of a 2 m deep, perfectly circular 10 m wide pockmark. In the bottom were some gravel and a few sea-anemones, otherwise only fine sand and clay. But, the ROV pilots had noticed a distinct sonar target out to port. After a few minutes *Scorpio* arrived at another pockmark, with a large square rock in its centre. The erratic rock was overgrown with sea-anemones and half-metre long serpulid tube worms (Figure 3.8). Their delicate feathery fan-like tentacles were sticking out of the upper end of each tube. When working under water, and finding new features, time passes extremely fast. During the last three hours, we had not only investigated the coral reef, but also some pockmark craters. Although this work was most interesting, our primary mission was to document a potential pipeline route through the entrance of the Lyngen and Ulls fjords. *Scorpio* was therefore recovered and the towed sidescan sonar re-deployed to continue our survey (Hovland *et al.*, 1994b).

Sec. 3.4] Pockmark craters   21

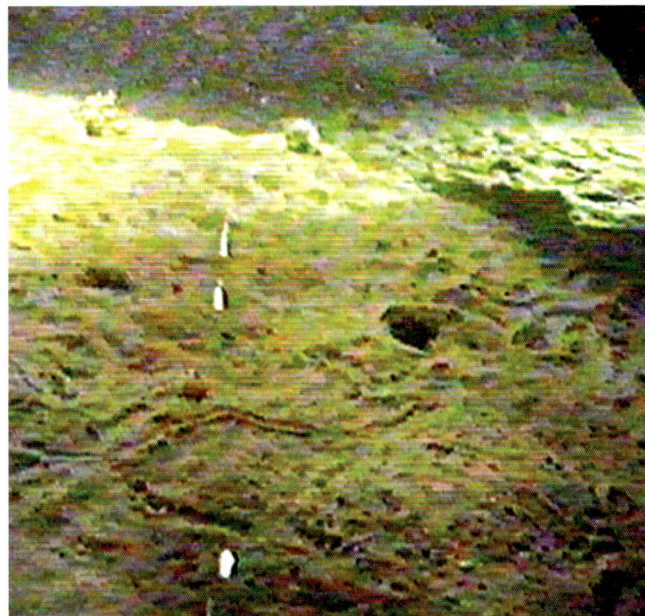

Figure 3.7. Methane gas bubbles are seen (mid-picture) emitting from the seafloor at Tommeliten, at 70 m water depth in the North Sea. The white objects on the left are plastic funnels used with success to collect gas samples into evacuated steel pressure cylinders above (see Judd and Hovland, 2007).

Figure 3.8. A large boulder (1.5 m high) located inside a pockmark near the Fugløy reef. The organisms are sea-anemones and serpulid tubeworms. It is rather surprising to find the serpulid worms in this area. However, if there is seepage of groundwater with dissolved gases, these organisms would be expected (see MacDonald et al., 1989).

22 A modern re-discovery

Although this occurrence of Norwegian coral reefs had been filmed and photographed with modern technology, a decade would pass before the real wave of re-discovery started. But in the meantime, we had several things to puzzle and ponder over ...

3.5 WHY NORTH OF THE POLAR CIRCLE?

Following this Fugløy reef discovery, there was a period of reflexion, and study of the old literature from Gunnerus (1768) to Naja Mikkelsen *et al.* (1982) on Norwegian coral reefs. During a drilling site survey over the Tommeliten field in the North Sea, a number of gas seeps had been noticed. Statoil funded a scientific study of these seeps, and in 1983, the *Scandi Ocean* was hired from Oceaneering Norway to do ROV, geological, and geophysical documentation of the Tommeliten seeps (Figure 3.7). Alan Judd was also a member of the research team, and he defended his PhD partly on the basis of this work. Small pockmarks, bacterial mats, and abundant other marine life, bioherms were found near the bubbling Tommeliten seeps (Hovland and Judd, 1988; Judd and Hovland, 2007; Wegener *et al.*, 2008).

During the mid- and late-1980s many major fields and pipelines were developed and constructed (Gullfaks, Statpipe, Sleipner, Zeepipe, Troll, etc.). Statoil also funded an R&D seep cruise to perform research work in 1985 with the vessel *Lador*. Using all the knowledge on seeps from these two main research cruises (1983 and 1985) the book *Seabed Pockmarks and Seepages: Impact on Biology, Geology and the Environment* (Hovland and Judd, 1988) materialized (Figure 3.9).

In 1985, Statoil unfortunately experienced a fatal blowout at the Midgard field on the mid-Norway shelf. This was called the "West Vanguard Blowout" (Hovland and Judd, 1988). A fast-working taskforce group was assembled to make sure that no

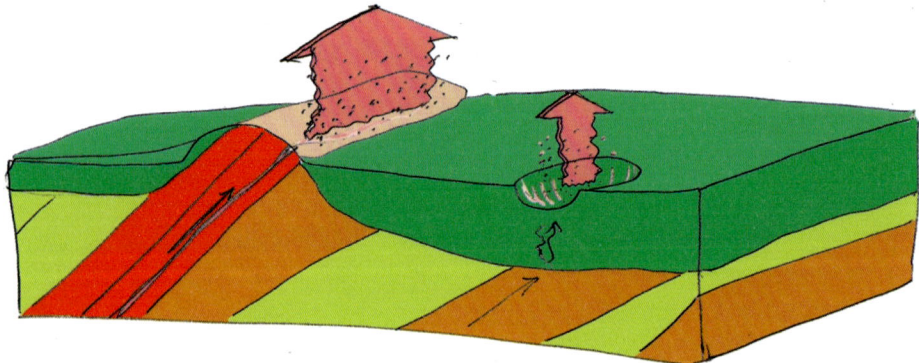

Figure 3.9. A sketch illustrating how I suggest light hydrocarbons (methane, ethane, and butane, the principal constituents of natural gas) dissolved in porewater seep up along dipping sedimentary rock layers through the seafloor in the Haltenbanken and other areas. The seep on the left occurs through the ridge (similar to the Sula Ridge and the escarpment at the Haltenpipe Reef Cluster, HRC). The seep on the right occurs through a layer of clay, causing a pockmark to form (from Hovland and Mortensen, 1999).

such disaster would happen again. This shallow-gas detection project started early in 1986 (Gallagher et al., 1991).

Completion of the "pockmark book" (1988) led to further reflexion over the mysterious aspects of deep-water coral reefs. Surely the reason they only occurred in certain areas of the Gulf of Mexico and North Atlantic was that they relied on local nutrient contributions. It could not be only by chance that pockmarks occur close by the Fugløy reef. The concept of a seepage link, not only to deep-water coral reefs—but also to other types of carbonate banks—was discussed in the *Terra Nova* article "Do carbonate reefs form due to fluid seepage?" (Hovland, 1990b). Later, one of the reviewers of this article came up and said: "Martin, because this idea is quite wild, I tried to find flaws and to vote against publication, but there were no flaws, neither in data nor reasoning . . .". Alan Judd also mentioned, on one occasion, "that is the best article you have ever written." Later, it partly led to the discovery of giant deep-water carbonate mounds with coral reefs in the Porcupine Basin, off Ireland (see Section 5.1.1).

Thus, there could be a striking similarity between hydrothermal vent systems and deep-water corals associated with pockmarks. Ever since they were first found, in 1977 (Ballard and Grassle, 1978), deep-ocean hydrothermal vents have fascinated researchers. Although it has been difficult to unravel how vent-associated life was sustained without sunlight, there was never any doubt as to where the energy fuelling the vent organisms came from: it must somehow originate from the vents themselves. Over the last 30 years, knowledge on an increasing number of species and lifeforms comes in the wake of that paramount discovery (Judd and Hovland, 2007).

3.6 SUSPECTED ORIGIN OF THE FUGLØY REEF

At the Fugløy reef location, there are two main factors suggesting the location of the reef is not determined only by a fast-flowing and generally nutrient-rich water current coming in from the southwest, the North Atlantic Current (the Gulfstream):

(1) It is located on the seaward side of a gravelly, morainic ridge,[1] with a slope angle of about 5°. This is in spite of ample possibilities to construct on adjacent crystalline bedrock at a similar or slightly shallower depth, where currents obviously tend to be stronger. Some theories say that reef animals prefer to build on a firm basement consisting of topographical highs with accelerated water currents (i.e., Freiwald et al., 1997; Wilson and Freiwald, 1998; Frederiksen et al., 1992). However, the Fugløy reef is located on sandy gravel, with clay, and not even on the summit of the ridge. This suggests there are other location determinants.

(2) The reef is located on a slope that faces a deeper basin at 320 m depth, with numerous pockmarks (Hovland and Mortensen, 1999). Because pockmark

---

[1] Morainic ridges are sediments that have been pushed up or deposited by moving ice sheets (glaciers and shelf ice). Morainic ridges in fjords are also called "thresholds", as they cross over the entire width of the fjord.

craters form as a consequence of focused fluid flow (gas and/or porewater, including groundwater, Judd and Hovland, 2007), this suggests the existence of overpressured underground fluids within the general area, which may also permeate and emit through the morainic ridge, where the reef is located. Another possibility is that the fluids come out through the pockmarks, which in turn signifies a locally enriched area of organic matter, partially fuelled by primary producers located in and near the pockmarks. The assumption, then, is that these nutrients would be transported by prevailing currents to the reef location in a higher-than-background concentration (as suggested by Lindberg *et al.*, 2007). In both cases there is a fluid flow (hydraulic) dependency (Figure 3.9).

Lindberg *et al.* (2007) performed a thorough modern mapping of the Fugløy reef area, using a multibeam echosounder on the vessel *Jan Mayen* (of Tromsø University) and explored the possibility that fluids seeping from the adjacent pockmark area could possibly be of beneficial influence to the coral reefs. Interestingly, they found many more large coral reefs perched above and to the south of the pockmarks, and what is more that the largest reefs are situated at the flanks of tidally active channels, where influx of water is continuous and currents are periodically high. They implied that there is a diurnal supply of water to the reef area from the pockmark basin. However, because there is no proof for any link, they concluded with the following remarks:

"It is tempting to link the Fugløy reef occurrences with seepage based on circumstantial evidences, but here, as for other published reefs, no *direct* evidence is found" (Lindberg *et al.*, 2007).

As with many other aspects of complex natural systems, I agree there is no direct evidence (*proof*), but, even so, their observations do not rule out that there may be a link of some type.

Pockmarks are indicators of fluid flow in whatever marine or lacustrine (freshwater) environment they occur, provided the sediments are suitable for their formation. Generally, they provide clear evidence of focused fluid flow (seepage), a common process in marine sediments. This has caused a developing scientific *paradigm shift*, which is slowly also gaining way into marine biological scientific thinking. Thus, before continuing the documentation of Norwegian coral reefs resident on the continental shelf and in fjords, two other very interesting fluid flow locations in deep waters of the northeast Atlantic will be reviewed. These are the Nyegga/Vøring Plateau and Håkon Mosby Mud Volcano (HMMV) locations, which are highly relevant in the context of enhanced organic life on the seafloor.

3.7 DEEP-WATER BIOHERMS AT NYEGGA

The Storegga slide is known as one of the world's largest submarine slides (Bryn *et al.*, 2003). Its name *Storegga* (named by fishermen) means a "large (long and high) scarp" or "edge" and refers to the abrupt termination of the continental shelf off mid-Norway. It represents the shoulder of the continental slope. From this shelf break, the water depth increases westwards to over 3,000 m in the Norwegian Sea basin. Storegga is the upper part of the slide scar and runs from south to north for a distance of about 500 km along the mid-Norway coast (about 100 km to 200 km offshore). At the northern termination of this long scarp, it turns east–west and becomes more distinct. This portion is named *Nyegga* (the "new scarp" or "edge"), and occurs at a water depth ranging from 600 m to about 800 m. To the north of the scarp, the continental slope is more or less undisturbed, at least by sliding. It also represents the southwest limit of the Vøring Plateau. In this area there are many fluid flow features. These have also been studied quite extensively by academic research (Bugge, 1983; Mienert *et al.*, 1998; Bouriak *et al.*, 2000; Gravdal *et al.*, 2003; Bünz *et al.*, 2005; Hovland and Svensen, 2006; Ivanov *et al.*, 2007). At Nyegga a new type of "complex" pockmark has been discovered. They contain up to 10 m high ridges of methane-derived authigenic carbonate (MDAC) rock. These pockmarks were investigated using ROVs first in 2003 and 2004 (Hovland *et al.*, 2005) (Figures 3.10 and 3.11).

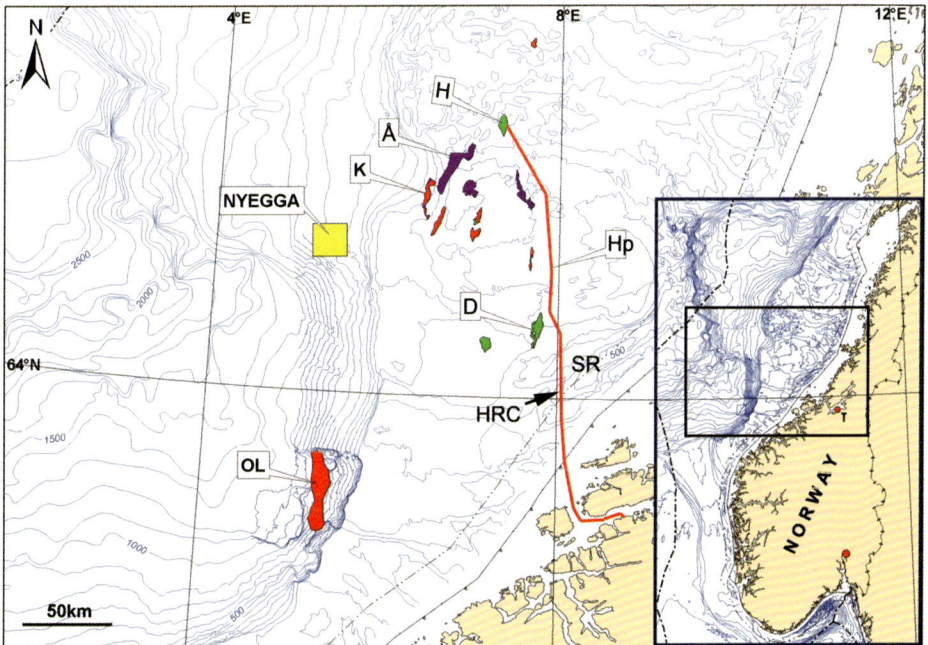

Figure 3.10. A bathymetric map, showing the locations of oil-fields (green), condensate-fields (purple), gas-fields (red), the Nyegga location (yellow square), and other locations mentioned in this book, off mid-Norway. OL = Ormen Lange, HRC = Haltenpipe Reef Cluster, SR = Sula Ridge, D = Draugen, Hp = Haltenpipe, K = Kristin, Å = Åsgard, and H = Heidrun. The Morvin field is not shown here, but it is located between the letters Å and H on the map. In the inset, T = Trondheim, shown as a red circle, Oslo is also shown as a red circle, farther south.

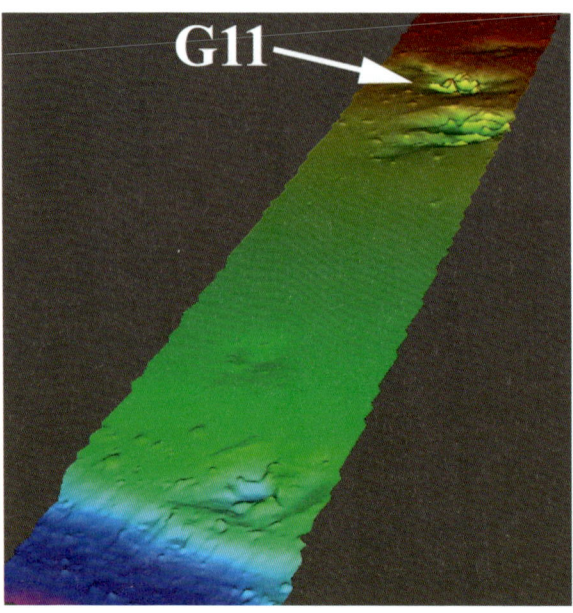

Figure 3.11. A perspective view of a 6 km long and 1 km wide section of the Nyegga seafloor area (see previous figure for location). The arrow points at complex pockmark G11, where authigenic carbonates, bacterial mats, gas hydrate pingoes, and a special animal community was found (Hovland *et al.*, 2005).

The complex pockmarks are located well within the gas hydrate pressure/temperature stability field, and are associated with a regional BSR (bottom simulating reflector). Furthermore, they occur above deeply rooted vertical pipe structures (Figure 3.12). They also have a distinct fauna with local bacterial mats, tubeworms (including small pogonophorans), stalked crinoids, and pycnogonids (sea spiders). The presence of light hydrocarbon gases (C_1–C_5) in clay-rich sediments, in which the methane $\delta^{13}C$ values range from $-54‰$ to $-69‰$ PDB (Pedee Belemnite standard),[2] suggests venting of both bacterial and thermogenic gases (Hovland *et al.*, 2005; Judd and Hovland, 2007). Only micro-seepage seems to occur presently (i.e., without ebullition or other visual fluid flow). Petrography and geochemistry of carbonate blocks from the pockmarks suggest precipitation of methane-derived aragonite within the sediments ($\delta^{13}C = -52‰$ to $-58‰$ PDB) (Mazzini *et al.*, 2006). The complex pockmarks probably formed by sudden (catastrophic) fluid flow, combined with long periods of subsequent micro-seepage, which is still active (Hovland and Svensen, 2006; Ivanov *et al.*, 2007).

Besides the outcropping large carbonate rocks, the fauna is very special. There are abundant up to 30 cm tall stalked crinoids and large (up to 1 m diameter) ophiurids (basket stars) on the carbonate rocks. These, together with hydrozoa, and numerous other unidentified sessile organisms also including some unspecified types of corals, are perched at specific locations on the carbonate rocks (see Figures 3.14 and 3.15). This biotope was only found inside the pockmarks, despite extensive visual surveys on the general seafloor of the region. The relatively high population

---

[2] The "Pedee Belemnite" is a marine carbonate rock used as a standard for carbon isotopes.

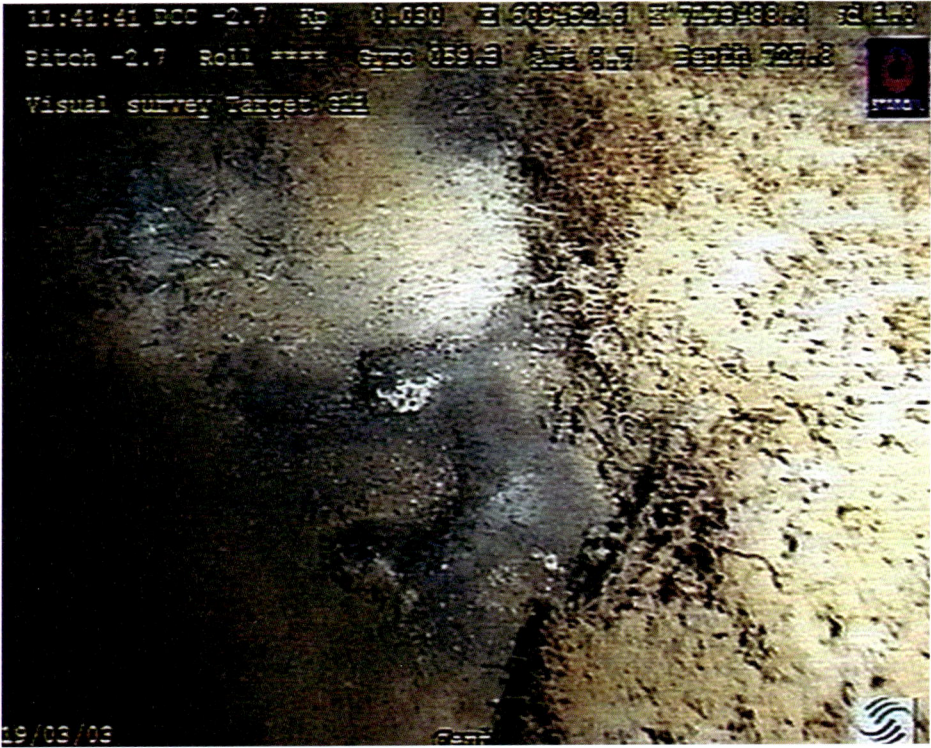

Figure 3.12. Gray and white bacterial mats on the sediment seafloor in the deepest, central part of complex pockmark G11 at Nyegga. A sediment core acquired at this very location demonstrates high concentrations of light hydrocarbons (Hovland and Svensen, 2006).

density and characteristic composition of this fauna is surprising when considering the sub-zero ($-0.7°C$) ambient seawater temperatures of the area. Specific microseepage areas are manifested by bacterial mats, but without any documented thermal effects.

In addition to the large carbonate ridges inside the complex pockmarks, some distinct organic-rich sediment mounds, hydrate pingoes, up to 1 m high and 4 m wide were found (Hovland and Svensen, 2006). A total of seven such mounds were visually inspected with ROVs. These have four characteristics in common:

(1) they have a positive topography (rounded mounds and cones);
(2) they are partly covered in bacterial mats (indicating ongoing fluid flow);
(3) they are partly covered in a carpet of small, living tubeworms (pogonophorans, which utilize methane); and
(4) they have distinct corrosion pits on their surfaces, indicating fluidisation and point source corrosion of the covering sediments (probably caused by localised sub-surface hydrate dissociation).

28 A modern re-discovery [Ch. 3

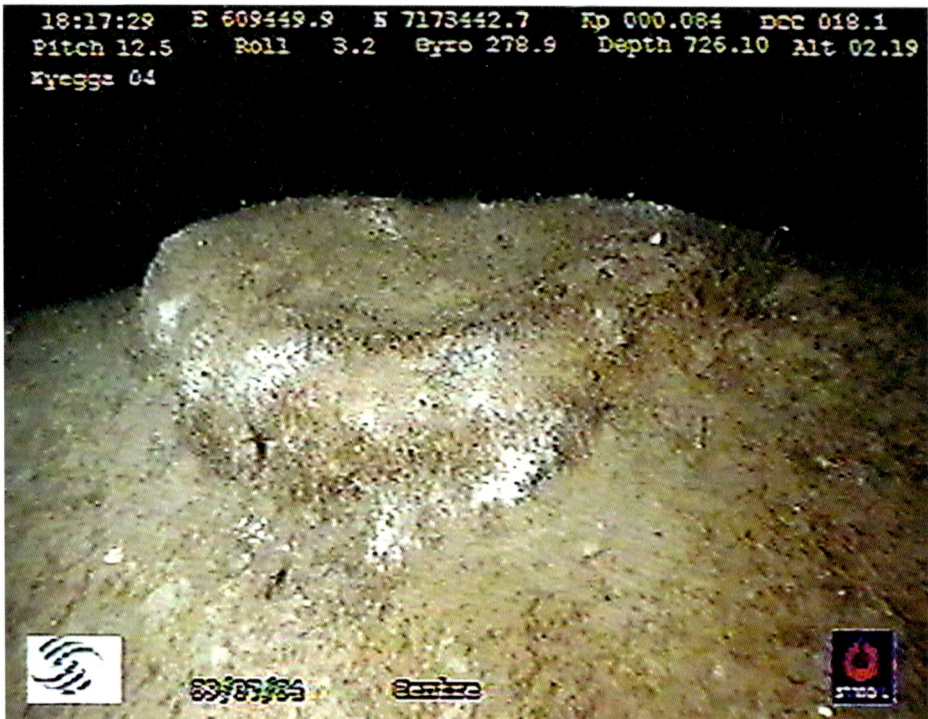

Figure 3.13. This is Pingo No. 1, located inside complex pockmark G11 at Nyegga. The characteristics of pingoes are (a) their elevated centre, indicating volume expansion due to hydrate ice formation below ground, (b) bacterial mats on their surface (white patches) indicating seepage of reduced fluids, (c) abundant small tubeworms (pogonophora), seen here resembling grass, (d) corrosion pits, where sediment has become fluidised by seepage, but not evident on this photograph.

The features were interpreted as true submarine pingoes, formed by the local accumulation of hydrate (ice) below the sediment surface (Hovland and Svensen, 2006). Recent work by Ivanov et al. (2007) proved that the mounds contain massive nodules or layers of gas hydrate occurring 1 m to 1.5 m below the surface:

"The discovery of hydrate in pockmarks of the Vøring Plateau has provided confirmation of their status as active methane seeps or vents. Whether the hydrate is sustained by methane in solution in pore water flowing up the chimneys and reaching saturation near the seabed or whether the hydrate is created during intermittent episodes of expulsion of free gas is a question still to be answered" (Ivanov et al., 2007).

Again, these particular fluid flow features demonstrate the very close relationship there is between primary producers (bacteria and archaea) residing within the surficial sediments and higher trophic organisms, seen either to reside within the

Sec. 3.7]	Deep-water bioherms at Nyegga 29

Figure 3.14. The white arrow indicates the location of a pingo (No. 2) beneath the carbonate rock overhang (see next figure). Note all the live organisms accumulating at this location. The two lights in the background are from another ROV. Water depth: 720 m, G11 complex pockmark. Water temperature = $-0.7°C$.

surficial sediments or perched somewhere near to the focused fluid flow location (Figure 3.14). Another interesting observation from this environment: despite there being an ample, firm sub-stratum and nutrients, for some reason it is out of bounds for *Lophelia* or *Madrepora*.

In 2006, researchers from IFREMER, France, also sampled the pingoes inside complex pockmarks at Nyegga. They found that the pingoes in G11 contained dense clumps of partly buried tubeworms, including pogonophorans (Ivanov *et al.*, 2007). Below is a general account of the association between such worms and seepage (based on Judd and Hovland, 2007).

3.7.1 Significance of the tubeworm

The tubeworm or pogonophore is one of the most fascinating organisms associated with seabed fluid flow; indeed, it is amongst the strangest of our planet's wildlife. They represent a diverse and geographically widespread phylum and occur at many cold seep sites, at depths ranging from 300 m to the deepest cold seeps (Peru Trench,

30 A modern re-discovery [Ch. 3

Figure 3.15. Arrow points at a pingo located between two large carbonate rocks (see previous image). Note the sea-anemones and other organisms hanging from the roof of the cavity above the pingo. Some of these organisms resemble and may be corals.

Monterey Fan, Laurentian Fan, and Aleutian Trench), as well as at hydrothermal vents.

Pogonophorans are divided into three sub-phyla (Sibuet and Olu, 1998):

- the *Perviata* live in reducing sediments;
- the *Monilifera* live in decaying wood, in reducing sediments (shallow water), and near cold seeps (deep water);
- the *Obturata* (or *Vestimentifera*) live near hydrothermal vents and cold seeps.

Some early work on marine chemosynthetic symbiosis was done on pogonophores from Scandinavia. Southward *et al.* (1981) analysed several species (including *Siboglinum ekmani* and *S. fjordicum*) from fjords near Bergen (Norway) and (*Siboglinum poseidoni*) sampled from the Skagerrak. Flügel and Langhof (1983) worked on specimens dredged from a depth of 294 m in the southern Skagerrak. These two teams found the animals living in normal marine sediments in symbiosis with large numbers of endosymbiotic bacteria, which were found to be strikingly similar to free-living bacteria that utilize methane (Völker *et al.*, 1977; Southward *et al.*, 1981). Stable carbon isotope ($\delta^{13}C$) values ranged from $-45.9‰$ to $-35.30‰$, at that time the lowest values reported for contemporary marine organic material (Southward et al., 1981).

After new sampling of *Siboglinum poseidoni* from the central Skagerrak, Schmaljohann and Flügel (1987) and Schmaljohann et al. (1990) performed various culture experiments on the pogonophorans and on their endosymbionts. They managed to keep the animals alive for up to 3 months in seawater under an atmosphere of air:methane (4:1), confirming that *S. poseidoni* takes up and oxidizes methane. The investigators had demonstrated the need for a flux of methane at the sites their samples came from. To my knowledge, there is yet no report of pogonophorans being found at *Lophelia* reefs.

3.8 HÅKON MOSBY MUD VOLCANO (HMMV)

Since its discovery in 1989–1990 the HMMV (Vogt et al., 1991) in the northeastern Atlantic has attracted considerable attention from American, European, and Russian researchers. It is the first true *chemosynthetic oasis* discovered in Nordic waters, more than 800 km from the Nyegga area, the nearest known cold seeps (Judd and Hovland, 2007). Currently, it is one of the most intensively studied marine mud volcanoes in the world. A special issue of *Geo-Marine Letters* was devoted to some of this work (Vogt et al., 1999). HMMV is located at 1,270 m water depth (Lat Lon), on the Bear Island Fan Slide complex (Laberg and Vorren, 1993). It is about 1 km in diameter, but has a height (above normal seabed) of only 5 m–10 m, and was discovered by chance as a slight feature on sidescan sonar (Figure 3.16).

Besides its production of fluids (methane and water) and solids (liquefied muds), it is characterized by several physical and biological aspects:

(1) very high temperature gradients;
(2) methane hydrate at 2 m sub-bottom;
(3) chemosynthetic tubeworms and widespread bacterial mats.

In 1996 the temperature in the centre of the mud volcano was found to be $15.8°C$, the temperature gradient an astounding $10°C\,m^{-1}$, and the heat flow $7{,}500\,mK\,m^{-1}$; this is one of the highest heat flows measured in the ocean, away from plate boundaries. Egorov et al. (1999) modelled the stability of HMMV gas hydrates, providing valuable clues about the migration of fluids to the seabed, gas hydrate production, and a distinct methane water column plume. Hydrates occur as nodules and thin layers at a distance of 200 m–830 m from the centre of the volcano. A methane concentration of $100\,\mu L \cdot L^{-1}$ was found in the water column 60 m above the mud volcano, whereas the residual gas content in sediments recovered from the central zone was as high as $100\,\mu L \cdot L^{-1}$–$200\,\mu L \cdot L^{-1}$ (millilitre per litre) of wet sediment. The extensive bacterial mats located inside the main crater are 0.1 cm–0.5 cm thick and are dominated by large bacteria whose filaments are up to 100 mm long and 2 mm to 8 mm thick.

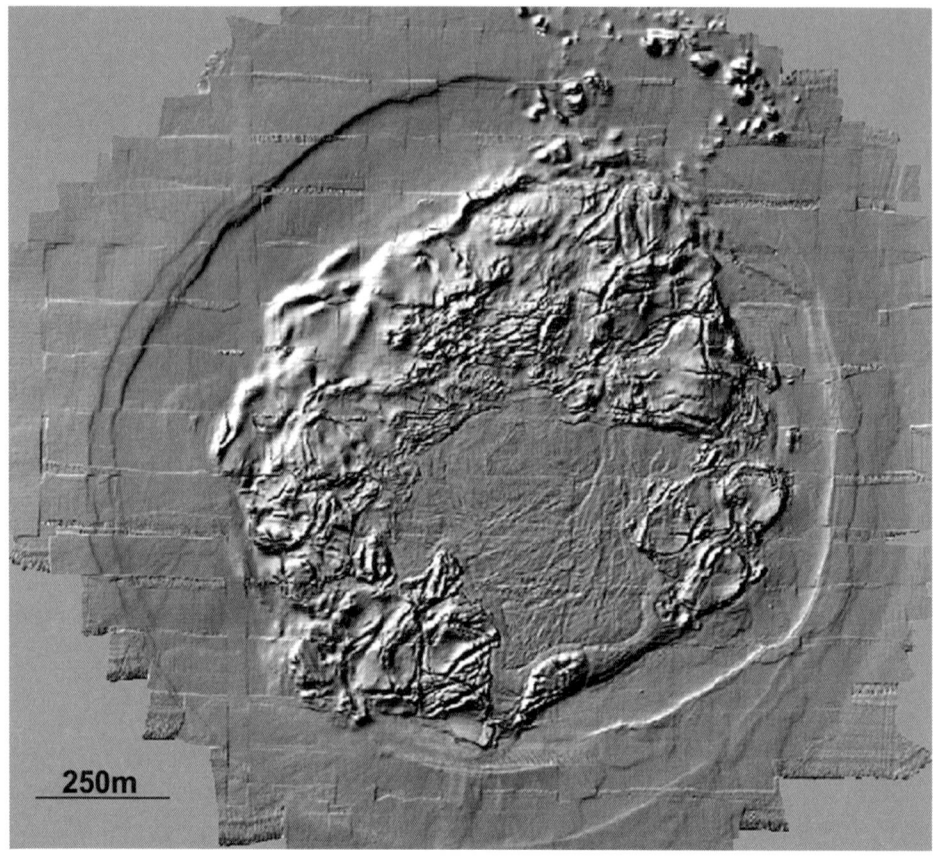

Figure 3.16. A shaded relief map showing the detailed topography of HMMV (modified from Jerosch *et al.*, 2007). The diameter across the entire mud volcano (outer circular ring) is about 1 km.

The fauna associated with HMMV includes tubeworms (at least two species, *Sclerolinum* sp. and *Oligobrachia* sp.) which form gigantic colonies spanning tens of metres. Pimenov *et al.* (2000) reported that *Sclerolinum*, which is the more abundant, lives in the oxidized sediments of the caldera, whilst bushes of *Oligobrachia* live in more reduced sediments. The fish population is also quite remarkable. Eelpout (*Zoarces viviparus*) occurs at a density of one per m^2, more than two orders of magnitude greater than normal. However, the most remarkable aspect of the varied Håkon Mosby fauna is that it occurs despite the ambient water temperature being consistently around 0°C! The fact that no corals have been found on or near HMMV, even though there is a firm sub-stratum (carbonate rock), provides a hint on their limiting factors, here believed to be low temperature.

Large portions of the central region of the HMMV are covered in bacterial mats. Seeps are often characterised by bacterial (*Beggiatoa*) mats, described as white or orange in colour. The significance of finding bacterial mats on the seafloor is discussed below. However, a recent publication shows a coverage map over the HMMV *Beggiatoa* coverage. This detailed map, reproduced in Figure 3.16 after Jerosch *et al.* (2007) shows how a seepage area is colonised, and it clearly demonstrates the very complex nature of seepage rate and coverage. This map is also a very good demonstration of how complex sub-surface seepage hydraulic processes are.

The bacterial mats and pogonophorans on the HMMV were interpreted as indicators of methane oxidation, or mud flowage. It was shown that only a fraction of the methane that migrates from below is exported to the above seawater column:

"This is mainly due to chemoautotrophic communities oxidizing methane and providing a 'microbial filter' reducing CH_4 fluxes" (Jerosch *et al.*, 2007).

They also found that the region with the highest methane discharge (16% of the area mapped) proved to be devoid of any benthic communities:

"*Beggiatoa* and pogonophorans use primarily hydrogen sulphide resulted from the process of the AOM (Anoxic Oxidation of Methane). If the zone of the AOM is near to the sediment surface, the rate of the sulphide synthesis and the sulphide gradients increases. The highly different vertical sulphide gradients below different communities strongly indicate the important role of sulphide availability in structuring community composition. Thus, the *Beggiatoa* density points on high AOM rate while pogonophorans indicate AOM in smaller extent; the tube worms are able to acquire sulphide from deeper sediment layers than the *Beggiatoa*" (Jerosch *et al.*, 2007).

In other words, they have found that the *Beggiatoa* grows best where there is a relatively high methane flux (in accordance with our own results from Tommeliten), whereas the pogonophorans grow at locations where methane occurs deeper down (probably in areas with gas hydrate pingoes, in accordance with our own results from Nyegga).

Only a total area of 5% of the mapped area is densely inhabited by *Beggiatoa*. The pogonophorans grow on "the hummocky outer part of the mud volcano" and cover 37% of the mapped area. Just as at Nyegga, I believe the hummocks may represent pingoes (i.e., locations where methane hydrate exists underground, Hovland and Svensen, 2006; Ivanov *et al.*, 2007).

3.8.1 Bacterial mats

Bacterial mats of sulphide-oxidising bacteria (thiotrophs) are perhaps the most obvious visual evidence of seepage, as they seem to occur wherever there are focused and prolific methane seeps. They have been reported from just about every cold seep site around the world, from inter-tidal sites to the deep oceans. They are also found at other places with sulphide-rich sediments at the seabed; for example, at the Florida Escarpment groundwater seeps around sulphide-rich hydrothermal vents, and in

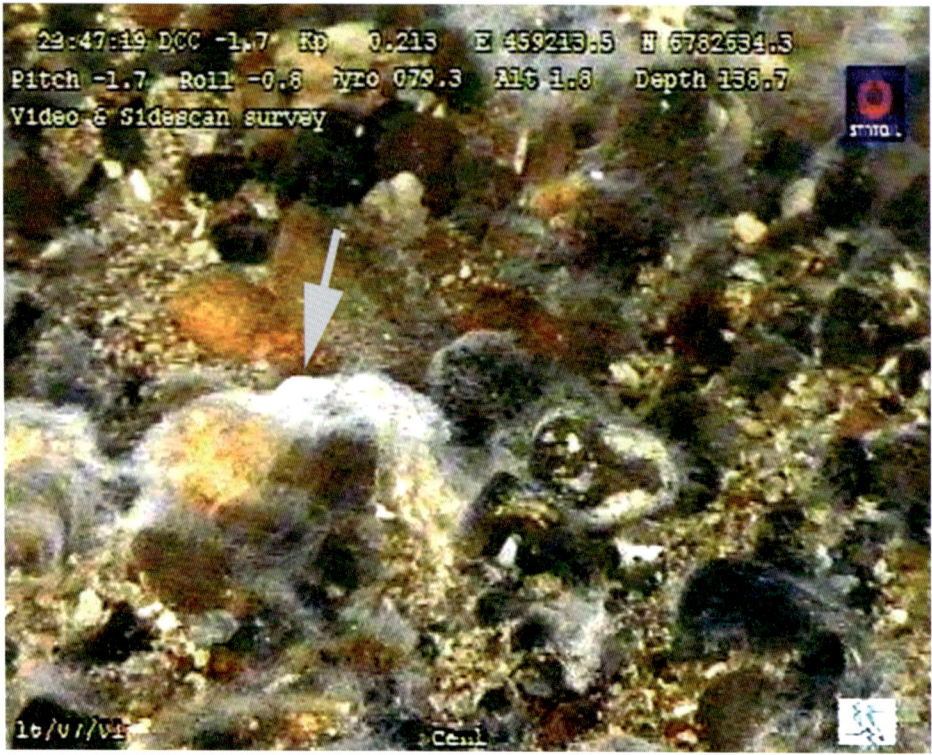

Figure 3.17. Close-up of bacterial mat at Gullfaks. Arrow points at where methane bubbles continuously emit from below (Hovland, 2007).

enclosed anoxic basins (Judd and Hovland, 2007). They are described as filamentous, white to orange fluffy mats lying on the seabed at the interface between anoxic, sulphidic sediments, and aerated seawater (Figures 3.17 and 3.18). Generally, the mats are less than 1 cm thick, but some may be as thick as 10 cm. On muddy seabeds they may extend *rootlike structures* several centimetres into the sediment (Roberts and Carney, 1997).

The most commonly reported bacteria are of the genus *Beggiatoa* (which is actually one of the largest known bacteria, in size), as described above for HMMV, but other sulphide oxidisers (e.g., *Thiothrix* and *Thioploca* spp.) are also known. These microbes oxidise sulphide using a variety of oxidants, usually oxygen or nitrate stored in vacuoles inside the microorganism (storing helps eliminate the problems of molecular diffusion in these huge bacteria). The process is summarised by the following equations:

$$HS^- + 2O_2 \rightarrow SO_4^{2-} + H^+$$

or

$$O_2 + 4H_2S + CO_2 \rightarrow SO_4^{2-} + H^+$$

Figure 3.18. Bacterial mats disturbed by an ROV manipulator, Gullfaks (Hovland, 2007). Note the hermit crabs; they were seen fighting over a piece of bacterial mat. They clearly eat bacterial mats and are, therefore, acting as bacteriovores!

Some of the most substantial mats of *Beggiatoa* have been found on sediments and amongst tubeworms at hydrocarbon fluid flow locations in the Guaymas Basin. Here, *Beggiatoa* filaments are up to 120 µm wide and 10 mm long, and the mats can be *thick enough to ladle* (Van Dover, 2000). Although *Beggiatoa* is characteristic of anoxic sediments, they are unlikely to occur in stagnant water as they need steep, opposed gradients of sulphide and oxygen (Møller et al., 1985).

One of the remarkable aspects of bacterial mats is their ability to modify the local ambient environment so that extreme chemical micro-environments occur beneath them. When they form, they represent the first step in what is called the "self-sealing of seeps" (Hovland, 2002). However, they do not only form in the ocean. They can also form in lakes where seepage or venting occurs. Thus, they occur at brine seeps at the bottom of the 594 m deep Crater Lake, Oregon. It is located in a caldera of Mt Mazama in the Oregon Cascades, which last erupted 6,850 years ago (Bacon and Lanphere, 1990). During nine dives with the submersible *Deep Rover*, bacterial mats up to 3 m in diameter and 2 cm–15 cm thick were observed. Measurements made beneath mats with a temperature probe found temperatures up to 6°C higher than the typical bottom water, and the concentrations of the major ions in water sampled from within and immediately below the mats were approximately twice those of any water sample collected within the caldera; Mn concentrations were about 1,000 times higher than in deep lake water (Dymond et al., 1989).

At oceanic hot vents and low-temperature seeps and vents, bacteria are ubiquous and often cause the shimmering venting warm fluids to appear milky. According to Van Dover (2000), elevated microbial biomass concentrations were first observed by Winn et al. in 1986. Naganuma et al. (1989) reported that plumes over the North Fiji vents contained a microbial community with both heterotrophic (80%) and chemoautotrophic (20%) members, and these had much shorter generation times (hours to days) than most deep-sea microbes (about 6 to 9 days). The microbial biomass can be four times that of the overlying surface waters, and suspended cell counts have been reported to be four orders of magnitude greater than background deep-sea microbial populations (Karl, 1995).

The largest deep-ocean blooms of bacteria, however, occur during submarine volcanic eruptions. Then, huge quantities of white, fluffy, flocculent material (floc) can be emitted in so-called "megaplume" events. Van Dover (2000) summarized the effects of an eruption of the East Pacific Rise in 1991. Even up to weeks after the eruption, microbial floc was blown from seabed cracks to an altitude of about 50 m. There was enough floc material to generate white-out conditions, obscuring visual navigation by the submersible pilots and requiring sonar guidance (Van Dover, 2000). Over an area several kilometres long and 50 m to 100 m wide there was a white floc blanket up to 5 cm thick (Haymon et al., 1993).

In general, bacterial mats do not need seepage or hydrothermal vents in order to grow, they can also form in other types of geological settings where sharp chemical gradients occur. However, the plumes of exotic fluids generated via seabed fluid flow processes seem to overshadow all other mechanisms, as mentioned by Kurt Konhauser (2007), University of Alberta, Canada:

"In O_2-depleted niches, chemolithoautotrophic sulfur-reducing and methane-generating bacteria respire using emitted H_2, and either S^0 or CO_2, respectively, while the organic substrates provided by the autotrophic communities serve as food for a variety of heterotrophs. Significant chemical energy to support biomass production is further available in the form of mineral substrates. This energy can be harnessed from the oxidation of seafloor hydrothermal sulfide deposits or from particles of elemental sulfur, metal sulfides, Fe^{2+}, and Mn^{2+} entrained within, or settling out of hydrothermal plumes. In fact, the latter study estimated that the primary productivity potential in the plume may represent a significant fraction of the organic matter present in the deep sea" (Konhauser, 2007).

Even though bacterial mats are good seep indicators, they do not seem to occur at many deep-water coral reefs. I have only found one small mat at the base of a coral, so far (Figure 4.36). It has just been reported that extensive bacterial mats were found upstream of the Floholman reefs in Vesterålen, northern Norway (Godø, 2007, see §4.4.2). The reason for the low amount of bacterial mats adjacent to coral reefs may be that the flux of migrating gas is too low. If this is so, then the gas is likely utilised by methanotrophs within the sediments and the gas is consumed by primary producers before it emits at the surface.

3.9 SUMMARY

The re-discovery of deep-water coral reefs was done with modern visual, *in situ* imaging and documentation, pioneered by John Wilson, off Ireland in the 1970s and by Statoil, in 1982, off Norway at Fugløy, north of the Polar Circle (Hovland *et al.*, 1994b). The fact that the Fugløy reef was found on the top of a morainic ridge, located adjacent to a deeper sedimentary basin containing numerous pockmark depressions, suggested there could be a link between the three features: reef, ridge, and pockmarks. The latter crater features were known to form by the focused expulsion of porewater (also groundwater) and gases. When the fluids emerge out on the seafloor and enter the water column, primary producers (bacteria and archaea) utilise the chemical energy and start off a local foodchain, which is independent of seasonal (Sun) variations. Similar gas (chemically) based ecosystems are described from the Nyegga complex pockmarks (750 m water depth), where gas hydrates occur in the sub-surface sediments. Another such example is also briefly reviewed, the Håkon Mosby Mud Volcano, where water, gas, and mud (clay) rises up within the sedimentary basin to the surface from 1,250 m water depth. The reason *Lophelia* and other common deep-water corals do not occur at these two latter locations is unknown, but perhaps it is the consistently cool water (around and less than 0°C) that is the cause (i.e., both locations lie outside the physical living envelope of *Lophelia*). This chapter has discussed and explained some of the reasons for originally suggesting why deep-water corals may partly rely on nutrients based on local seabed hydraulic conditions and explains why and how the hydraulic theory (see end of Chapter 8 for explanation) for deep-water coral reefs was first formulated.

4

Scandinavian coral reefs

The subseafloor biosphere constitutes the least explored part of our global environment, yet geobiological research reveals that extremely energy-poor environments deep beneath the ocean floor are home to most microorganisms on Earth.

Jørgensen (2007)

The camera pod had just hovered above some beautiful coral-oases, full of fish and other organisms. Only another few metres further along the sandy seafloor plain, and then, suddenly—the bacterial mats appear.

Godø (2007)

4.1 INTRODUCTION

To understand the true nature and variability of deep-water coral reefs, it is necessary to review as much detailed information as possible. In this chapter I have tried to review as many first-hand observations as possible and to discuss their relevance to understanding the whereabouts of reefs in Scandinavian waters. The chapter starts with a description of fjord-dwelling reefs and continues with reefs on the continental shelf, especially off mid-Norway. Of particular significance are the reefs associated with hydrocarbon fields, like the Kristin and Morvin fields, and corals found inside pockmarks, even in the North Sea, near the Troll gas-field. The chapter ends with a short review of coral colonies found on human-made structures in the North Sea (Figure 4.0).

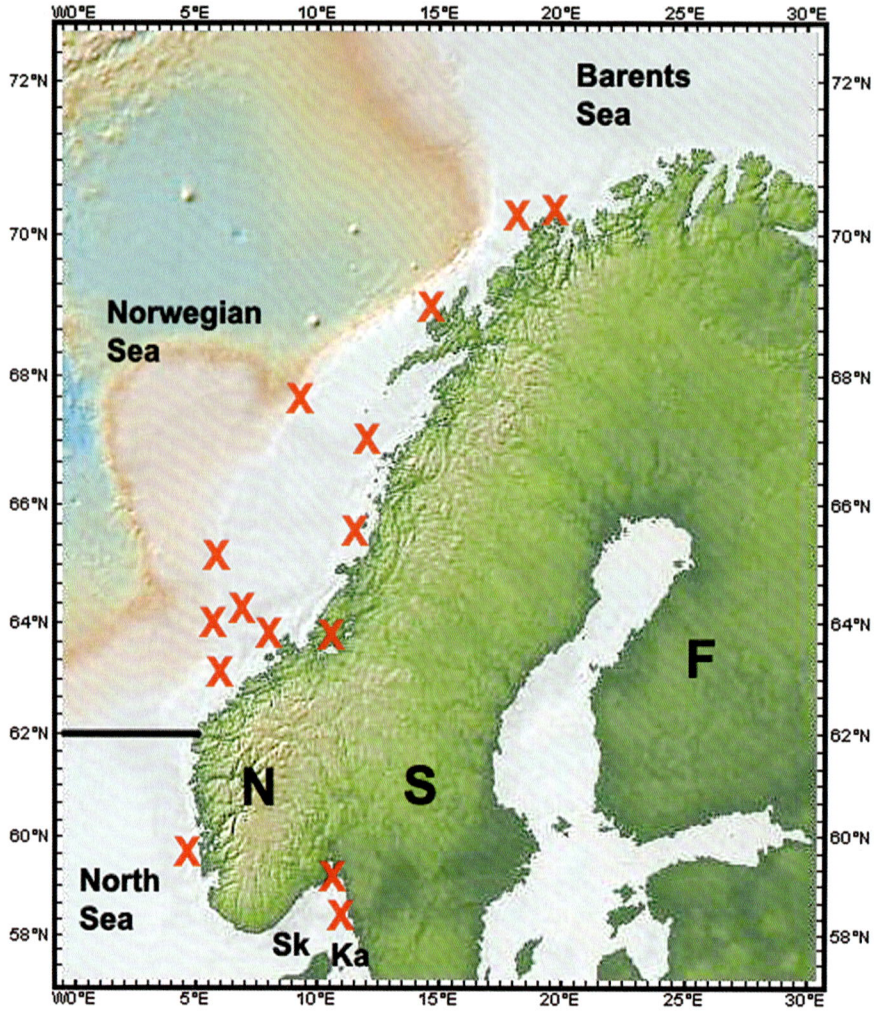

Figure 4.0. A general map of parts of Scandinavia where deep-water (and cold-water) coral reefs are found (red X's show main locations). N = Norway, S = Sweden, F = Finland, Sk = Skagerrak, Ka = Kattegat. The black horizontal line (the 62nd parallel) indicates the boundary between the North Sea (south) and the Norwegian Sea (north). See text for further details.

4.2 REEFS IN NORWEGIAN FJORDS

4.2.1 Agdenes and Tautra reefs

During a reconnaissance pipeline route mapping using a multibeam echosounder between Tjeldbergodden and Skogn in the Trondheimsfjord, the previously known Agdenes and Tautra reef complexes were also mapped (Dons, 1944; Hovland and

Figure 4.1. This collage of images has been assembled from many different Norwegian coral reefs. The artificial image was made in order to provide a wide-angle view of the biodensity and biodiversity that can be experienced on some Norwegian deep-water coral reefs. It illustrates the abundance of species and prolific life that exists on parts of these reefs. The ROV on the left (*Hirov 6*) is about 1.5 m high.

Mortensen, 1999; Mortensen and Fosså, 2001). Both structures are located on top of morainic threshold deposits, consisting of gravel, sand, and clay (Figure 4.2). These very distinct "terminal ridges", which cross over glaciated fjords at specific locations, were built at locations where the fjord glacier had a temporary termination. These deposits often resemble small deltas, with dipping sandy layers at the "delta front", facing away from the glacier towards the ocean side of the fjord. Towards the inland, where the glacier still occupies the fjord, the fjord bottom is much deeper than oceanward.

At Agdenes the ridge is about 150 m deep, but the fjord basin to the east (inland) of the ridge is up to 500 m deep. Below the Agdenes reef complex there is clear evidence of sub-surface dipping layers which come to the surface immediately below some of the reefs. Some of these layers are acoustically enhanced, signifying that they may contain minute gas bubbles and transport fluids (gas and groundwater).

Figure 4.2. An artist's impression of the *Seaway Commander* located above the prominent *Lophelia* reefs on the Agdenes morainic ridge at the mouth of the Trondheimsfjord. These reefs were mapped during a route survey between Tjeldbergodden and Skogn in 2000. The width of the "digital terrain model" of parts of the Agdenes reef and ridge shown here is about 100 m.

The Tautra reef is the world's shallowest deep-water coral reef, at only 39 m water depth. Because it has become a target for scuba diving and other activity, it has now become a protected area, by Norwegian law, as of June 2000 (location, Figure 4.0). Below the Tautra reef complex no dipping layers were detected. The morainic threshold on which the Tautra reef occurs, was constructed during the Younger Dryas period, about 12,000 YBP.

4.2.2 Stjernsund and other Norwegian fjord reefs

Farther north, in the Stjernsund area, Freiwald *et al.* (1997) have studied reefs located on similar morainic thresholds, but so far there has been no investigation of possible dipping sub-surface layers or enhanced acoustic reflections. Reefs are also found in some fjords near Bergen, southwest Norway. At one of these locations (Osterfjord), there are reports of smelly sediments (H_2S) when sediment samples are taken near their base. Although anaerobic sediments are not only associated with porewater seepage, it may signify the existence of anaerobic conditions caused by seepage of porewater (Hovland and Judd, 1988; Boetius *et al.*, 2000).

Because fjord threshold moraines typically cause local current acceleration, it is suspected that the reefs preferentially build up there because of higher-than-background nutrient concentrations associated with increased nutrient flux. But, even so, a groundwater seep relationship cannot be ruled out. There are three reasons to suggest that seepage may be a factor which determines their location in fjords.

Figure 4.3. Photograph of resting *Sebastes* sp. (about 15 cm long) on the Malangen reef (from www.mareano.no, 2007, Lene Buhl-Mortensen, IMR, Bergen).

These are

(1) The reefs are based on sediments, rather than on adjacent, protruding crystalline bedrock. By visual inspection of the Tautra reef complex, it was determined that non-reef-building species, like the octocorals (*Arborea* sp.) were located on bedrock, whereas *Lophelia* colonies were always based on sandy and gravelly sediments (unpublished data from visual inspection, 2000).
(2) Dipping and surfacing, enhanced acoustic layers suggesting migration of gases (methane, carbon dioxide, and/or hydrogen sulphide) and porewater towards the base of the Agdenes reefs.
(3) The reported H_2S smell of surface sediments at the base of a fjord reef near Bergen.

4.3 REEFS ON THE CONTINENTAL SHELF

4.3.1 The Haltenpipe reefs

The Heidrun oil and gas-field located about 200 km off the coast of mid-Norway was one of the first fields to be found on the continental shelf in the Norwegian Sea, north of the 62nd parallel. In order to produce and export the gas from the Heidrun field, a pipeline route had to be found from this field to shore. However, the rugged coastline along mid-Norway proved to have less shore approach opportunities than farther south, in the North Sea, where several pipelines had been constructed by 1988. Abundant reconnaissance mapping was therefore performed from 1985 onwards. Thus, large mapping campaigns were run from the Heidrun field in eastern, southeastern, and southern directions. This was the reconnaissance mapping for the "Haltenpipe" (i.e., the gas transport pipe from Heidrun to shore). Although many new features were noticed on the seafloor, some of which resembled the Fugløy reef, it was at first not appropriate to interpret them as coral reefs (Figures 4.4–4.7). The reason: they were larger and occurred in great abundance. We could not imagine such features being coral reefs of the Fugløy type. They were therefore dubbed "haystacks" until further documentation. If they were coral reefs, the consequences would be enormous.

There also seemed to be a correlation between abundance of haystacks and suspected seep features, such as shallow "bright-spots" (strong underground acoustic reflections typical of shallow gas), suspected diapiric mud, and pockmarks (Hovland, 1990a), see Figure 4.4.

In 1990 the long-awaited opportunity to dive with an ROV on a haystack came, as a significant obstacle occurred in the selected main route for the Haltenpipe. This obstacle was about 200 m long, 50 m wide, and 15 m high. The *North Sea Surveyor* was used for the inspection, and the obstacle *indeed* turned out to be an impressive deep-water coral reef.

Because of the isolated occurrence of such reefs it was no problem to re-route the Haltenpipe round the reef. The following summer, a new mapping campaign was run

Sec. 4.3] Reefs on the continental shelf 45

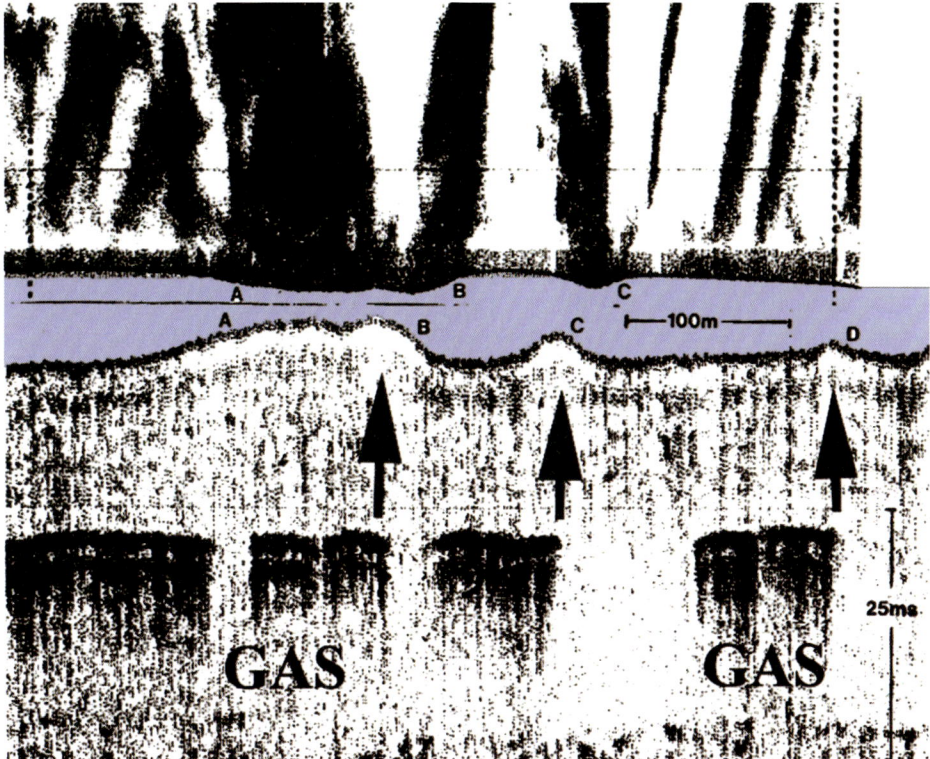

Figure 4.4. Bright-spots (black, acoustically highly reflective horizontal bands in the lower image) caused by gas accumulations located about 10 m below the seafloor (Hovland, 1990a). The upward-pointing arrows show how the gas drains upwards to the seafloor surface and causes a doming of the seafloor (domes marked with A, B, C, and D). The upper image shows the effect on the sidescan sonar record of the seafloor surface. Where gas drains through the seafloor and into the water column, it becomes acoustically highly reflective (from Hovland, 1990a, 1992).

with a newly developed Simrad multibeam hull-mounted echosounder. A strip 3 km in width was mapped with this system for the whole length of the route: about 200 km. This mapping included the westernmost portion of the so-called "Sula reef". On the basis of the splendid maps produced by these first ever multibeam echosounder systems, we managed to find a new and improved route. In particular, the route managed to bypass a dense cluster of reefs on the Sula ridge, later to be named "The Haltenpipe Reef Cluster" (HRC) consisting of nine live reefs (Hovland *et al.*, 1998).

In 1993 a baseline survey of the HRC was performed by Statoil. The objective was to improve our very meagre understanding of these biohermal occurrences, which could potentially be affected by our pipeline construction activities. Numerous high-definition colour still photographs (with a pressure-housed ROV-mounted Hasselblad camera) were taken and a few markers (four) were placed out on the seabed for reference. Sediment samples were taken close to the coral reefs and also at

Figure 4.5. This is how the haystacks appeared on sidescan sonar (thin arrows). It was not before 1990 that we visually ground-truthed (checked using ROVs) what they were. On this record, white represents acoustic non-reflectivity (acoustic shadow), and black represents strong reflectivity. Scale: the length of the reef at the far left is 80 m.

control sites for geochemical analysis. Over 200 very good underwater colour photographs were released to relevant researchers (Hovland and Thomsen, 1997). Thus, in September 1993, a group of marine biologists were invited to join a seminar on Statoil's findings for the Haltenpipe. A handful of marine biologists from Trondheim, Bergen, Stavanger, and Copenhagen came to this seminar, and were impressed by seeing, for the first time, live video footage and high-resolution photographs from the Norwegian deep-water coral reefs.

4.3.2 HRC

The Haltenpipe Reef Cluster (63°55′N, 07°53′E) is located about 75 km north of the town Kristiansund, on the western coast of Norway (Figures 4.8–4.12). It lies about 1 km west of the Haltenpipe gas pipeline, which was constructed in 1996. HRC consists of one dead and nine live individual reefs, each up to 25 m high and 200 m in length (Hovland et al., 1998; Hovland and Mortensen, 1999; Hovland and Risk, 2003). The cluster is confined to a relatively small (500 m by 500 m) seabed area. The age structure of the HRC was investigated by sediment coring and radiocarbon dating. The deepest and oldest *Lophelia* remnants were found in a sediment layer about 2.4 m below the seabed adjacent to the living reefs. The oldest skeleton remnant is 8,150 calendar years old; several others were grouped around 6,000–6,600 YBP and 2,000–3,000 YBP (Hovland and Mortensen, 1999). The HRC grows on a gentle ridge on the seabed that delineates the transition from sub-cropping sedimentary rocks of Palaeocene[1] age, in the west, to rocks of Cretaceous[2] age in the east.

---

[1] The Palaeocene was the first epoch after the Tertiary period of geological time. It started following the dramatic Cretaceous extinction (i.e., indicated geologically by the K/T boundary) in which the dinosaurs became extinct. The Palaeocene lasted from 65 million YBP until 56.5 million YBP.

[2] The Cretaceous geological period lasted from 144 YBP until 65 million YBP.

Sec. 4.3] **Reefs on the continental shelf** 47

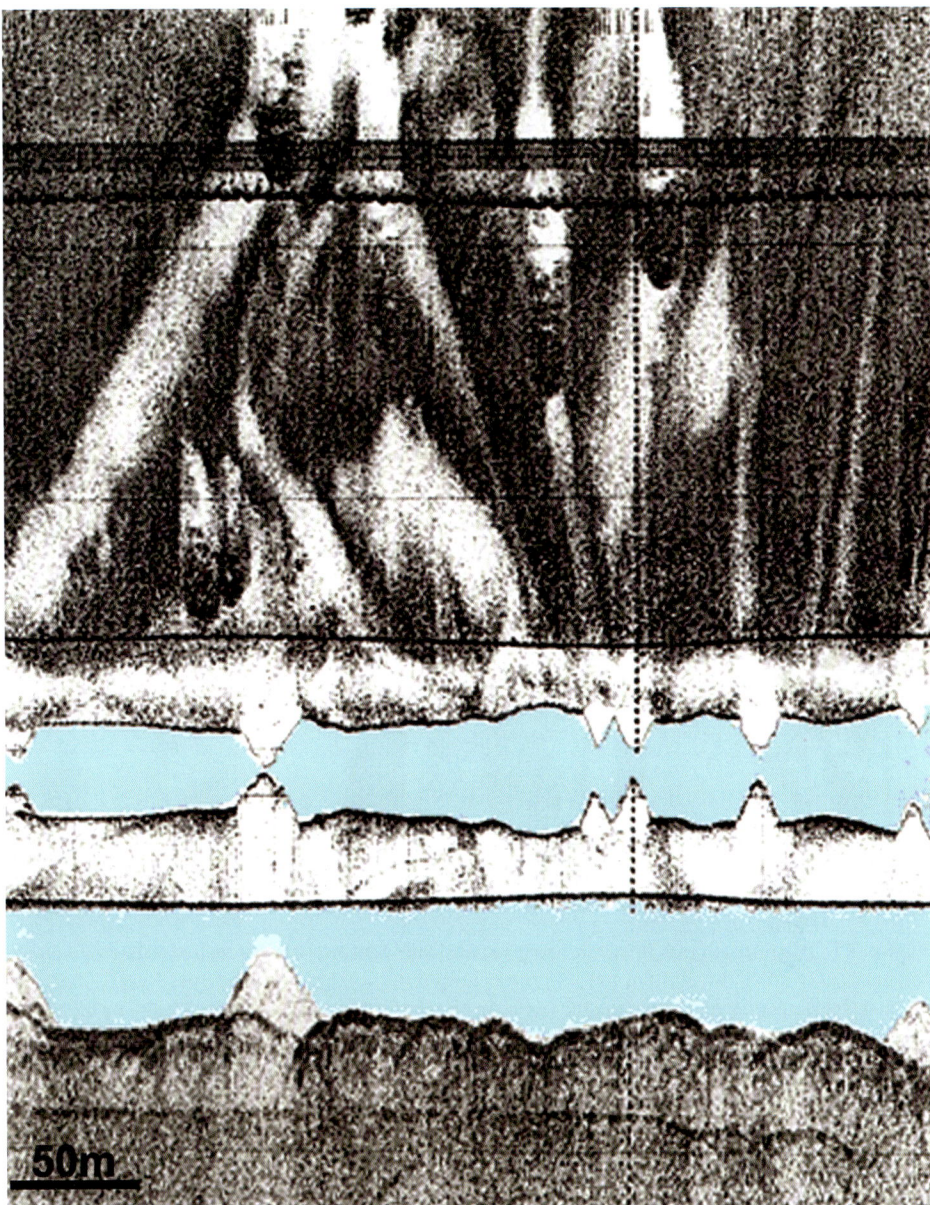

Figure 4.6. More haystacks (proved to be deep-water coral reefs) off mid-Norway, here obviously in high density. Sidescan sonar, upper image, and simultaneous deep-towed boomer, lower image (recorded during reconnaissance surveys off mid-Norway between 1985 and 1987). The reefs range between 80 m and 100 m across and 10 m–15 m in height. One of the sidescan sonar channels (starboard) is shown in full-range recording (i.e., 500 m width, upper portion), whereas the other one (port, lower portion) only shows the near-field record (50 m). A total of 12 reefs can be counted on the full-range field of view (upper portion).

48 Scandinavian coral reefs [Ch. 4

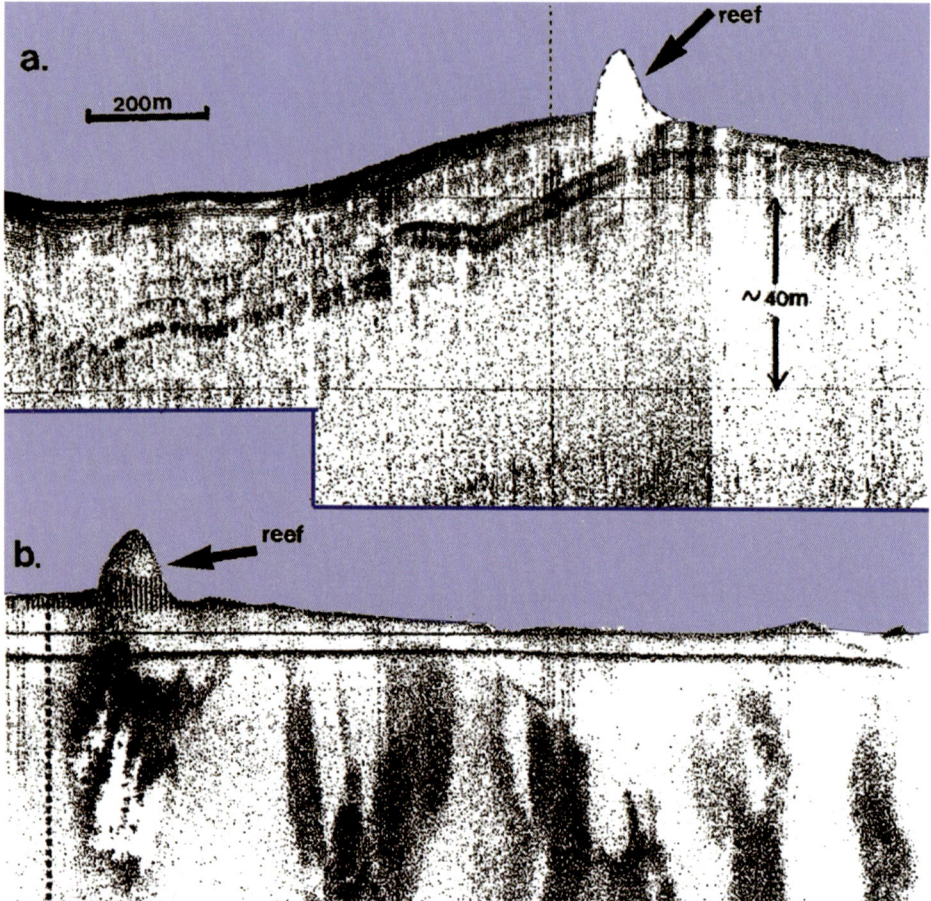

Figure 4.7. A simultaneous sidescan sonar (a) and sub-bottom profiler record (b) of a *Lophelia* reef mapped during the Haltenpipe-mapping campaign (northwest of the Sula Reef). There are at least five *Lophelia* reefs in a cluster (seen on the sidescan sonar record). The 12 m high reef, which is directly intercepted by the sub-bottom profile is seen to be located where a steeply dipping band of reflectors comes near the surface. There are no reefs at the iceberg ploughmark berms, seen to the right. The dipping, acoustically strong underground reflector (sedimentary bedding plane), is suspected to be gas-charged. This is one of the images that sparked off the idea that coral reefs are closely associated with seeping gas (see Hovland, 1990a, b) (from Hovland and Thomsen, 1997).

4.3.3 Bivalve association near HRC

Less than 1 km to the southwest of HRC, there is a 15 m high reef, which is perched on a ledge over a 2 m deep pockmark crater. In 1997, we had the opportunity to visually document parts of this reef. To our amazement, there were abundant *Acesta excavata* lucinid bivalves attached near the base of the reef. They all seemed to be

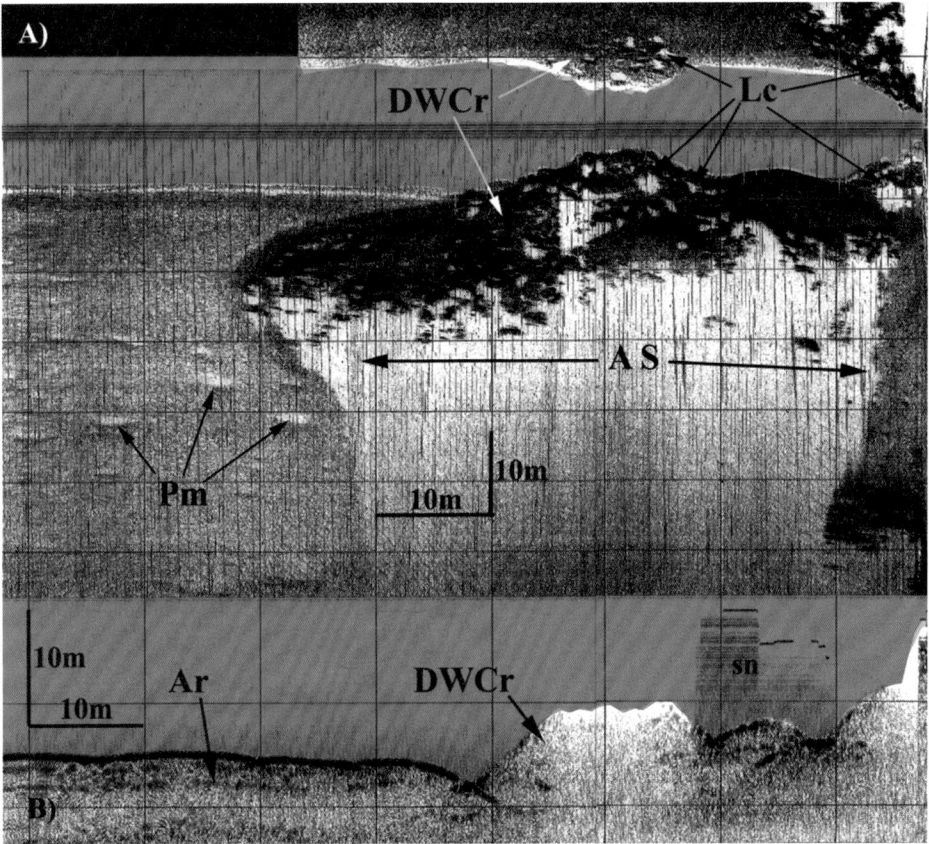

Figure 4.8. A sidescan sonar (A) and sub-bottom profiler (B) record over one of the HRC reefs (from 1997). The two images provide a good representation of the acoustic characteristics of DWCr (deep-water coral reefs) in general. The two records were acquired in one ROV pass over the reef, both acoustic systems running simultaneously. DWCr = live coral reefs, PM = pockmarks, Ar = coherent acoustic reflector, AS = acoustic shadow, Lc = individual live *Lophelia* colonies, Sn = system noise. The sidescan sonar nadir is along the upper horizontal line in (A) (for more information on sonar interpretation, see Blondel and Murton, 1997).

thriving just above the floor of the pockmark (Figure 4.13). Although we had noticed a few individuals of this species on HRC, and especially a lot of dead shells, such abundances of live ones had never been seen. Later, it would be proven that they also live in a similar setting (pockmark) at the Kristin reefs. They are also associated with coral individuals occurring at the Troll field in the North Sea. Below is a review of the known associations between bivalves and seeps. In some cases, it is known that the bivalves represent a manifestation of ongoing seepage through the sub-seafloor.

Bivalves, like the pogonophores, were part of the first hydrothermal vent and cold seep communities discovered. The mussels found during the East Pacific Rise (Galapagos Rift) discovery dive in 1977 were described as filter feeders. It was soon

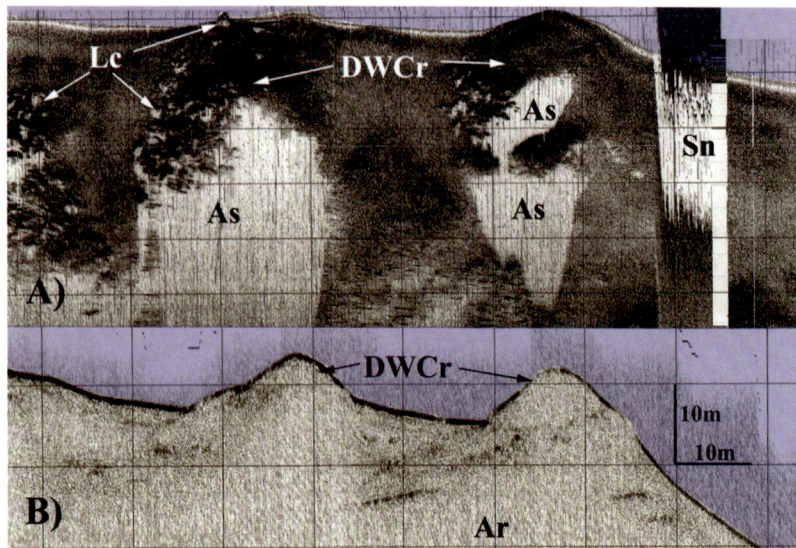

Figure 4.9. Same as in the previous figure, but across two other reefs of the HRC. (A) One channel (port) of the sidescan sonar record. Note that the live *Lophelia* colonies (Lc) occur on one side of the mounds. The actual summits of these two reefs are smooth and consist of sediment, dead *Lophelia* material, and sponges (confirmed from visual ROV inspection). The abbreviations are as in the previous figure. (B) This sub-bottom profiler image shows that there are upward-dipping, coherent, discontinuous acoustic reflections beneath the surface (Ar). Some of these reflections may represent gas (methane) charged layers (from Hovland and Risk, 2003).

realised that bivalves could play host to symbiotic microbes (Childress *et al.*, 1986). In general, some of these bivalves grow near locations where venting is active. As such, this is a "manifesting species".[3] Like tubeworms, the bivalves often contain endosymbiotic chemolithoautotrophic bacteria that reside within their tissue:

> "At sulfide-rich vents, the host tube worm provides shelter and gases to sulfur-oxidizing bacteria (e.g., O_2 and H_2S via an unusual hemoglobin), while the bacteria supply organic carbon. This activity leads to elemental sulfur and iron sulfide mineral precipitation within the organic tissues of the animals, in effect fossilizing part of the tube. In areas where methane is abundant, a symbiotic relationship exists between giant mussels and methane-oxidizing bacteria. Still other bacteria act as epibionts, growing attached to the outer surface of the host (Nelson and Fisher, 1995)" (Konhauser, 2007).

In their review of deep-water cold-seep communities, Sibuet and Olu (1998) identified five families of bivalves known to inhabit cold seeps: the Vesicomyidae,

---

[3] Often, the only visible evidence of "micro-seepage" of fluids through the seafloor are various kinds of organisms called "manifesting species".

Figure 4.10. A photograph showing the top of one of the reefs in HRC. Up to 2 m wide live *Lophelia pertusa* colonies are seen (right side).

Mytilidae, Solemyidae, Thyasiridae, and Lucinidae (to which family *A. excavata* belongs).

Vesicomyidae

The vesicomyid family includes more than 50 species (including the genus *Calyptogena*) of clams that are found nearly exclusively in sulphide-rich habitats such as cold seeps, hydrothermal vents, and accumulations of organic debris (e.g., whale carcasses) (Barry and Kochevar, 1998). They live partly buried in soft sediment, extending a foot down into the sulphide-rich sediment, but provide themselves with oxidants through a short siphon (Roberts and Carney, 1997). They host endosymbiotic sulphur-oxidising bacteria, although different species seem to inhabit locations with different sulphide concentrations.

Mytilidae

Bathymodiolus, the single genus of this family that inhabits cold seeps, is a genus of mussels. Unlike the burrowing clams, these mussels require a hard substrate to live on. They have been reported from water depths ranging between 400 m and 3,270 m (Sibuet and Olu, 1998). Childress *et al.* (1986) showed that members of this family are dependent upon methanotrophic bacterial symbionts, but Roberts and Carney (1997) noted that family members may use either methane or sulphide, depending upon

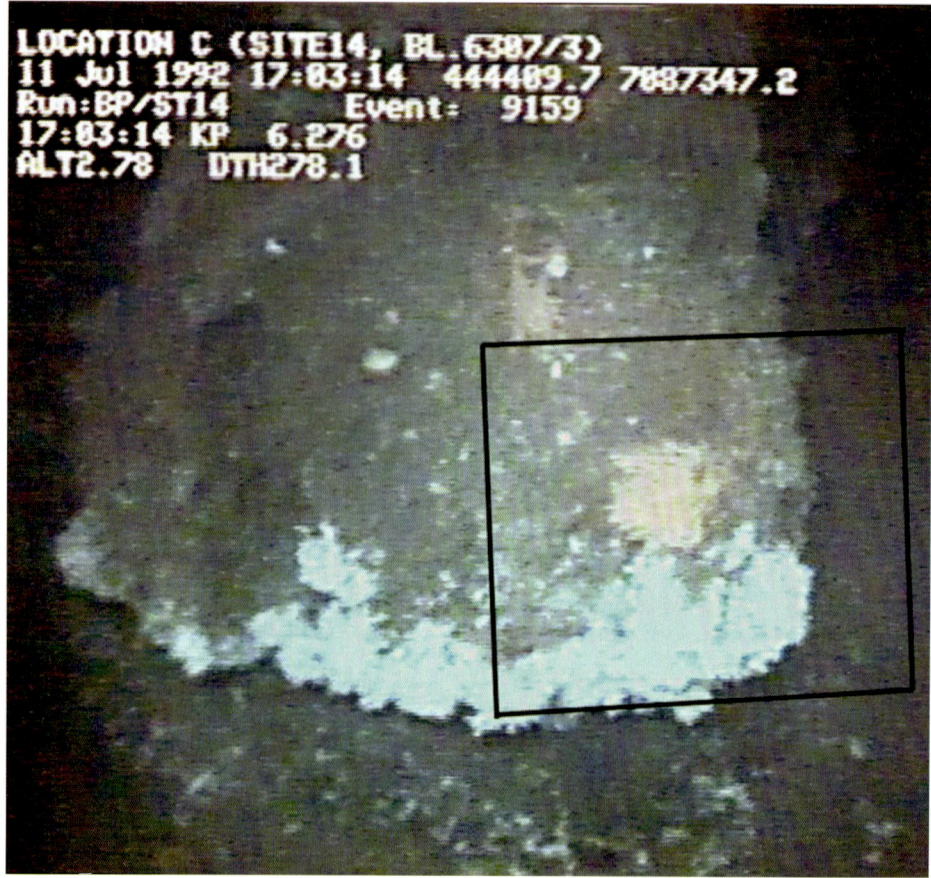

Figure 4.11. This video-grabbed (low-resolution) image from the lower steep slope of one of the largest HRC reefs shows a large naturally displaced *Lophelia* colony. Note that it is upside-down (i.e., it has recently toppled over from its original living position higher up on the reef slope). This signifies a natural process, whereby large *Lophelia* colonies have grown upwards and outwards to such an extent that overhangs have developed. Further growth then caused mechanical failure and toppling due to gravity.

the location, species, and symbionts; van Dover *et al.* (2003) identified both methanotrophs and thiotrophs in *Bathymodiolus heckerae* collected from Blake Ridge.

Solemyidae

Two genera of the deep-burrowing Solemyidae, *Solemya* and *Acharax*, have been descibed from cold seeps in deep and shallow water, and from hydrothermal vents (Sibuet and Olu, 1998). They have been found in the western Atlantic, the Gulf of Mexico, and on both sides of the Pacific Ocean. These are thought to be most

Figure 4.12. A high-resolution still photograph of the framed portion of the previous image of the upside-down coral colony at HRC. Note the still live (white) *Lophelia* in the lower portion, beneath the red rice-coral branch.

common at sites with relatively low flux rates, and to be dependent upon sulphide-oxidising symbionts (Sibuet and Olu, 1998). Dando and Southward (1986) explained that they burrow into the sediment to obtain sulphide-rich porewater, and extend a siphon upwards to the seabed to inhale oxygen-rich water. They reposition their inhalant siphon frequently in order to partially oxidise the overlying sediment. Their bacterial symbionts are located in their gills, which are enlarged as a consequence.

Thyasiridae

The thyasirids are well known from the sulphide-rich sediments of shallow-water seeps and other "polluted" sites, but they have also been found in the deeper waters of the Atlantic, and Pacific cold-seep sites (Sibuet and Olu, 1998; Kamenev *et al.*, 2001). They are suspension feeders with endosymbionts which utilise reduced sulphur. Like the lucinids, they burrow into the sediment, but the thyasirids make an extensive network of mucus-lined tunnels into which sulphide-rich porewater presumably seeps.

Figure 4.13. Clusters of *Acesta excavata* bivalves hanging down from dead and live *Lophelia pertusa* colonies at the very base of a large deep-water coral reef soouthwest of HRC. The reef is perched on the western shoulder of a large pockmark depression, the floor of which can be seen here.

Lucinidae

This family is poorly represented at clear-cut seepage sites compared with the others mentioned above. However, the coral reef-associated giant limid bivalve *Acesta excavata* belongs to this phylum. Sibuet and Olu (1998) listed only two genera (four species of *Lucinoma* and one of *Myrtea*) from sites in the Western Pacific, the Gulf of Mexico, offshore West Africa, and the Eastern Mediterranean Sea. The known geographical range was extended further when a new species, *Lucinoma kazani*, was described from the Anaximander (mud volcanic) mountains in the Eastern Mediterranean (Salas and Woodside, 2002). A shallow-water species, *Lucinoma borealis*, has been described from the North Sea (Dando, 2001). When it comes to a link between the *Acesta excavata* and possible seepage, there is no documentation available, yet. However, as will be mentioned later (Section 8.8.1), Statoil-supported research has found some interesting indications and evidence.

4.3.4 Suspected origin of the HRC

Because very little was known at that time about the ecology and robustness of deep-water coral reefs, the Haltenpipe Reef Cluster became the object of further detailed studies (Hovland et al., 1998). It was soon found that both the Sula Reef Complex and this cluster were located on top of a geological structure termed a "flat-iron" feature, which forms when an eroding agent fails to erode the most competent (hardest) rocks. In this case, the eroding agent was a large expanse of grounded shelf-ice, similar to those currently found on the Antarctic continental shelf (Bugge et al., 1984). The grounded but slowly moving shelf-ice off mid-Norway during the LGM rendered a surface where the most competent rocks remained as highs and the least competent rocks were eroded into troughs or basins. Thus, when the shelf-ice off mid-Norway started floating during post-glacial sea-level rise, sedimentary rocks of Palaeocene age were exposed as highs and the adjacent (less competent and landward) sedimentary rocks of Cretaceous age were exposed as basins. Subsequently, a 0 m–15 m thick layer of sub-glacial clayey till and soft, layered muds were deposited on top of these exposed sedimentary rocks (Figure 4.14). Although the till became ploughed by grounding icebergs during the de-glaciation (between 15,000 YBP and 10,000 YBP), this surface has remained largely unaltered since then (Rokoengen et al., 1995).

Underlying the HRC are sedimentary rocks of Palaeocene age. They "subcrop" below the thin layer of overlying Weichselian clays. Immediately to the southeast, the)less competent Cretaceous sedimentary rocks subcrop. Both Palaeocene and Cretaceous formations have beds (strata) dipping westward at an angle of 10°. These strata clearly bear reflection seismic evidence of being gas-charged (Hovland et al.,

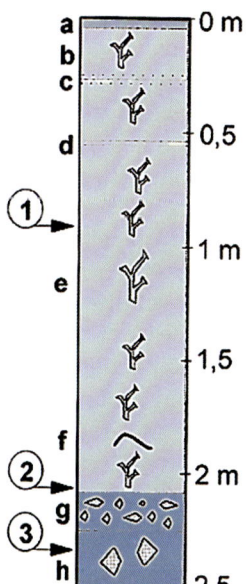

Figure 4.14. Drawing illustrating where dead *Lophelia* skeleton pieces were found inside a 10 cm wide sediment core acquired from the base of HRC, Reef A (from Hovland and Risk, 2003). It spans the ages: 5,000 YBP at 1 m depth (1), 8,000 years at 2 m depth (2), to 56 million years at 2.2 m depth (3).

Table 4.1. Geochemical results (occluded, free hydrocarbons, in ppb by volume) from Statoil/BP geochemical sampling performed during a survey off mid-Norway in 1992. The selected samples are from different locations near coral reefs (within 1 km of sample location). C_1 = methane, C_2 = ethane, C_3 = propane, iC_4 = iso-butane, nC_4 = normal-butane, nC_5 = normal-pentane.

	Hbk6	Hbk7	Hbk8	Hbk11	Hbk12	Hbk13	Hbk10	Hbk17	Hbk21
C_1	5.868	5.312	3.967	21.341	7.592	13.762	24.249	26.792	25.251
C_2	2.116	1.552	0.851	5.881	2.513	4.501	8.461	3.218	7.065
C_3	1.359	0.747	0.336	2.990	1.233	2.876	5.181	1.626	4.596
iC_4	0.066	0	0	0.233	0.002	0.284	0.408	0.151	0.589
nC_4	0.531	0.227	0	1.306	0.331	1.192	2.396	0.726	2.135
nC_5	0.091	0	0	0.405	0.056	0.311	0.781	0.282	1.038
nC_6	0	0	0	0.028	0.096	0.168	0.186	0.098	0.454
Sum	10.031	7.838	5.154	32.184	11.823	23.094	41.661	32.893	41.128

1998). This is further documented by the numerous large pockmarks that formed in the soft clays overlying the Cretaceous subcrop basin. The largest pockmarks occur to the southeast, within a distance of less than 500 m of the HRC (Figure 4.15).

Sidescan sonar, sub-bottom profiler, and multibeam echosounder data acquired in 2000 show that there are numerous small pockmarks not only inside the Cretaceous depressions, but also on portions of the seabed where the Palaeocene dipping layers subcrop and where the reefs grow (Figure 4.8). These pockmarks are between 3 m and 10 m in diameter and up to 0.5 m deep. "Unit pockmarks" are most dense close to the base of the three westernmost coral reefs in the HRC, strongly suggesting a link between these reefs and seabed fluid flow (Hovland and Risk, 2003; Judd and Hovland, 2007).

In order to document baseline geochemical conditions of seafloor sediments, prior to pipeline construction, Statoil also conducted a geochemical sampling and analysis investigation over the Palaeocene and Cretaceous sub-cropping rocks. Some of the samples targeted the HRC as well as one of the large pockmarks located about 500 m east of the HRC (arrowed in Figure 4.15). These results prove that the upper sediments contain light hydrocarbons (methane, C_1, ethane, C_2, propane, C_3, butane, C_4, and pentane, C_5) which have migrated from the Jurassic[4] source rocks below.

It was found that light hydrocarbon concentrations increase well above the background value inside the pockmark and also near the two largest reefs of the HRC (Tables 4.1 and 4.2). Whereas the background value of the sum of adsorbed methane to pentane concentrations in the near-surface clay (at about 1.5 m below the

---

[4] The Jurassic lasted from 206 million YBP until 144 million YBP.

Table 4.2. Geochemical results (adsorbed, sediment-bound hydrocarbons in ppb by volume) from Statoil/BP geochemical sampling performed during a survey off mid-Norway in 1992. Note that the selected samples in this and Table 4.1 are not all the same.

	Hbk18	*Hbk12*	*Hbk13*	*Hbk10*	*Hbk17*	*Hbk21*
C_1	11.112	42.308	50.512	143.546	560.878	222.520
C_2	1.314	6.652	7.597	26.792	83.350	36.182
C_3	0.736	2.896	3.518	17.543	44.852	20.395
iC_4	0.652	1.254	1.491	7.896	18.723	7.589
nC_4	0.461	1.482	1.861	9.232	20.541	9.434
iC_5	0.433	1.359	2.003	7.929	16.863	7.070
nC_5	0.301	0.494	1.158	3.405	7.725	3.602
nC_6	0	0	0.251	1.018	2.120	0
Sum	15.009	56.445	68.391	217.161	755.052	306.792

surface) is 100 ppbv (parts per billion by volume), the value inside the large pockmark is about nine times higher, and at the reef base about six times higher than background (Hovland et al., 1998). This geochemical investigation, therefore, confirms that both the large pockmark and at least one of the reefs are associated with seepage of light hydrocarbons (Hovland and Risk, 2003) (Figures 4.16–4.17).

Thus, at the HRC there are four separate factors indicating a seep association with *Lophelia* reefs:

(1) Shallow seismic records over the HRC show enhanced (strong) dipping reflections under the reefs, suggesting the presence of gas.
(2) Geochemical results document higher-than-background concentrations of light hydrocarbons in the sediments at the base of HRC reefs (Tables 4.1 and 4.2).
(3) The presence of large pockmarks adjacent to the HRC area, where dipping Cretaceous strata sub-crop. These features strongly indicate seepage of hydrocarbons in the area (Figure 4.15).
(4) Small, unit, pockmarks occur at the base of at least two individual reefs of the HRC (Figure 4.15).

4.3.5 The Sula Reef Complex

One of the largest known *Lophelia* reef complexes occurs on the Sula Ridge, some 15 km northeast of the HRC (Freiwald, 1995; Freiwald et al., 1997). This ridge has a

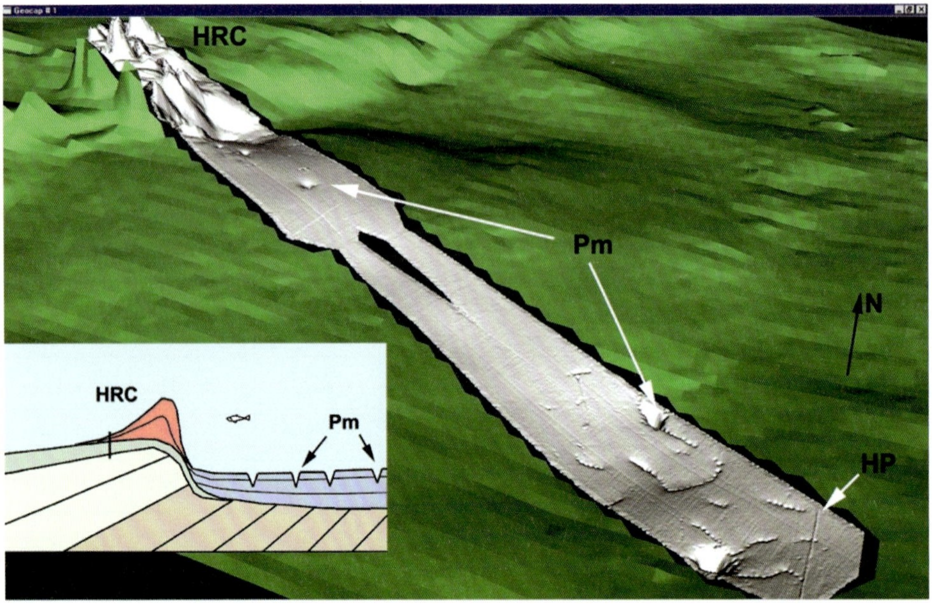

Figure 4.15. This digital terrain model, shown in artificially shaded perspective view, has been constructed using two datasets of different resolution. The green, low-resolution background is from hull-mounted multibeam echosounding. The white and grey, high-resolution strip has been acquired using an ROV-mounted echosounder (two lines merged together). The black "hole" in this latter dataset is due to the ROV being close to the seafloor. The high-resolution strip crosses Cretaceous sub-cropping rocks, from the Haltenpipe pipeline (HP, seen at the lower right), to the Haltenpipe Reef Cluster (HRC, seen upper left), which is located on sub-cropping Palaeocene rocks. Pm = pockmarks. The inset drawing is a simplified geological profile along the strip.

more pronounced flat-iron structure than the HRC. Here, individual *Lophelia* reefs grow to heights of 45 m. According to Freiwald *et al.* (1999), the length of the main reef complex is about 13 km, and the depth range for *Lophelia* reefs is from 233 m–330 m (Figure 4.18).

Sula Ridge sedimentary rocks are covered by a relatively thin layer (0 m–10 m thick) of sub-glacial till, which was ploughed by icebergs during the deglaciation (Lien, 1983). Steeply dipping strata of Palaeocene age underlie the Sula Ridge in the same way as at the HRC. Also in this region there are large pockmarks in the adjacent surface sediments to the southeast, where Cretaceous rocks sub-crop. During 1992, Statoil and BP conducted a combined geophysical and geochemical survey of some locations off the mid-Norway continental shelf, including the northern part of the Sula Ridge. Here it was confirmed that "High-amplitude reflectors in the Palaeocene probably indicate shallow gas" (Thrasher *et al.*, 1996).

Geochemical sediment samples acquired from the eastern slope of the Sula Ridge also proved to contain light hydrocarbons (methane to hexane) of above background values. Whereas background values of occluded gases in the sediments off mid-

Figure 4.16. A conceptual diagram explaining how locally produced nutrients, originating from seeping fluids, indicated by thin, sub-surface black arrows, feed into one of the HRC reefs and into the water column. The "hydraulic theory" suggests that such "fertilisation" by reduced organic and inorganic chemical species is a necessary prerequisite for permanent (long-lived) 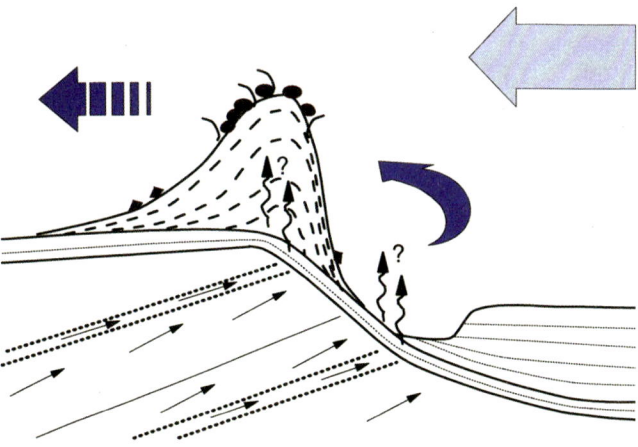 deep-water coral reef construction. This conceptual model is clearly inspired by the HRC situation shown in the previous figure.

Figure 4.17. An artistic photo-montage showing a fossil deep-water Devonian reef imaged in its suspected growth position on the seafloor. The situation greatly resembles the Haltenpipe Reef Cluster (HRC). Notice the Landrover, for scale. See Figure 7.3 for the non-manipulated photograph (photograph by Wendt et al., 1998).

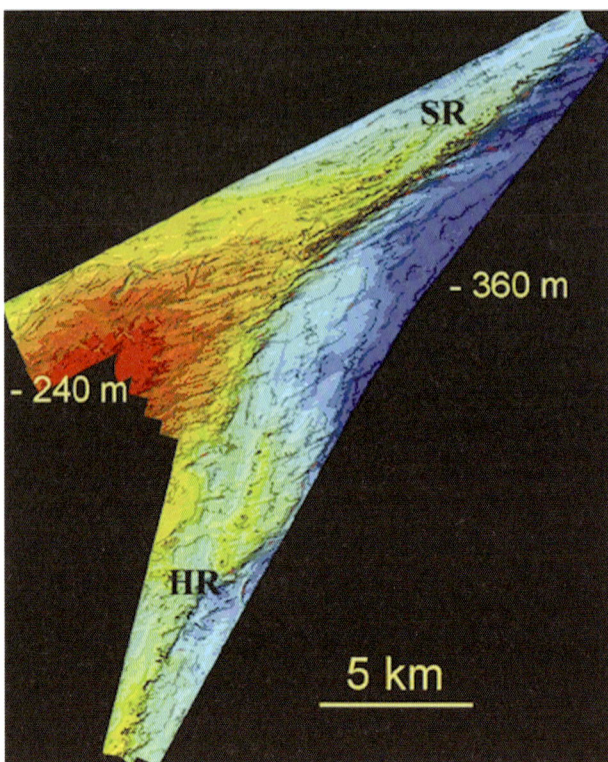

Figure 4.18. The Sula Reef (SR) and the HRC reefs (HR) shown on a shaded digital terrain model (DTM). The water depths are colour-coded, and specific depths are indicated with negative numbers on either side of the DTM (courtesy Dag Ottesen, NGU, 2002).

Norway were typically found to be lower than 10 ppb (see Table 4.1), the values found in the eastern slope of the northern Sula Ridge (i.e., below the large coral reefs) were up to three times greater (32 ppb). For adsorbed gases the values were notably higher. Whereas background values of adsorbed gases in the sediments off mid-Norway were typically found to be lower than 15 ppb (see Table 4.2), the values found in the eastern slope of the northern Sula Ridge were up to four times greater (68 ppb). Methane concentrations in the water column near these reefs are found to be relatively high (Michaelis, 2000), although they decrease near the bottom at the locations of seawater sampling. Even so, this water column methane profile suggests a source of methane from nearby the Sula Reef.

It is also worth mentioning that some of the highest concentrations of light hydrocarbon in surface sediments were found adjacent to a solitary coral reef (Hbk10 in Tables 4.1 and 4.2), located a couple of kilometres north of the HRC reefs. The occluded gas values were here 4 times greater than background (42 ppb), and the adsorbed values 15 times greater (217 ppb) than background.

In a cruise report from a marine geological and geophysical survey over the Sula Ridge, onboard the *Jan Mayen* (of Tromsø University), the researchers noted:

"Moreover, a band of steeply dipping, high-amplitude reflections exists underneath the Sula Ridge. Following the stratigraphic interpretation of Rokoengen *et al.* (1995) we assign this unit a Palaeocene age. In spite of the good resolution of the subsurface, none of the profiles exhibit faulting that can be directly related to reef growth. This supports the hypothesis of Hovland *et al.* (1998), who propose that the reefs are fed by hydrocarbons that seep through the Palaeocene package" (Laberg *et al.*, 2000, p. 19).

Although these conclusions have never been supported by Freiwald *et al.* (2002), who after performing some geophysical and biological work at the Sula Ridge complex concluded that the *Lophelia* reefs are associated with iceberg ploughmarks, a concept and an interpretation that will be discussed in Section 8.5.1, the following indicators at the Sula Reef Complex support a seep relationship (Figures 4.19–4.20):

(1) Enhanced dipping acoustic reflectors beneath the reefs, suggesting not only the presence of gas, but active migration (advection) of gas towards the surface.
(2) Documented higher-than-background light hydrocarbon (methane to hexane) concentrations in the near-surface sediments adjacent to the reef complex.
(3) Documented higher-than-background methane concentrations in the water column adjacent to the reef complex.
(4) The presence of large pockmarks east of the Sula Reef Complex area, where dipping Cretaceous strata sub-crop.

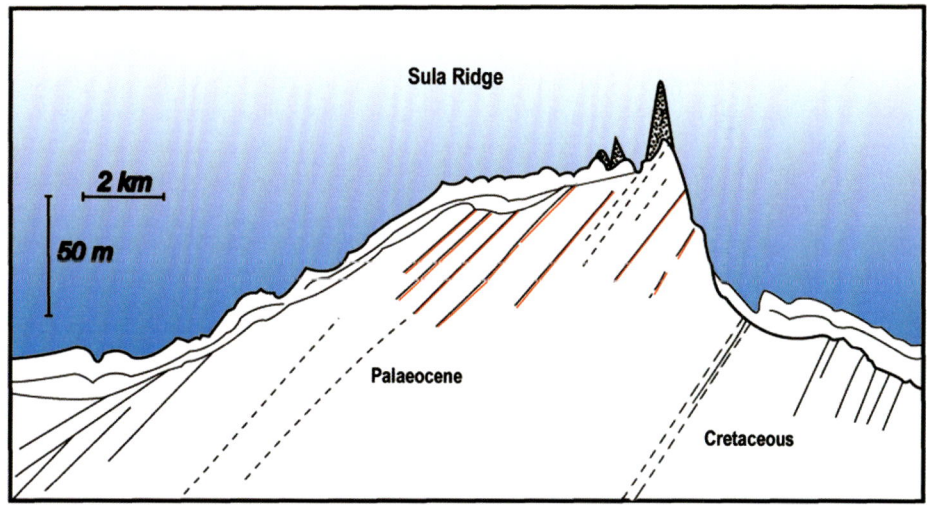

Figure 4.19. Interpreted image of a 2-D seismic line running across the Sula Ridge flat-iron formation. The dipping rock layers immediately below the coral reefs (seen as pyramids on top) are of Palaeocene age. The dipping sedimentary rocks immediately to the east (right) are less competent and are of Cretaceous age. Numerous pockmarks occur in the seabed to the east (over the Cretaceous rocks).

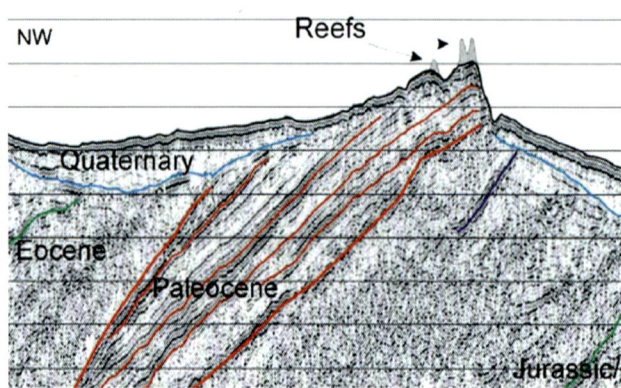

Figure 4.20. An interpreted seismic image (modified from Freiwald *et al.*, 2002), over the Sula Reef. It clearly shows the varying acoustic reflection strength in the up-dipping layers, beneath the Sula Reef, which indicates migrating hydrocarbon gases. Note the small depression (sometimes termed a "moat") in the seafloor on the right-hand side (east): this may well represent a pockmark depression or a gully induced by seeping fluids.

4.3.6 The Husmus reefs, Draugen field

Perhaps the strangest *Lophelia* reefs are those occurring only a few kilometres due east of the Draugen field, off mid-Norway. They were found when Statoil performed a detailed seabed mapping using ROVs in 1994 (64°20′N, 07°59′E). The Haltenpipe had to be re-routed past the small Husmus hydrocarbon reservoir. When coming from the south, along the Haltenpipe route, the seafloor consists of soft clay. It is even and smooth, at a general depth of 290 m. There are numerous up-to-100-m-wide and 10 m deep pockmarks in the seafloor (Figure 4.21). About 4 m sub-seafloor, however, there is an anomalously strong acoustic reflector. But, whenever a pockmark is crossed, this reflector "disperses" (Figure 4.22), just as if whatever causes the reflector "dissolves". Continuing northwards a couple more kilometres, the terrain changes, and a distinct seascape of ridges and troughs takes over, at the same time as the general water depth decreases to 280 m (Figure 4.23). Also, beneath these ridges the strong acoustic reflector dissolves. The Husmus *Lophelia* reefs occur on the tops and sides of some of these strange ridges (Figure 4.21). Only two of the reefs were visually documented, as little time was available for such work.

Based on water depth, sediment type, and the seismic records this strong acoustic reflector is interpreted as one representing the boundary between free gas below and "frozen" gas hydrates above the reflector, an interpretation which is in agreement with previous work (Hovland, 1990a, 1991; Hovland and Svensen, 2006; Judd and Hovland, 2007; Ivanov *et al.*, 2007). In effect, this reflector represents a true BSR (bottom-simulating reflector), caused by thermobarically stabilized gas hydrates (Minshull *et al.*, 1994; Flemings *et al.*, 2003; Tréhu *et al.*, 2006). The reason the reflector dissolves beneath the pockmarks and the ridges is that the gas hydrates "melt" or dissociate, and their contents of bound gas (methane and ethane) and fresh (ice-bound) water is released to the sediments and the above water column. The reefs,

Sec. 4.3] Reefs on the continental shelf 63

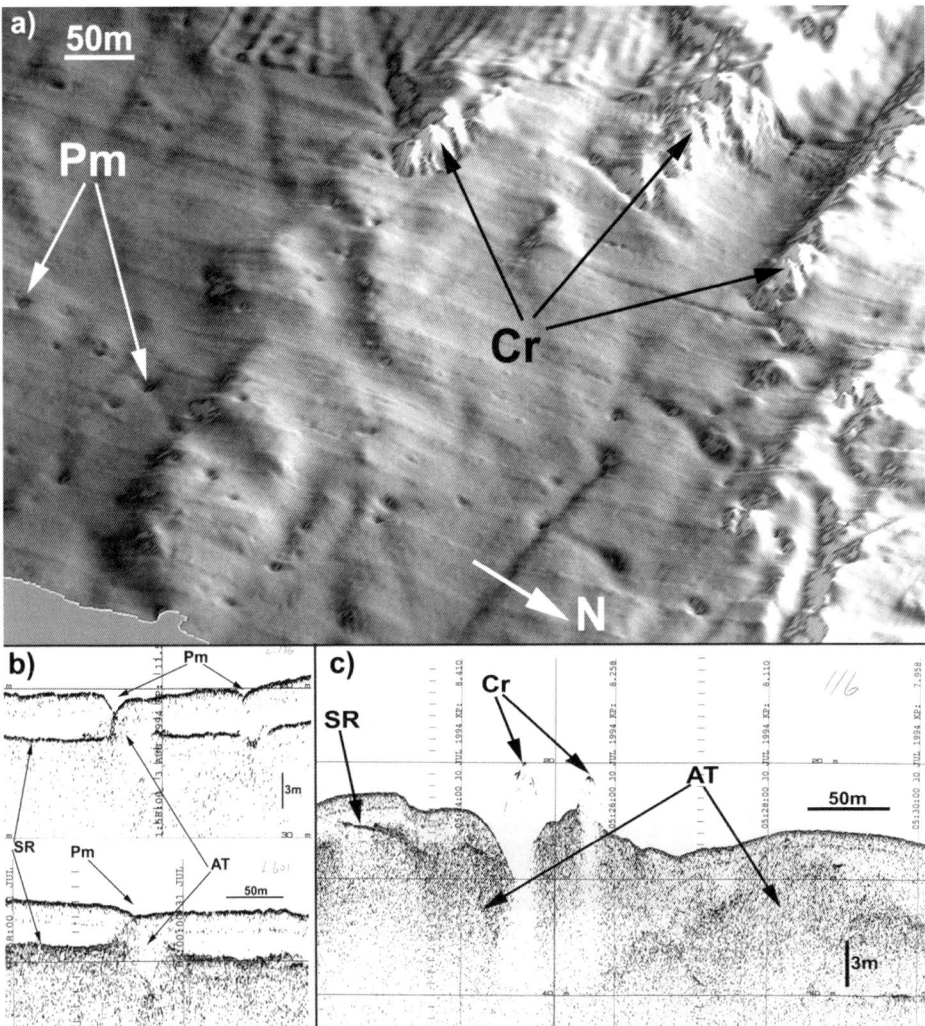

Figure 4.21. (a) A digital terrain model (DTM) showing the location of ridges, with coral reefs (Cr) in the Husmus area (see text), relative to the pockmarks (Pm) of the area. (b) A high-resolution seismic record across two of the pockmarks, showing a prominent (strong), shallow bottom-simulating reflector (BSR), which "disintegrates" (AT = acoustic turbidity) immediately beneath the pockmarks. (c) A high-resolution seismic record across one of the coral reef-associated ridges shown in (a). From this image two reefs are seen (Cr) on either side of the ridge.

therefore, are inferred to occur at the locations in the seafloor where these fluids (gases and low-salinity water) enter the seawater, and they are thought to be directly associated with seepage. However, unlike at the Kristin field (see Section 4.3), the reefs only form on the ridges and not inside the pockmarks.

64 Scandinavian coral reefs   [Ch. 4

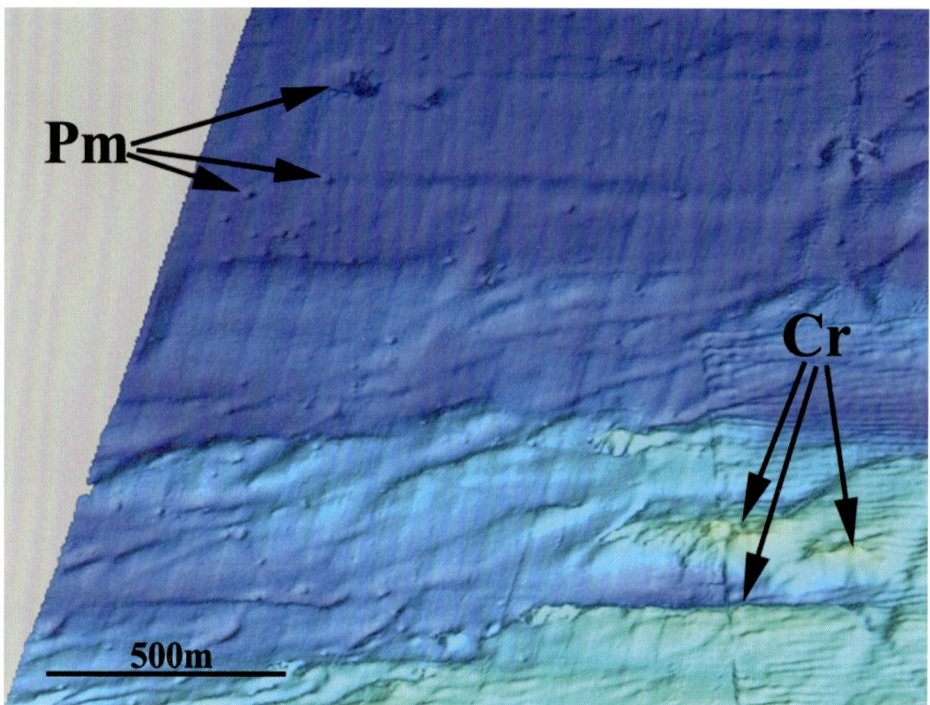

Figure 4.22. Same as Figure 4.21(a); however, here with a wider field of view and in colour. See text for further information.

According to theoretical calculations of gas hydrate stability and porewater pressures (Nabil Sultan, pers. commun., 2006), methane hydrates are at their stability limit at the water depths (280 m–290 m) and temperatures (5°C–7°C) occurring in this area. Therefore, it is envisaged that hydrocarbons migrating upwards through the sediments, perhaps originating in the Husmus reservoir, are trapped in the shallow sediments as free gas below a boundary of upper gas hydrates. Because these bounding gas hydrates are at their thermobaric stability level, very small perturbations, such as annual temperature changes of less than 1°C, and/or lunar tidally induced pressure cycles, are sufficient to induce the formation ("freezing") and dissociation ("thawing") of near-surface gas hydrates (Figure 4.21). It is hoped that more geobiological work can be performed at this exciting location in the near future.

4.3.7 The Træna Deep Reef Complex

An area with large *Lophelia* reefs was found by Statoil during a pre-drilling site survey (location, Figure 4.1) (66°58′N, 11°07′E) in Norwegian Concession Block 6610/3 about 100 km west of the island Træna, at 300 m to 330 m water depth off mid-Norway (Hovland and Mortensen, 1999; Hovland and Risk, 2003). These reefs are constructed on top of an Oligocene deltaic sandy fan deposit with arcuate

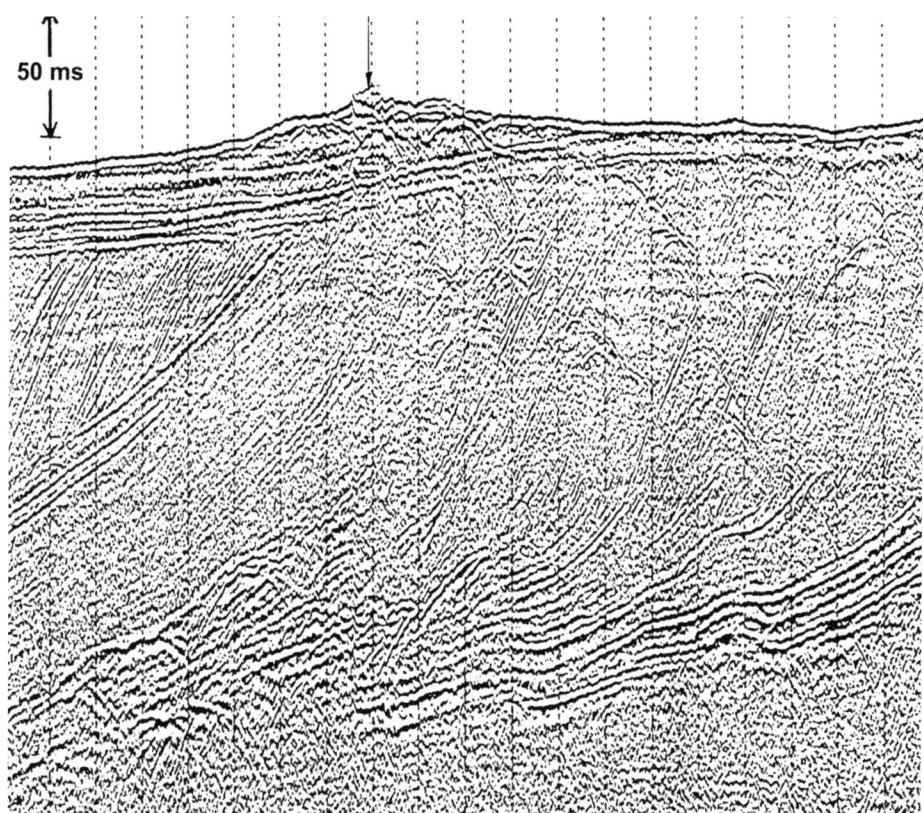

Figure 4.23. Seismic line through the Træna Deep reefs. Steeply dipping sand layers of Oligocene age can be seen. However, clay (morainic and glacimarine) layers of younger, Quaternary age cover them. These latter sediments form a seal for migrating gas from below. Enhanced reflections in this image bear evidence of migrating gases (see also Figure 4.24).

bedding, dipping towards the west (Figures 4.23–4.25). The largest reef within the 5 km by 5 km wide mapped area is about 15 m high, 700 m long, and 150 m wide. There are 265 individual coral reefs within this area, each wider than 15 m. The density of reefs is very high, at about ten reefs per square kilometre. On seismic records there is strong acoustic evidence that some of the layers below the reefs are gas-charged and may represent hydrocarbon migration pathways to the surface (Figures 4.22–4.24), also documented by Lindberg et al. (2004). Sidescan sonar and echosounder data collected during the site survey also contain clear evidence of acoustic plumes in the water column, probably representing free gas emitting directly to the water column from some of the dipping layers (Hovland and Mortensen, 1999).

A couple of more recent geophysical and visual surveys were performed by the University of Tromsø, Norges Geologiske Undersøkelser (NGU, Norwegian Geological Survey), and IMR in this area (Fosså et al., 2000; Fosså and Alvsvåg, 2003,

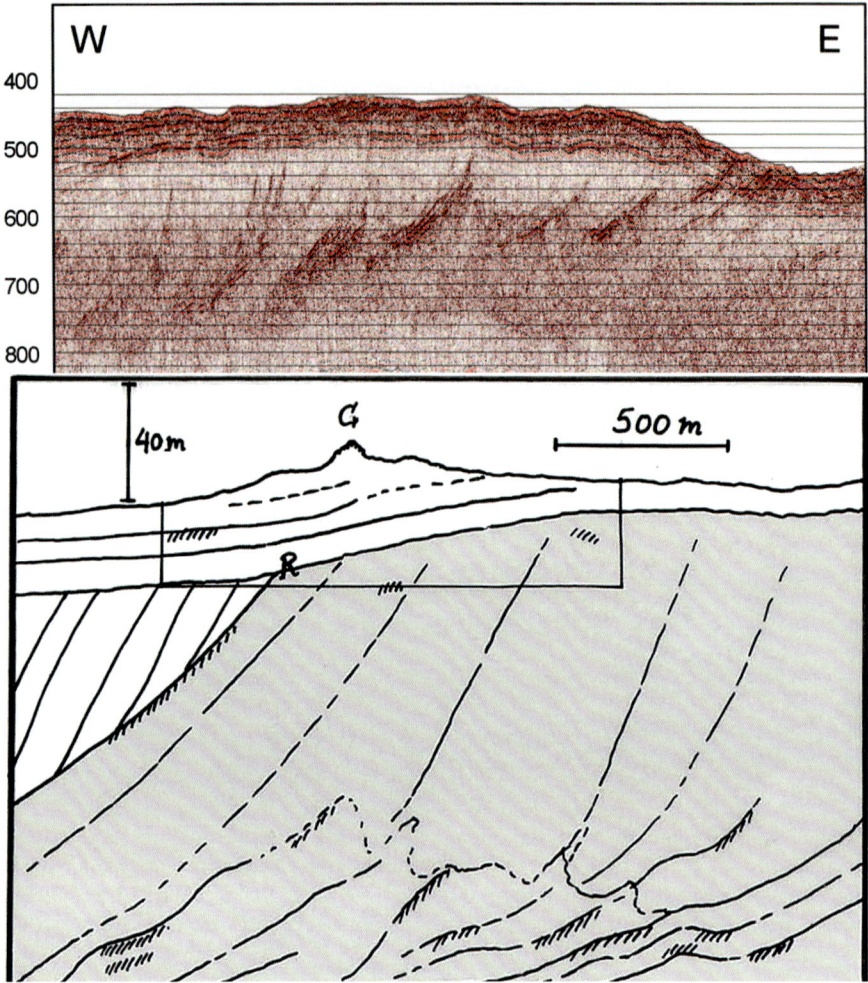

Figure 4.24. Dipping strata underneath the Træna Bank reefs. The reflections are enhanced, indicating the sediments are partly gas-charged. The upper seismic record is courtesy of Björn Lindberg (Lindberg et al., 2007). The lower interpreted seismic is based on Hovland and Mortensen (1999) and is not at exactly the same location as the upper record. Cr = coral reef; R = strong coherent acoustic reflector. Hatching indicates scattered and diffracted reflections caused by the presence of (migrating?) free gas in the sediments.

2004; Lindberg et al., 2004). Although their conclusions as to the alignment and origin of the Træna Deep Reefs differ from mine, their findings are of interest. The descriptions and assumptions made by Björn Lindberg (2004) are very interesting:

> "On a more local scale, the reefs display a slight tendency to settle on topographic highs, being found in somewhat higher densities on the flanks of the mega-scale lineations . . .

Sec. 4.3]   **Reefs on the continental shelf**   67

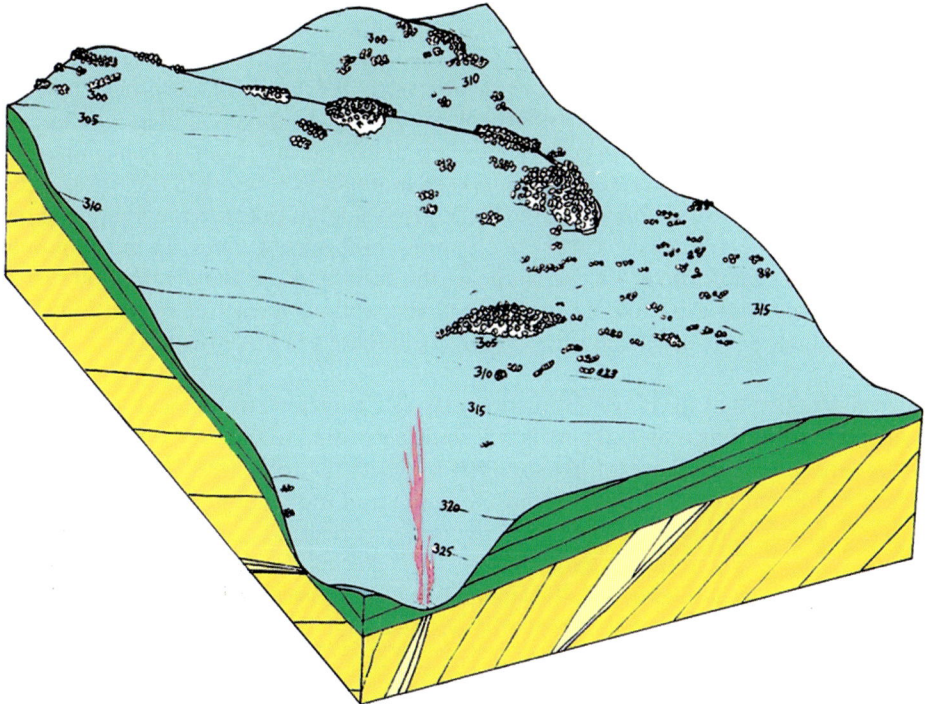

Figure 4.25. A perspective view of the Træna Deep Reef Complex, based on an integrated sidescan sonar, bathymetric and high-resolution shallow seismic interpretation. This block diagram covers an area of approximately 4 km by 5 km and shows the coral reefs drawn onto the topography of the area. North is towards the top. Note the acoustic "flare" (red) suspected to represent a gas plume in the water column at lower left. It is located immediately above a steeply dipping sub-cropping sedimentary bedding plane, which is acoustically enhanced (Judd and Hovland, 2007). The numbers drawn onto the block are water depths.

On a local scale, the orientation of the individual reefs also display a clear trend with parallel longest axis, oriented 300° (120°) in the southeast, 270° (90°) in the north and gradually turning towards 240° (60°) in the south-western part. Thus it seems that the distribution and orientation of the reefs are governed by several environmental factors. In total, 1.2% of the survey area is covered by the mounds, but locally the mounds cover up to 10% of the seafloor along the western and northwestern flank of the depression" (Lindberg, 2004).

Lindberg et al. (2004) thus try to find support for the iceberg ploughmark concept of Freiwald et al. (1999) as well as the general current direction in the area. They argue as follows:

"On a large scale the reefs need sufficient hard substrate to settle on favourable topography—there is a tendency that the reefs settle on highs such as ridges and ploughmarks. The orientation of the reefs' longest axes does not follow that of any seafloor feature seen on our data, and we can assume that other factors cause the pronounced parallel orientation observed. We know of no other environmental factor that has the possiblitiy of structuring the direction of growth except for the current direction, and base our hypothesis on this. Food is transported to the reefs with the current, and it is thus advantageous to grow in the front of the reef, creating a situation where the front (stoss side) of the reefs are living and the back (lee side) of the reefs experience waning coral growth. Thus, we infer a purely oceanographic control on the shape of the reefs, but for the *distribution* of the reefs, we believe that the morphology of the seafloor (large-scale lineations and ploughmarks) is a contributing factor" (Lindberg *et al.*, 2004, p. 9).

I certainly agree that the current direction plays a significant role in local shape of the reef and that the long axis of the reefs will settle according to the local direction of the governing current. However, when it comes to where the reefs grow in the first place, this is a different matter altogether. In the paper by Lindberg *et al.* (2004), there are two interesting figures, which tell their own tale, as these are observations, and not interpretations. Their fig. 2 shows where the "glacial large-scale lineations", the distinct seafloor "escarpments", "ploughmarks", and "N–S broad lineations" occur relative to the "selected reefs". To me, there is no correlation whatsoever between any of these features and the localization of the reefs.

However, on their fig. 5 (Figure 4.23), which is a high-resolution 2-D seismic line shot from the *Jan Mayen*, there is clear evidence of acoustic bright-spots along steeply dipping layers beneath the projected location of the reefs. This confirms my own interpretation: that there is upward-migrating gas in the Oligocene deltaic sandy sediments below the reefs and that the gas emits along linear trends on the seafloor. Therefore, the indicators at the Træna Deep Reef Complex supports a seep relationship (Figure 4.25).

4.4 REEFS ASSOCIATED WITH HYDROCARBON FIELDS

Since work started in the early 1980s with the planning and construction of large export gas trunklines across wide swaths of the North Sea, Statoil has mapped and visually documented tens of thousands of kilometres of seafloor. These transects stretch from the Barents Sea in the north, through large portions of the Norwegian Sea (off mid-Norway) and through much of the North Sea (south of the 62nd parallel). Even though one of the areas in the North Sea is called "Korall banken", there are no corals there, only boulders. Except for the Fedje Reefs, which are hugging the coast of Norway, there are no known coral reefs on the North Sea continental shelf. Thus, more than 90% of the Norwegian offshore coral reefs occur off mid-Norway and northern Norway.

During the mapping and route documentation for trunk pipelines between the Norne field, the Heidrun, and the Åsgard fields off mid-Norway, coral reefs up to 3 m high were found at varying densities. Many of these visually inspected reefs were not as regular and large as the larger ones previously studied along the Sula Ridge and along the Haltenpipe route. One of the reasons seems to be that these reefs, which all occur farther than 150 km from the coastline, are founded on a seabed with slightly softer clay than at the Sula Reef and Haltenpipe reefs. There is a strong possibility that the reefs develop into relatively small (low) structures because they start sinking into the soft clay when they grow too big and become heavier than the bearing capacity of the clay surface. At this distance from the coast, the upper clay layer is thick and there are no underlying shallow sub-cropping sedimentary rocks to support heavy structures. It is therefore inferred that the reefs in this area do not attain heights greater than 3 m–4 m before they sink into the soft substratum, due to their increased weight.

4.4.1 Pockmark reefs at the Kristin field

The Kristin hydrocarbon field is located on the shoulder of the continental shelf off mid-Norway, at water depths ranging from 310 m to 385 m (65°N, 06°32′E). The field was fully developed in November 2005 with several sub-sea and one floating production units (Figure 4.26). During the detailed mapping period (prior to field development) it became evident that this location contains numerous large pockmarks and, curiously, pockmark-dwelling coral reefs. This was *not* expected according to previous expectations on where to find coral reefs (see Chapter 5). The up-to-3.5-m-high coral reefs were actually living some metres below the general seafloor level, inside local pockmark craters!

Within the 14 km^2 area mapped over the Kristin field, there is a medium density of pockmarks of approximately six per square kilometre. They are typically 130 m wide and 10 m–12 m deep. A total of 120 reefs were also mapped within the same area. About a quarter of these (33 reefs) occur inside or along the inside rim of pockmarks. Most of the other 87 reefs are located within 200 m of a pockmark. The largest reefs at Kristin are up to 3.5 m high and 90 m in length (Hovland, 2005).

4.4.2 The "bio-expectations" for DWCRs

According to Freiwald *et al.* (2002), "all known occurrences" of deep-water coral (DWC) reefs share a set of common, general characteristics. These are as follows:

(1) They form on locally elevated hard substrates.
(2) They grow below the local storm wave base.
(3) They are associated with strong currents "that prevent the settling of fine-grained detritus".
(4) They prefer stable physical oceanographic parameters (temperature and salinity).
(5) They prefer to settle within the most saline water mass in a given area, "which often coincides with the seasonal oxygen minimum zone" (Freiwald, 2002).

I call this listing of expected conditions (as summed up here) "bio-expectations", as they clearly rely on a biologist's logical expectations rather than on direct observations.

Figure 4.26. The semi-floating platform at the Kristin field off mid-Norway. The platform acts as a hub, collecting the gas and condensates from four underwater wellhead templates, each located about 6 km from the platform. On the nearest leg, four anchor chains can be seen. To the right of this leg, a "forest" of risers can be seen. These are the flexible pipes through which the products flow (gas and condensates, etc).

However, neither the pockmark-associated reefs at Kristin, nor those at the Morvin field (Section 4.4.5) attach themselves to a locally elevated substrate. The substrate they grow on is diametrically opposite from that expected. They choose to grow inside relatively deep pockmarks, even though there are locally elevated ridges nearby! Furthermore, there is little evidence that they are associated with "strong currents that prevent the settling of fine-grained detritus". Exactly the opposite seems to be the case. At the times we have been performing visual inspection at Kristin, there has hardly been any currents at all, and there is ample evidence of fine-grained

Sec. 4.4] **Reefs associated with hydrocarbon fields** 71

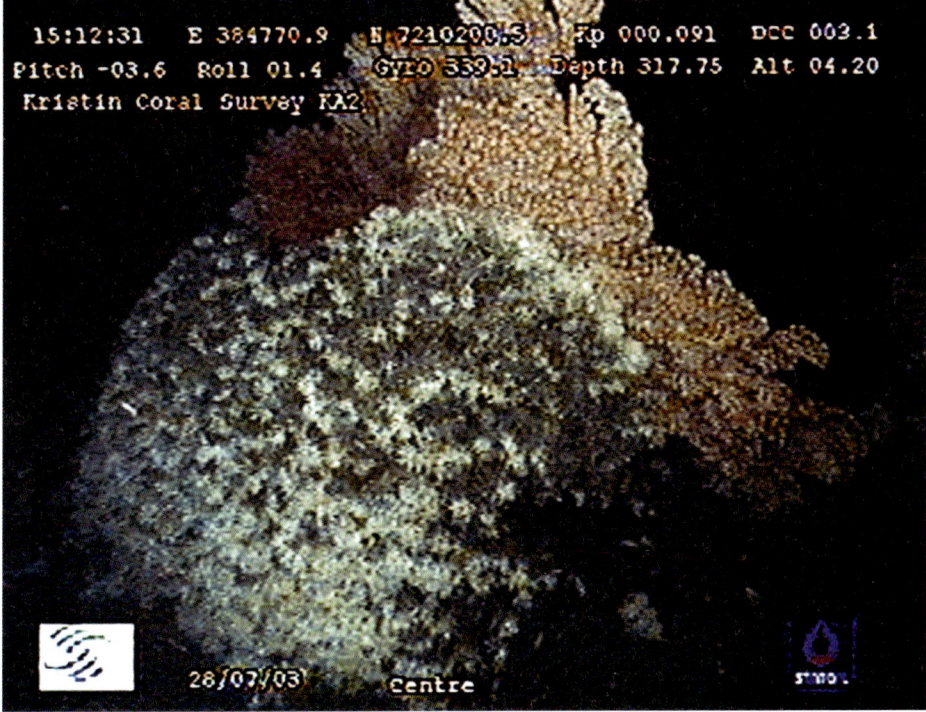

Figure 4.27. A small portion of the coral reef KA2 inside the pockmark KA2 at the Kristin field. The large white, near-circular *Lophelia* colony shown here in the foreground is about 2 m wide.

detritus settlement. Thus, at least 40% of the listed bio-expectations turn out to be wrong.

4.4.3 Some relevant emerging results

As part of the seafloor documentation prior to development of the Kristin field, three of the pockmark-dwelling coral reefs were surveyed visually in July 2003 (pre-construction) and also in November 2006. None of the reefs has been damaged during the field development and initial production phase (i.e., post-May 2006), as anchor patterns and flowline routes have been chosen such that they avoid the reefs. The three inspected reefs are designated KP1, KA1, and KA2 (Figures 4.28–4.30). The size of the reefs, their depth span, and size of their associated pockmarks are given in Table 4.3. Because detailed mapping of the seafloor has been performed with ROV-mounted multibeam echosounders and sidescan sonar, the exact locations and extents of the coral reefs are well documented.

The reefs at Kristin provided a good opportunity for Statoil to support some very interesting research. As operators of the Kristin field, Statoil is obliged to look after

Figure 4.28. An example, from HRC, of various types of sponges at coral reefs. The two round ones shown in the upper image are *Geodia* sp. They are each about 30 cm in diameter.

Sec. 4.4] Reefs associated with hydrocarbon fields 73

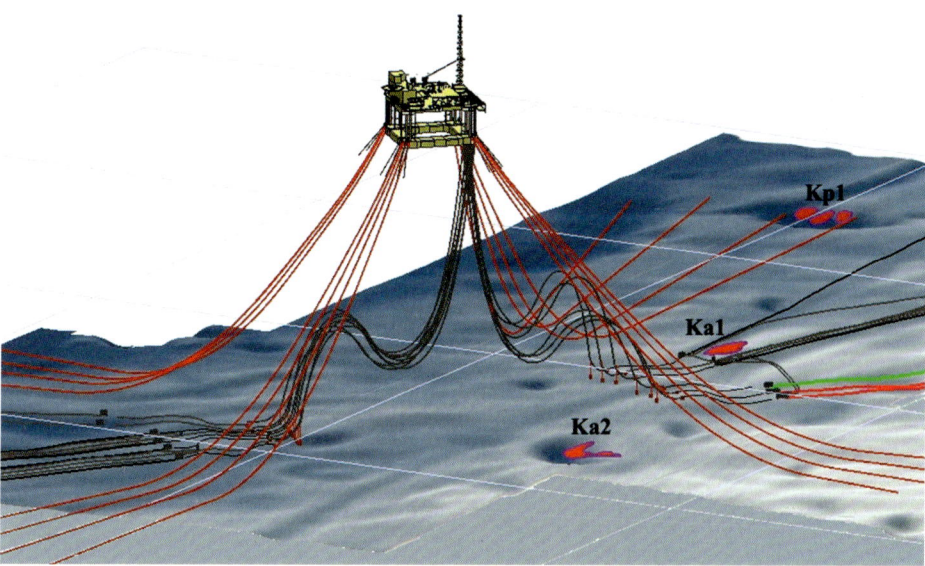

Figure 4.29. This 3-D perspective image shows how the Kristin semi-floating platform is located above the seafloor. The water depth below the platform is 305 m. Red lines indicate anchor chains and wires. The black lines indicate flowlines and risers piping products (gas, condensates, etc.). The chains and wires that hold Kristin in place are attached to suction anchors, inverted domed steel cylinders sucked into the seafloor clays. The coral reefs, Ka2, Ka1, and Kp1, were inspected twice before installation and twice after installation of the platform and the other infrastructure on the seafloor. No damage was found on the reefs (image courtesy of Leslie Austdal, Statoil).

the reefs and protect them from damage, during field development and during oil and gas production. Inspection of the three main reefs is done on a yearly basis, whereby they are visually documented by video "transects", which are repeated every time. In addition, samples of the specific epifaunal organisms have been collected for scientific analysis. Thus, we have collected samples from water, sediment, and epifauna (bivalves, sponges, and *Lophelia* corals). These have been used for microbial analysis undertaken at the Biological Institute, University of Bergen (UiB).

Two interesting results stem from this work:

(a) Sigmund Jensen (UiB) and his colleagues have found a new bacterium living inside the gills of the reef-associated *Acesta excavata* bivalve (Jensen et al., 2008b). Although this seems to be a symbiont, it has no known relatives, and its function is still unknown.
(b) No less interesting, however, DNA analyses of bacteria living in the water, sediments, and sponges adjacent to the KA2 (pockmark) reef at Kristin have provided some intriguing preliminary results: comparative analyses show that

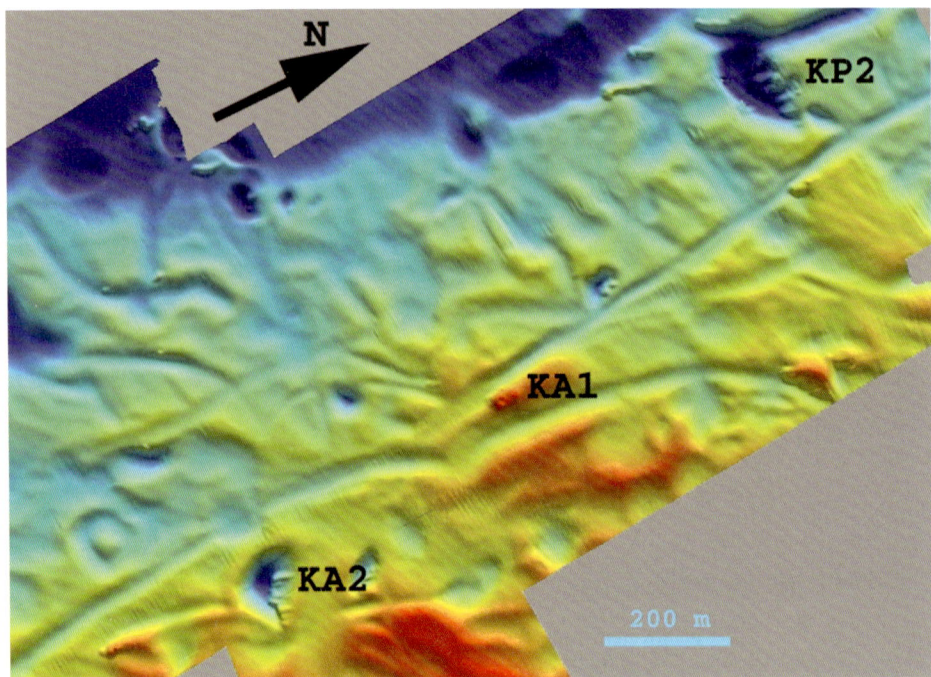

Figure 4.30. An artificially shaded relief map of parts of the Kristin field, off mid-Norway (Hovland, 2005). Black arrows point at coral reefs, some of which occupy the interior slopes of pockmark craters. Note the relict iceberg ploughmarks (long, straight, and curvilinear troughs). The named reefs, KA1, KA2, and KP2, are regularly inspected by Statoil (now StatoilHydro). This mapping was performed using an ROV-based MBE (multi-beam echosounder), with the ROV flying about 20 m above the seafloor. The grid size is 0.5 m by 0.5 m.

Table 4.3. Characteristic size and depth spans of coral reefs and their associated pockmarks, at three locations in the Kristin field. The reef height is measured from the sloping pockmark bottom.

	Reefs			*Pockmarks*		
Depth span (m)	319–327	323–329	316–326	316–328	317–329	316–327
Height/Depth (m)	3.0	3.5	3.0	12	12	11
Length (m)	90	60	85	150	180	130
Width (m)	30	50	40	120	75	120
Reef or pockmark id	KP1	KA1	KA2	PmP1	PmA1	PmA2

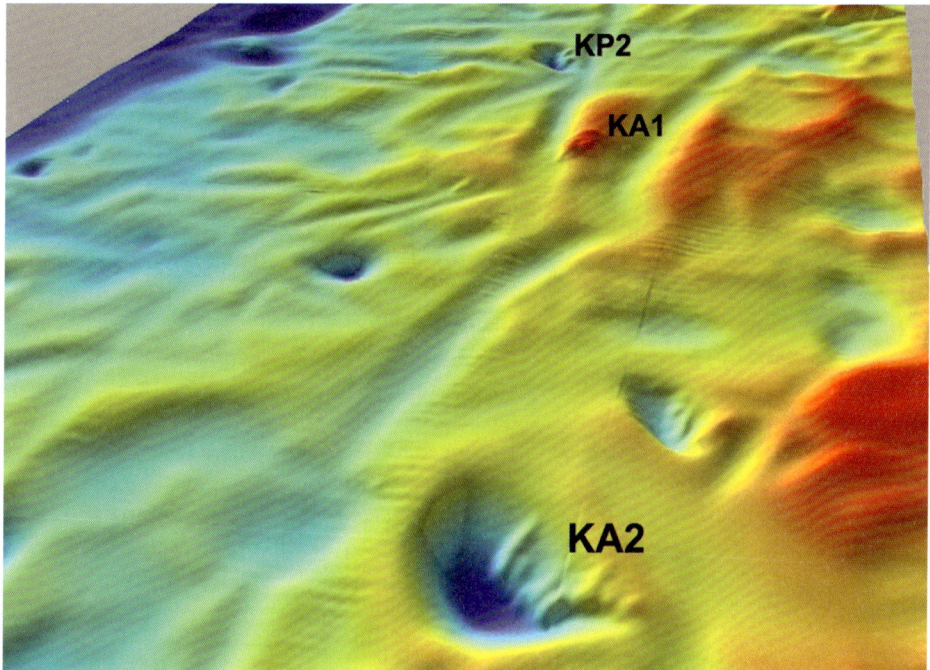

Figure 4.31. Oblique view of parts of the image shown in Figure 4.29. Note how the coral reefs represent an extra roughness in the otherwise relatively smooth seafloor scape. Arrows point at named and un-named reefs. Abbreviations and scale as in Figure 4.30.

some of the bacteria found here are associated with *chemosynthesis*: "Phylotypes taken to represent this deep-water coral reef are related to octocoral, sponge and methane associated sediment bacteria" (Jensen *et al.*, 2008a).

By comparing where similar DNA sequences have been found elsewhere in the world, Jensen and his colleagues found that:

"The retrieved *pmoA* gene fragment (found in a sponge at KA2) clustered *pmoA* from a type I methanotroph. This is in agreement with the stacked membranes visualized within cells of another *Demospongiae* of the same order *Poecilosclerida*, the unusual genus *Cladorhiza*, which thrive near methane sources in deep sea mud volcanoes in the Barbados accretionary wedge (Vacelet *et al.*, 1996). To our knowledge, the *pmoA* we detected is the first methanotrophic sequence from a sponge. The closest cultured relatives to this *Desmacidon* (sponge) associated methanotroph are thermophilic species of the genera *Methylococcus* and *Methylocaldum*" (Jensen *et al.*, 2008a).

Thus, it seems that the bacteria and the archaea living in association with the coral reefs and inside some of the associated epifauna may hide the key to understanding what really fuels deep-water coral reefs.

4.4.4 Sponge-associated bacteria

The seepage link between sponges and bacteria is very interesting. In a different context, at the Atalante mud volcano (Barbados) accretionary prism (water depth 4,700 m to 4,900 m) there are dense bushes of carnivorous sponges of the genus *Cladorhiza*. Although it is thought that, like other members of this genus, they trap and eat small swimming prey such as crustacea, Vacelet *et al.* (1996) found that sponge tissues contain at least two, and possibly three, separate types of bacteria. Carbon isotope values ($\delta^{13}C$ of −48.4‰ to −48.8‰) show that these are methanotrophs, and it seems they provide a significant proportion of the sponges' nutrition. Although the microbes seem to live outside the sponge cells, they are absorbed and digested by certain sponge cell types. It also seems that some sponge cells harbour bacterial embryos; so, these sponges look after the developing bacteria, then eat them!

From analysis of the video footage, it can be determined that the dominating sessile macro-organisms and reef builders on the three inspected reefs are (in order of apparent abundance): the stony coral *Lophelia pertusa*, various types of large sponges (including *Geodia* sp.), the octocorals *Paragorgia arborea* and *Primnoa resedaeformis*, and the bivalve *Acesta excavata*. The most common fish living on these pockmark reefs are *Sebastes* sp. and *Brosme brosme*.

4.4.5 Silted reefs at the Morvin field

Recently, numerous reefs (up to 4 m in height) have been mapped at the Morvin field, which lies about 20 km northwest of Kristin, in the Norwegian Sea (65°09′N, 6°28′E). Many of the reefs at Morvin cluster at the bottom of slight ridges, local depressions (some of which may be pockmarks), and along iceberg ploughmarks (see Figure 4.31). Furthermore, they seem to be aligned and often form long chains of small coral reefs. According to Christian Lysholm (pers. commun., 2007), who interpreted video information from the field, they seem to grow in the direction of the incoming prevailing bottom current (running from the southwest). Only the outer tips of the *Lophelia* branches are white and healthy, suggesting that there is a relatively high degree of sedimentation (silting) (i.e., that these *Lophelia* colonies are stressed by, or adapted to?, natural silting processes).

4.4.6 Origin of Kristin and Morvin reefs

From Statoil's extensive seabed features database from the Barents Sea in the north, via the mid-Norway continental shelf (Norwegian Sea) to France, in the southern North Sea, it is possible to say that the "normal" (dominating) seafloor topography in northern European waters is rather "uneventful" (i.e., there are very few significant features to see). In order of occurrence, the most common significant natural seabed features are

— relict iceberg ploughmarks (Lien, 1983), typically 50 m wide, 5 m deep, and up to 10 km long;

Sec. 4.4] **Reefs associated with hydrocarbon fields** 77

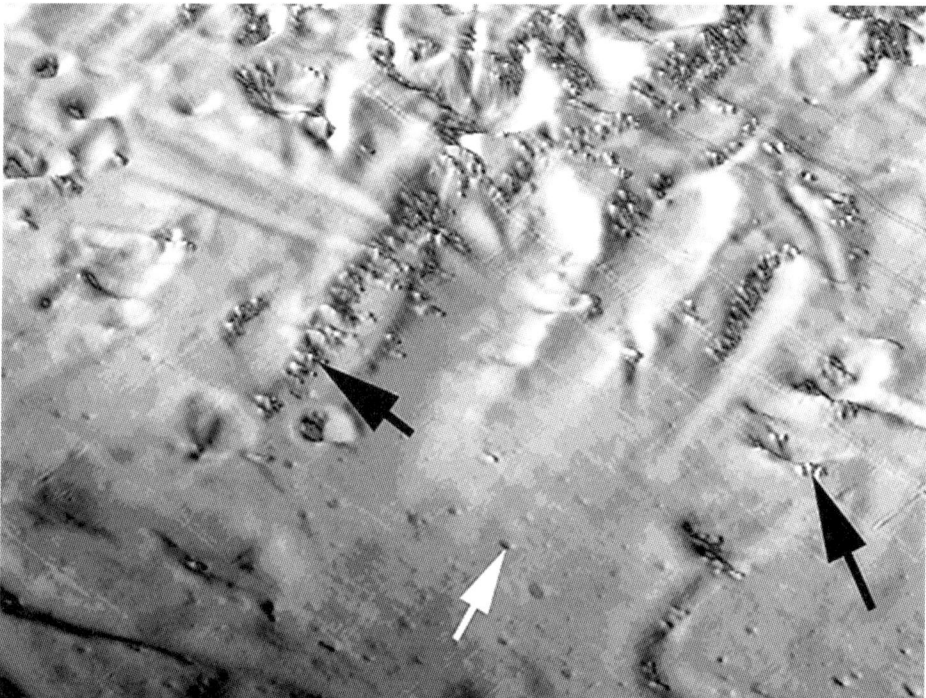

Figure 4.32. A section of the seafloor mapped using ROV-based MBE. The water depth is between 340 m and 320 m in this perspective view of the digital terrain model. Artificial lighting is from the left (north). Black arrows point at small *Lophelia* coral reefs, whose size is about 1.5 m high and 10 m in diameter. The white arrow points at a small pockmark crater. Note how the coral reefs are clustered inside pockmark and iceberg ploughmark depressions.

— normal sized (>30 m wide, up to 15 m deep) pockmarks (Hovland and Judd, 1988; Judd and Hovland, 2007);
— ahermatypic coral reefs (>20 m diameter, up to 30 m high) (Dons, 1944; Hovland and Mortensen, 1999; Freiwald *et al.*, 2002);
— sandwaves, normally 1 m–2 m high, but up to 15 m high and 300 m long, in areas of strong tidal currents, such as in the southern North Sea.

Relict iceberg ploughmarks tend to occur within large regions (i.e., off mid-Norway and in large parts of the Barents Sea). Pockmarks also occur regionally, such as in the Barents Sea and the Norwegian Trench (Hovland, 1981; Hovland and Judd, 1988; Judd and Hovland, 2007). It is known that pockmarks are generally dependent upon two factors:

(1) sub-bottom hydrology (hydraulically active conditions, due to the presence of shallow gas, or pressurized porewater);

Figure 4.33. Morvin *Lophelia* reefs. Note that the live *Lophelia* (white "rind") only occurs in the outer few centimetres of the colonies (the white portions).

(2) a seabed that is "pockmarkable" (i.e., consists of fine-grained sand, silt, and/or clays).

However, ahermatypic coral reefs occur only patchily, and their whereabouts are impossible to predict before detailed surveys have been conducted (Mortensen *et al.*, 2001). The causative formation mechanism for Norwegian ahermatypic coral reefs is still unknown, mainly because of lacking long-term detailed and dedicated (current, nutrient, microbial, and hydraulic) measurements. It is believed that they are generally dependent on two factors:

(1) relatively hard and stable seabed sediments; and
(2) a stable supply of nutrients (i.e., possibly originating locally from the primary and secondary trophic level).

Because pockmarks tend to destabilize fine-grained sediments, it is therefore a paradox to find ahermatypic coral reefs inside and along the inside rims of pockmarks, as is the case at Kristin. However, some large pockmarks are known to erode the fine-grained sediments right down to a firm substratum (Judd and Hovland, 1992,

2007), and on rare occasions to exhume sediment-embedded rocks, subsequently found inside the pockmarks. Furthermore, authigenic methane-related carbonate slabs and nodules occur inside some pockmarks (Hovland *et al.*, 1987). But, at Kristin, no such hard substrates have been found so far.

Based on this database of seabed conditions, it is fair to say that pockmarks only occur in special areas, and that large (more than 2.5 m high) ahermatypic coral reefs are also relatively rare seabed features. If the two seabed features occur together, then, statistically, there must be a common connecting factor. From numerous submarine seep sites worldwide, including pockmarks in the Tommeliten area of the North Sea and areas of the Gulf of Mexico, we know that seeps stimulate the growth of primary producers (bacteria and archaea), which rely on the chemical gradients always present at seep locations (MacDonald and Leifer, 2002). The only reasonable unifying connecting factor expected to exist, therefore, between pockmarks and ahermatypic reefs, is that gas and porewater emanating from pockmarks stimulate primary organic production and provide a stable nutrient source to the near-seafloor porewater and the downstream water column (Hovland and Risk, 2003). Thus, the Kristin pockmark-associated reefs strengthens the validity of the hydraulic theory (Hovland, 1990b; Hovland and Thomsen, 1997; Hovland *et al.*, 1998; Hovland and Mortensen, 1999; Hovland and Risk, 2003).

Because pockmarks are formed by episodic removal of fine-grained sediments by focused gas and porewater expulsion, the pockmark-dwelling, reef-building animals (i.e., *Lophelia*, *Paragorgia*, *Primnoa*, and *Acesta* spp.) especially must be able to tolerate periods, probably lasting for several hours at a time, of very turbid water ("silting episodes"). Thus, it is believed that ahermatypic coral reefs, in general, are more robust against silting episodes than previously suspected. The conspicuous finding of dense aggregates of the giant limid bivalve *Acesta excavata*, inside and at the base of *Lophelia* corals of the Kristin reefs, suggests that it is especially seep-prone. This means that it not only tolerates silting episodes, but that it perhaps depends on chemicals dissolved in the porewater, emanating from pockmarks. To follow this reasoning through, it may mean that *Acesta excavata* may host endogenic bacterial or archaea symbionts, as is known to be the case with giant bivalves living in similar seep-prone settings in the Gulf of Mexico (MacDonald *et al.*, 2002). The bacterium Sigmund Jensen *et al.* (2007) found in the gills of *Acesta excavata* also suggests this.

4.4.7 Seep-associated corals and sponges over a leaky field

In Concession Block 6407/4, Statoil drilled a prospect and found only gas in the upper section of the sediments. There were no commercial accumulations at depth, so the field was judged to be "dry". However, the reason for it not to be commercial seems to be prolific seepage through the seafloor. This fluid flow is still ongoing as can be seen from the image in Figure 4.34.

In 1992 we had the opportunity to dive here, using an ROV, and check for any sign of corals or other organisms. Although the shallow seismic data suggest prolific ebullition through the seabed, no bubbles were found during the ROV dive. However,

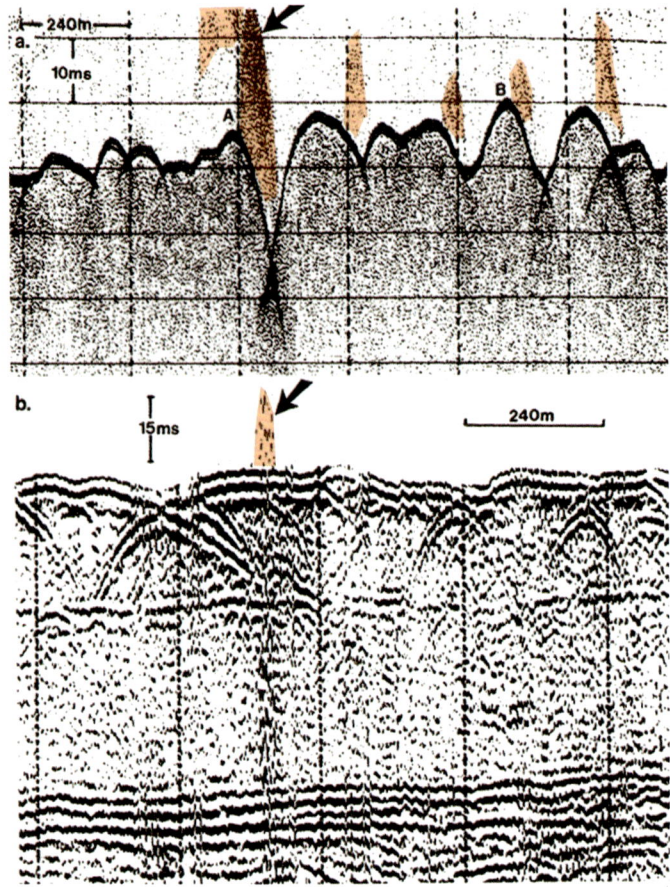

Figure 4.34. Acoustic evidence of prolific hydrocarbon seepage out of the seafloor in Norwegian Concession Block 6407/4. The arrows in these two seismic records point at so-called acoustic flares, coming from bubbles in the water column (see Judd and Hovland, 2007). Patches of coral and sponges were found together with bacterial mats on the seafloor at this location (from Hovland, 1990a).

there were large patches of corals, prolific sponges, and also evidence of bacterial mats, of which one very small one is seen in Figure 4.36. One geochemical sample was taken in the area, which confirmed that there were high concentrations of hydrocarbons, including ethane, propane, butane, pentane, and hexane (i.e., thermogenic hydrocarbons). Unfortunately, no more work has been performed in this exciting area, to date.

4.5 OTHER REEF OCCURRENCES OFF MID-NORWAY AND NORTHERN NORWAY

4.5.1 Reefs in Breisunddjupet, near Ålesund

During the planning of the Langeled pipeline, between Norway and the UK (in 2003), a potential route was mapped just west of Ålesund, on the mid-Norway

Figure 4.35. Accumulation of sponges found at the site seen in the previous figure. It is uncertain if this represents a dead coral reef or not. From video and photographs it is not possible to discern any coral debris (Concession Block 6407/4).

coast. The route had to cross over a 170 m deep and 5 km wide gulley striking east–west, the Breisunddjupet. Inside this submarine valley, which has a water depth of about 280 m, the seafloor consists of sandy clay, and there are patches of large coral reefs. Unfortunately, some of the reefs have been damaged due to trawling. However, Statoil documented some of the reefs (Figures 4.37–4.39), before this portion of the pipeline route was found to be unsuitable for a large-diameter trunk pipeline. The Langeled pipeline has now been constructed farther west, and in deeper water. Although there are some strange local depressions in the seafloor of the Breisunddjupet, it is uncertain if these represent pockmarks. Therefore, the reason for the patchy occurrence of reefs inside this trough is suspected to be current strength and funneling of water masses, possibly combined with upwelling of deep, nutrient-rich water that flushes into the underwater valley from the west.

4.5.2 The Floholman reefs, with associated bacteria and gas!

Inside a local downfaulted block (or ice-gouged trough), just west of Skogsøy, Vesterålen, there is a group of impressive coral reefs, called the "Floholman reefs".

Figure 4.36. This photograph was acquired in 1991 from the Block 6407/4 location, off mid-Norway above a leaky hydrocarbon reservoir (same location as shown in Figure 4.34). Normally, a bacterial mat suggests seepage of anoxic fluids from the subsurface (Hovland, 2002). Note the small arrowed bacterial mat located at the base of the *Paragorgia arborea* in the background. This evidences the flow of anoxic fluids from the seafloor. A small white *Lophelia pertusa* colony is seen in lower centre.

They are located at a water depth of more than 200 m. A recent survey onboard the *G.O. Sars*, of the University of Bergen, Norway, has found some very interesting features upstream of these reefs:

> "The researchers stumbled over mats of bacteria indicating seepage of light hydrocarbons on the seafloor" (Godø, 2007).

Some additional information came from the chief scientist onboard, Pål Buhl Mortensen:

> "There are between 200 and 300 coral reefs in the area. Their shape is similar to the ones in the Træna Deep, with a 'live front' against the incoming current and a tail of dead coral a couple of hundred metres in the current direction, towards the west. The reefs are surrounded by sand, and therefore resemble oases in the desert. There does not seem to be trawling in the area, but frequent finds of lost long-lines (fishing-lines) witness of abundant [non-destructive] fishing. From previous visual

Sec. 4.5] Other reef occurrences off mid-Norway and northern Norway 83

Figure 4.37. A beautiful *Lophelia* colony in the Breisunddjupet area. Note the pink shrimps sitting on the top of the colony.

inspections in this area we knew that there were living corals here, but not that the mounds actually represented deep-water coral reefs. After having inspected about 20 of the structures scattered over the whole area, we can now conclude that this is the case" (*www.mareano.no*, October 2007).

In the *Harstad Tidene* (a newspaper) of October 15, 2007, this bacterial mat association with coral reefs is documented for the first time:

"The whole area is just like a formidable coral park. The camera pod had just hovered above some beautiful coral-oases, full of fish and other organisms. Only another few metres further along the sandy seafloor plain, and then, suddenly— the bacterial mats appear. The researchers sitting in the small video control room of 'G.O. Sars' burst into loud, excited discussion, as the camera pod continued at a speed of 1 knot. They became aware of the fact that these life forms appear as a result of natural hydrocarbon seepage. There were five or six marine biologists and geologists in the room, intensely following the camera recording. Chief

Figure 4.38. *Lophelia* colonies and a sponge in the Breisunddjupet area. Note the pink shrimps sitting on the white sponge.

Figure 4.39. A pink version of a *Lophelia* sp. at Breisunddjupet.

Sec. 4.5] Other reef occurrences off mid-Norway and northern Norway 85

Figure 4.40. A perspective view of the Floholman reefs, based on swath bathymetry data. This seascape is quite stunning. It shows a group of large, up to 10 m high *Lophelia* reefs off Norway. The water depth is about 250 m. The vertical scale is enhanced by about five times. Note how sediments accumulate in drifts on the suspected lee side of the reefs. Note also that the reefs do not occupy particularly pronounced highs on the seafloor.

scientist, Pål Buhl Mortensen, will not call this a 'gas-find', but the large mats and cakes of bacteria and other organisms clearly tell their own story. 'From the oil sector, we have a hypothesis saying that natural gas seeps form this type of bacterial growth and that it can also stimulate formation of a high faunal biodiversity. This is what we have here' " (Godø, 2007).

Figure 4.41. An underwater photograph of bacterial mats (yellow arrows) and suspected methane-derived authigenic carbonate crusts (black arrows) at the seep site near the Floholman reefs, west of the Island of Skogsøy, northern Norway (P. B. Mortensen, IMR, *www.mareano.no*, October 2007).

86 Scandinavian coral reefs [Ch. 4

As this new information was further analysed, the following, more complete description appeared on the Mareano website (translated from Norwegian):

"Mareano's autumn cruise off Vesterålen and Troms has disclosed gas seepage from the seafloor. Material on the seafloor indicates that methane gas and dissolved chemical components stream out of the seabed. Now geologists want to find answers to what type of gas this is.

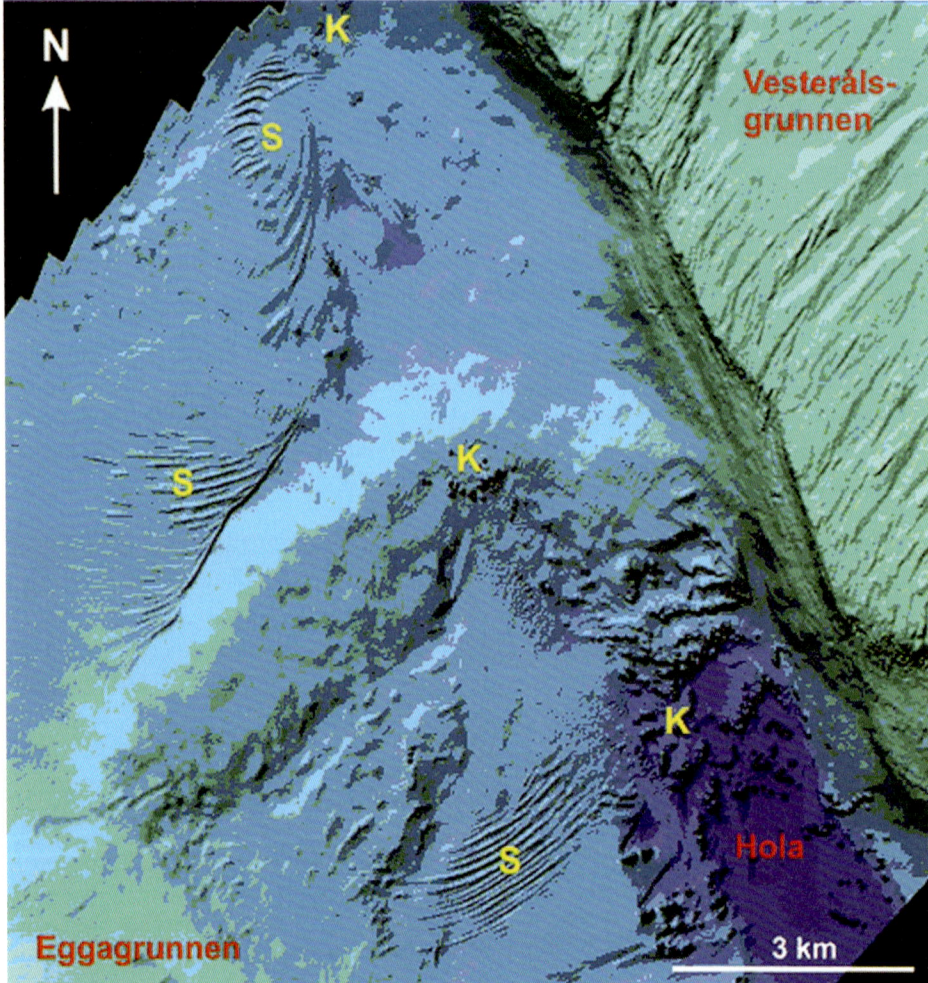

Figure 4.42. Terrain map from Hola. The occurrence of sediment waves (S) and coral reefs (K) on the seafloor are shown. The deepest areas (200 m–270 m) are in the darkest tones, whereas the areas on Vesterålsgrunnen (70 m–80 m) are shown in lighter tones. The map has been produced on the basis of depth data from Forsvarets Forskningsinstitutt (FFI) (the caption is translated from *www.mareano.no*, Bøe, 2007).

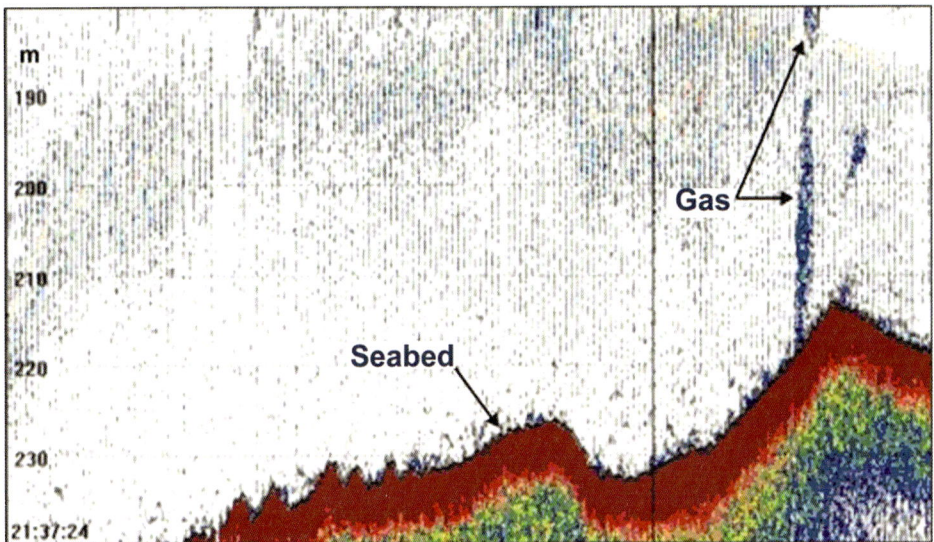

Figure 4.43. Echosounder data from Hola, collected during the Mareano autumn cruise onboard *G.O. Sars*, 2007. The image is from the area containing coral reefs and shows ascending gas bubbles in the water column (vertical column on the right, "gas"). The water depth is about 220 m. The gas possibly originates from Cretaceous layers beneath the seafloor, but further investigations including the study of seismic data are necessary in order to confirm this. Numbers on the left are water depths in metres (from www.mareano.no, Bøe, 2007).

On echosounder records, gas bubbles were observed in the water column, and it seems they come from the same area [as the carbonates and bacterial mats].

The area being investigated is in 'Hola', which is a deep portion of the seafloor, southwest of Vesterålsgrunnen. Hola is a trough between the two shallow banks Vesterålsgrunnen and Eggagrunnen, which was formed during the last ice age.

Between 200 and 300 coral reefs in the area between Vesterålsgrunnen and Eggagrunnen were mapped. They are seen as upward protruding mounds on the relatively flat seafloor in Hola. The coral reefs occur at water depths between 200 m and 270 m. Each individual mound is 20 m high and 150 m–200 m in diameter. The coral reef areas mainly lie in the northeastern section, near Vesterålsgrunnen, where the seafloor currents run towards the northwest. The area with sediment waves mainly lies in the southwest, near Eggagrunnen, where the dominating current component is towards the southeast.

We do not know the origin of the gas that leaks out of the seafloor in Hola. Geological maps show that the rock beneath the seafloor is shale and sandstone of the Cretaceous, i.e., 65–145 million years old. Some of the rocks have high organic content and may produce oil and natural gas. Large faults in these rocks have also been mapped. The gas is possibly leaking up along these faults, where open cracks may occur" (Bøe, 2007).

4.5.3 The Malangsreef

The most recent large-reef discovery off Norway was made in May 2007, by IMR of Bergen, onboard their research vessel *G.O. Sars* (see *www.mareano.no*). The 30 m high and over 1 km long, lush reef occurs on a ridge between two fishing grounds, Malangsgrunnen and Fugløybanken (Buhl-Mortensen, 2007). Its position is at 70°07.49′N, 18°04.84′E, which is only about 80 km southwest of the Fugløy reef, mentioned earlier. Whereas both the adjacent fishing banks are at water depths less than 150 m, the Malangsreef is located at water depths ranging from 290 m to

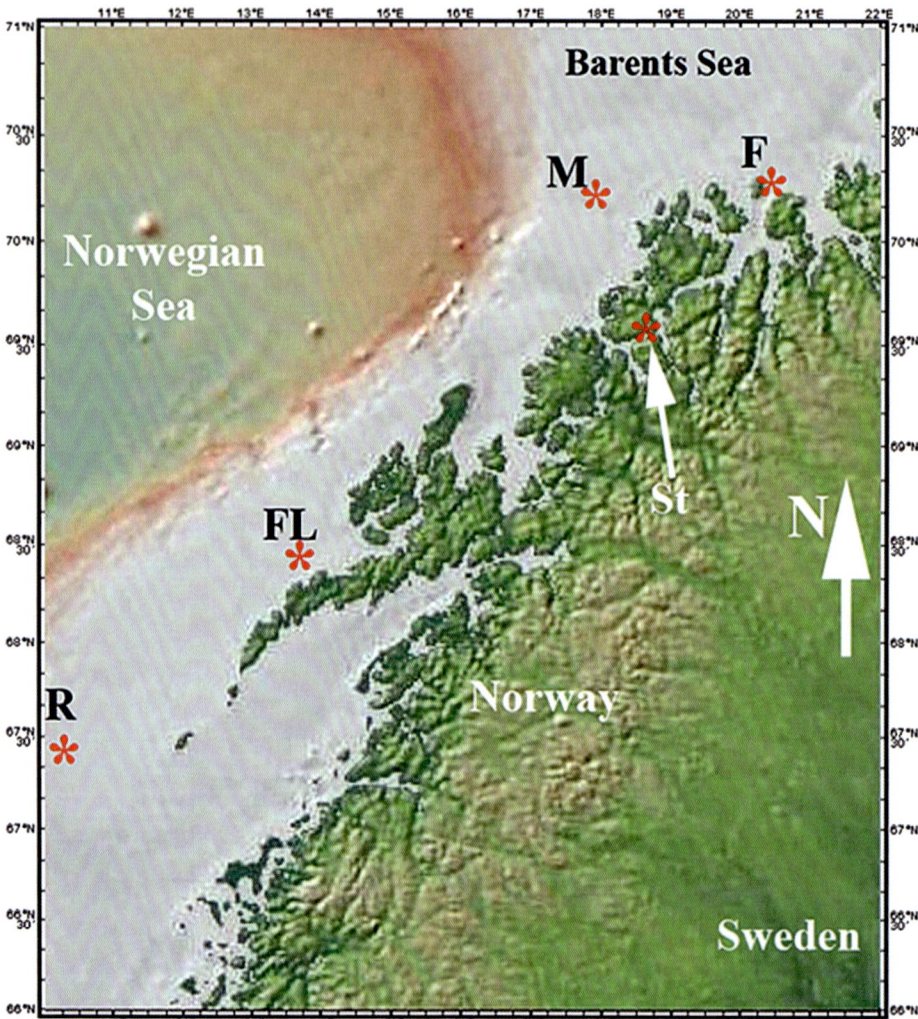

Figure 4.44. The location of Malangsreef (M). Other reefs shown are the Floholman reefs (FL), Fugløy reef (F), the Røst reefs (R), and the Stjernsund reef (St).

Sec. 4.6] Reefs on the continental edge 89

Figure 4.45. Photo from the lower portion of the Malangsreef. In the background is a pink *Lophelia* colony. The white variant is seen in the front, at right, together with many other species, including sponges, sea-anemones, and the red *Paragorgian* octocoral (gum-coral). The *Sebastes* fish at left is about 20 cm long. Two thin, red, parallel laser beams, 10 cm apart can be seen upper right and as red spots where the beams encounter the seafloor, centre (courtesy of Lene Buhl-Mortensen and *www.mareano.no*, 2007).

270 m:

> "On both sides of the ridge where the reef is located, there are deeper sediment basins, with soft seafloor where sea urchins, sea pens, and sea feathers (crinoids) are common" (Buhl-Mortensen, 2007).

The underwater photographs acquired from this reef are outstanding in their detail and clarity, because the researchers used a newly developed towed video-rig over the reef (see Figures 4.3, 4.45, and 4.46). There are, however, no published seismic sections across the reef on which to extract substratum information: basement rock, moraine, sand, or layered sediments?

4.6 REEFS ON THE CONTINENTAL EDGE

Lophelia reefs occur along many parts of the Storegga slope break. But these have only briefly been described scientifically (Fosså and Mortensen, 1998; Jung *et al.*, 2001). There are strong currents flowing along and over this extensive topographic feature. And there is also evidence of seepage in some areas. Bacterial mats and pockmarks have been found in the general Storegga slide area during a visual

Figure 4.46. A beautiful photo of a *Chimaera* sp., "flying" with its characteristic fins, above corals on the Malangsreef (courtesy of Lene Buhl-Mortensen and *www.mareano.no*, 2007).

inspection by an NR-1 US Navy nuclear submarine (Jung *et al.*, 2001). The location of the reefs on the edge of a major slide scarp either suggests nutrient upwelling by currents or that nutrients are charged regionally into the water column from disrupted slide material.

Another factor is also worth considering: the fact that numerous sedimentary bedding planes have been cut off and eroded, so that they become exposed to the water column after a slide disruption. This actually produces abundant possibilities for overpressured porewater to escape (i.e., fluid flow). Considering that the upper tens of metres of marine sediments have porosities of between 30% and 50%, their volume partially consists of interstitial fluids (fluids that occupy the internal pore-spaces). When large segments of layered sediments are removed by slides, there should be a considerable leakage of fluids out of the remaining head-wall scarp, where the sedimentary layers have been sheared off. The confining pressures have been removed, resulting in an "off-loading effect". The effect is analogous to that of cutting into a cactus plant, which will immediately start to sap or "sweat". The question is how long after disruption do the layers of sediment sap: tens, hundreds, or thousands of years? The process of such fluid sapping from sediments in gulleys is termed "spring sapping" (Orange *et al.*, 1999). Also, theoretical calculations show that sediments disturbed by landslides are prone to fluid flow (Huyen Bui, pers. commun., 2007).

4.6.1 The Røst Reef Complex

During spring 2002, IMR, Bergen, used acoustic techniques to map deep-water coral reefs along the shelf break. The largest ever *Lophelia* reef complex was discovered, measuring 35 km long and 3 km wide. It occurs at between 300 m and 400 m water depth along a steep and rugged part of the continental break. Because the closest island to the reef is Røst, at the very tip of the Lofoten archipelago, the reef was named the "Røst Reef" (Fosså and Alvsvåg, 2003).

4.7 CORALS AND REEFS IN THE NORTH SEA AND KATTEGAT

4.7.1 The Fedje reefs

The southern boundary of the Norwegian Sea is at the 62nd parallel, which is also the northern boundary of the North Sea. Compared with the Norwegian Sea continental shelf, the North Sea contains very few *Lophelia* reefs. The largest ones known to date occur about 8 km west of the island of Fedje (location, 60°45′N, 04°30′E, Figure 4.47). They are located on top of a few elongated ridges composed of Cretaceous or Tertiary sedimentary rocks (Rokoengen and Østmo, 1985; Hovland and Mortensen, 1999). Crystalline bedrock is exposed on the seafloor only 3 km to the east of the reefs,

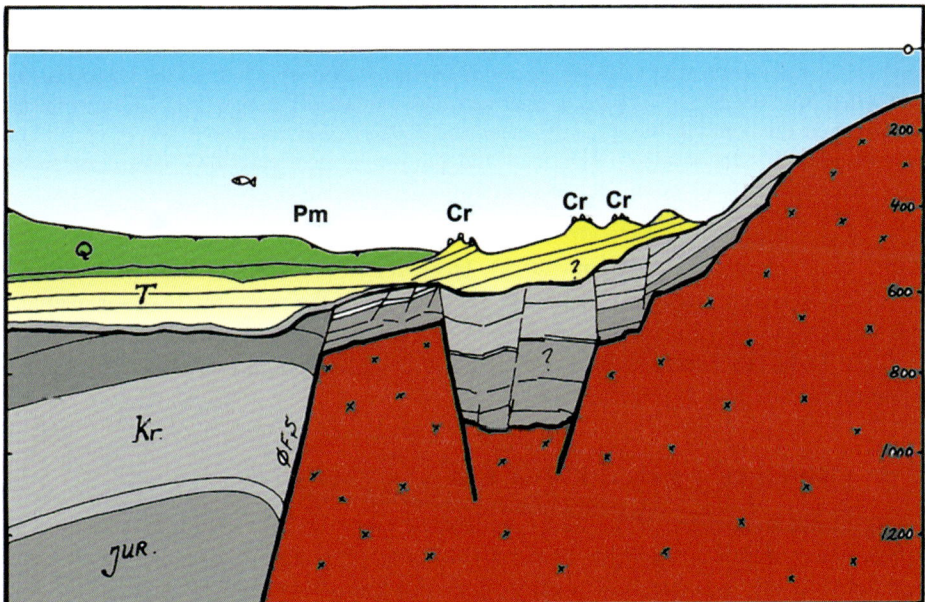

Figure 4.47. A geological and topographical section through the Fedje reefs (Cr). It is based on an interpreted 2-D seismic record across the coastline west of Fedje island, near Bergen, western Norway. ØFS = Øygarden fault zone, which forms a sharp boundary between Norwegian crystalline basement rocks (red), centre and to the right (east), and sedimentary rocks to the west. Cr = the deep-water coral reefs off Fedje island. Pm = pockmark craters, which indicate active leakage of gases through the seafloor. Kr = sedimentary rocks of Cretaceous age. Jur = sedimentary rocks of Jurassic age. Numbers on the right are depth in metres (from Hovland and Mortensen, 1999).

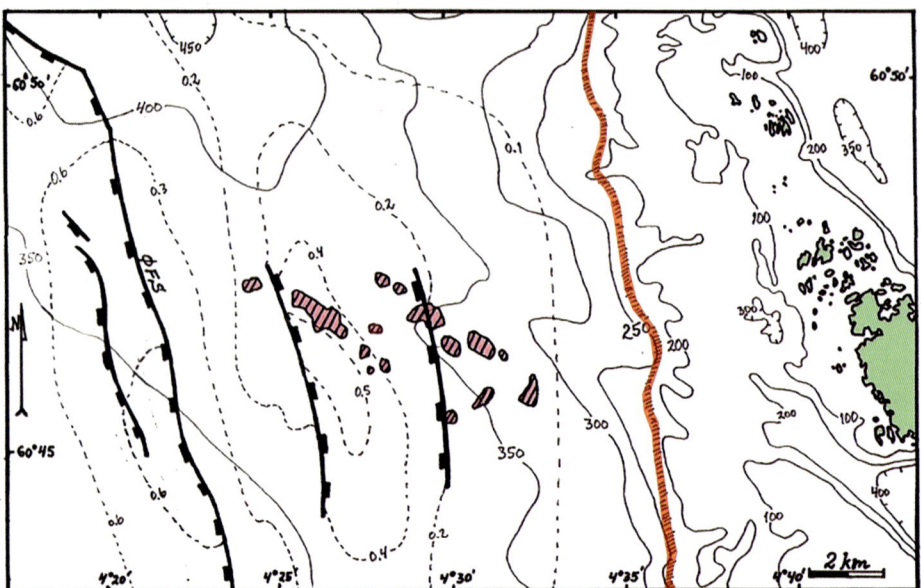

Figure 4.48. Map of the Fedje reefs (hatched "blobs"), located about 8 km west of Fedje island (seen on the right, east). The reefs live on top of a series of elongated hills on the seafloor in an area called *Nakken* (the neck) by local fishermen. The water depth ranges from 330 m to 380 m. The hatched line running north–south near the 250 m isobath indicates where crystalline bedrock intersects the seafloor. This map and the profile in Figure 4.47 have been partly constructed on the basis of Rokoengen and Østmo (1985).

at the same water depth (250 m–300 m). However, to our knowledge, there are no reefs located on the crystalline bedrock. To the west of the reefs, the seafloor deepens and pockmarks occur (Hovland and Judd, 1988; Judd and Hovland, 2007). Therefore, the Fedje reefs seem to be associated with sub-cropping sedimentary rocks and pockmarks (i.e., just like the HRC).

4.7.2 Corals inside a pockmark, Troll field

About 30 km northwest of the Fedje reefs is the large Troll gas-field. It is located at 310 m water depth inside the broad (>100 km) Norwegian Trench, which runs parallel with the southern and southwestern coast of Norway. There are numerous pockmarks in this area. They are up to 100 m in diameter and 8 m deep (Tjelta *et al.*, 2007; Judd and Hovland, 2007).

Statoil investigated some of the largest pockmarks over parts of the Troll field to find out more about the rate of natural sub-seafloor hydraulic activity (Forsberg *et al.*, 2007). On visual inspection using an ROV, they came across some large colonies of soft coral (*Paragorgia* sp.) inside one of these pockmarks. To our knowledge, this is the first time large corals (however, not reefs) have been documented in the Norwegian Trench. The find came as a surprise to the biologists as conditions

Sec. 4.8]　　　　　　　　　　　　　　　　*Lophelia* **colonies on man-made structures**　93

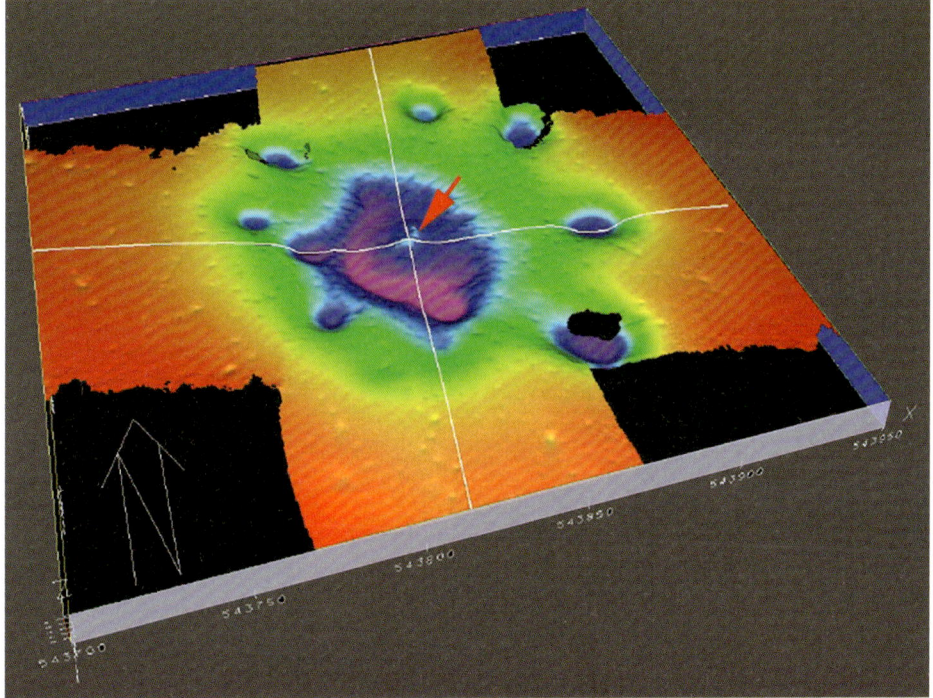

Figure 4.49. Map of a complex pockmark at Troll. There is one central (old) "mother pockmark" in the middle and seven smaller "satellite pockmarks" surrounding it. The two following photographs (Figures 4.50 and 4.51) were taken at the centre of the mother pockmark.

(or, rather, bio-expectations) are regarded to be far from perfect for such filter-feeding organisms. The two large *Paragorgia arborea* (one white and one red) individuals are perched inside an 8 m deep pockmark, which has a 1 m high layered, conical methane-derived carbonate rock protruding up from its centre (Figures 4.50 and 4.51). The corals are firmly based on this "natural concrete" substratum. Clusters of up to 30 *Acesta excavata* bivalves are also affixed to the same structure. Because these animals live at a depth of up to 6 m below the general seafloor, it is likely that they must tolerate frequent periods of heavy silting and sedimentation. This relatively prolific megafauna undoubtedly occurs here as a result of seepage-induced nutrient enrichment.

4.8　*LOPHELIA* COLONIES ON HUMAN-MADE STRUCTURES

4.8.1　The Brent Spar Buoy

That *Lophelia* corals build colonies on man-made structures as well became clear when the infamous Shell Brent Spar Buoy was recovered and salvaged instead of

Figure 4.50. These two large, white and red *Paragorgia arborea* (gum-corals) were found inside the Troll pockmark. A dense cluster of about 10 cm long *Acesta excavata* bivalves are attached to their stems and to the underlying rock (a methane carbonate).

being dumped onto the seafloor, as originally planned (Roberts and Gass, 2005; Skjærseth *et al.*, 2004). Perched on its base there were several *Lophelia* colonies:

> "No-one had found this cold-water coral living in the North Sea before, and no-one expected it would turn up on oil rigs. But the discovery wasn't a one off. More turned up in the Beryl oil field. And delving into the video libraries the oil and gas companies collect during their maintenance surveys, I've found *Lophelia* on 13 more platforms" (Roberts and Gass, 2005).

4.8.2 Statfjord

Similar colonies, up to 22 years old, have been found on pipes leading up to the giant concrete platforms at the Statfjord field, at 135 m water depth, northern North Sea. Figures 4.54 and 4.55 show photographs of these. According to Roberts and Gass (2005) and Gass and Roberts (2006), the rigs present perfect structures for the *Lophelia* to attach to, which are protected from trawling activity. Thus, *Lophelia*

Sec. 4.8] *Lophelia* **colonies on man-made structures** 95

Figure 4.51. This large, apparently layered carbonate rock, a methane-derived authigenic carbonate, MDAC (or methane carbonate) serves as the foundation (or substrate) for the paragorgians and *Acesta*. Actually, this feature resembles and may fit the description given to methane reefs by Han *et al.* (2008).

pertusa can proliferate, even in the North Sea. Furthermore, besides peace (refuge from trawling), there are plenty of currents and nutrients around the platforms.

These particular modern finds clearly demonstrate the true cosmopolitan nature of *Lophelia pertusa*. If it is granted a relatively firm substrate, relatively strong currents, a stable nutrient source, and is protected from mechanical disturbance, it will thrive even in "busy", noisy, and supposedly polluted industrial environments.

4.8.3 UK Block 22/4, blowout site

During exploration drilling in UK Block 22/4, on November 21, 1990, the operator lost control over the well, which led to a blowout situation ($57°55'$N, $01°38'$E). The well was shut down and abandoned. However, gas has been issuing out of this well ever since, and the location is marked on navigation charts with a warning to shipping. It was not before diving with the German research submarine *Jago* (Alkor Cruise 259) that the amount of currently emitting bubbles (about 10^3 m^3 per hour) was seen. The blowout crater is about 10 m deep, and on the flank of this crater some

Figure 4.52. Arrows point at living (articulating) *Acesta* individuals. They are apparently thriving even though they are partly covered by sediments. They seem to be embedded in the methane carbonate rock.

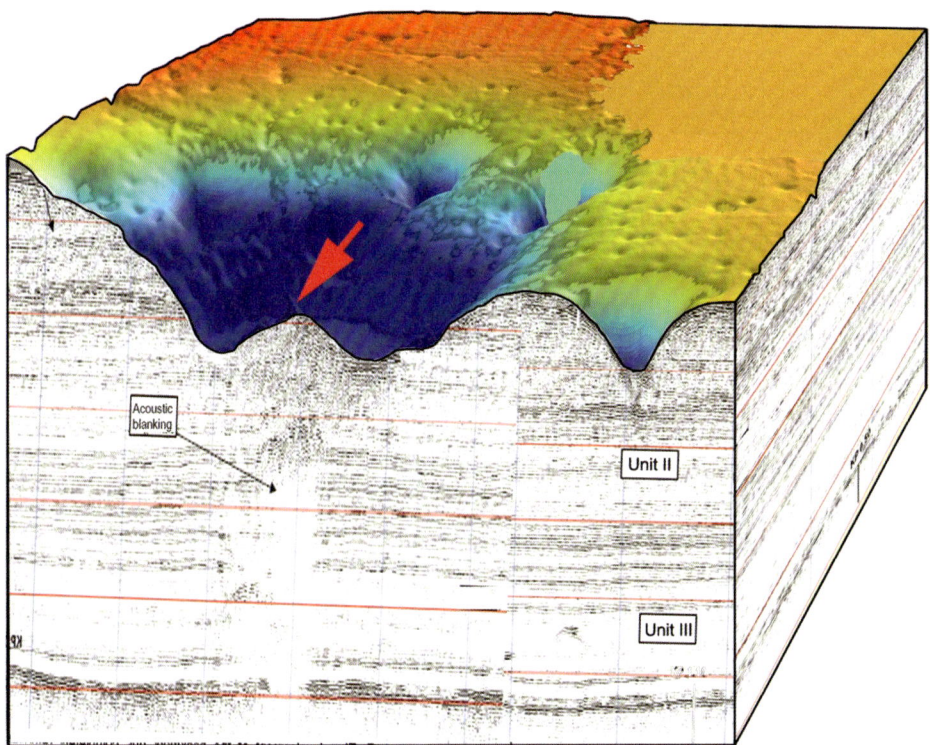

Figure 4.53. The location of the Troll pockmark corals and *Acesta* bivalves is indicated by a red arrow. This drawing is made as a composite image based on real sub-bottom profiler and bathymetric data from the actual location. Note the enhanced reflections and acoustic blanking below the pockmark, indicating gas charge and gas migration. The sides of this block diagram are about 800 m by 600 m by 30 m (the vertical scale is exaggerated).

abandoned trawler wires were found with some small (about 20 cm high) colonies of live *Lophelia* attached (Peter Linke, IFM-Geomar, Kiel, pers. commun., 2008). This is one of the very few sitings of *Lophelia* in the North Sea. More information about the project COMET (controls on methane fluxes) can be found at *www.ifm-geomar.de*

4.9 SWEDISH/NORWEGIAN REEFS

4.9.1 Kosterfjord reefs

Along the Swedish and Norwegian eastern coast of Kattegat (the strait between Denmark and Sweden/Norway), a series of relatively large *Lophelia* reefs have been documented (Lundälv and Jonsson, 2003). One of these, the Tisler reef was found by researchers at Tjärnö Marine Biological Laboratory (Stömstad, Sweden). It is located on the Norwegian/Swedish border as it sits on a sill in the narrow channel connecting

98 Scandinavian coral reefs [Ch. 4

Figure 4.54. A *Lophelia* colony (lower right) on a human-made structure at Statfjord, Norwegian sector of the North Sea. The colony cannot be older than about 22 years (age of structure).

the Kosterfjord with the rest of the Skagerrak water mass, north of Tisler (small island) in Yttre Hvaler (Norway). It measures 1.2 km in length and is about 200 m wide. On June 11, 2003, the Tisler reef was protected by law against trawling activity. Other living corals occur at between 75 m and 160 m water depth. Here, yellow varieties of *Lophelia pertusa* were documented for the first time. There are several more reefs in the Kosterfjord/Yttre Hvaler area, centred at 58°58.7′N, 11°01.6′E. Two reefs occur on a morainic sill at Säcken (Sweden), north of Strömstad, at 82 m–87 m depth. One of these covers an area of about 250 m^2, the other 50 m^2 (Nilsson, 1997; Jonsson and Lundälv, 2001). Both reefs have also been protected against fishing.

4.10 DANISH "BUBBLING REEFS"

Although there are no known *Lophelia* reefs in Danish waters, there is a very interesting "reef" type, the so-called "bubbling reefs" in the Kattegat, off

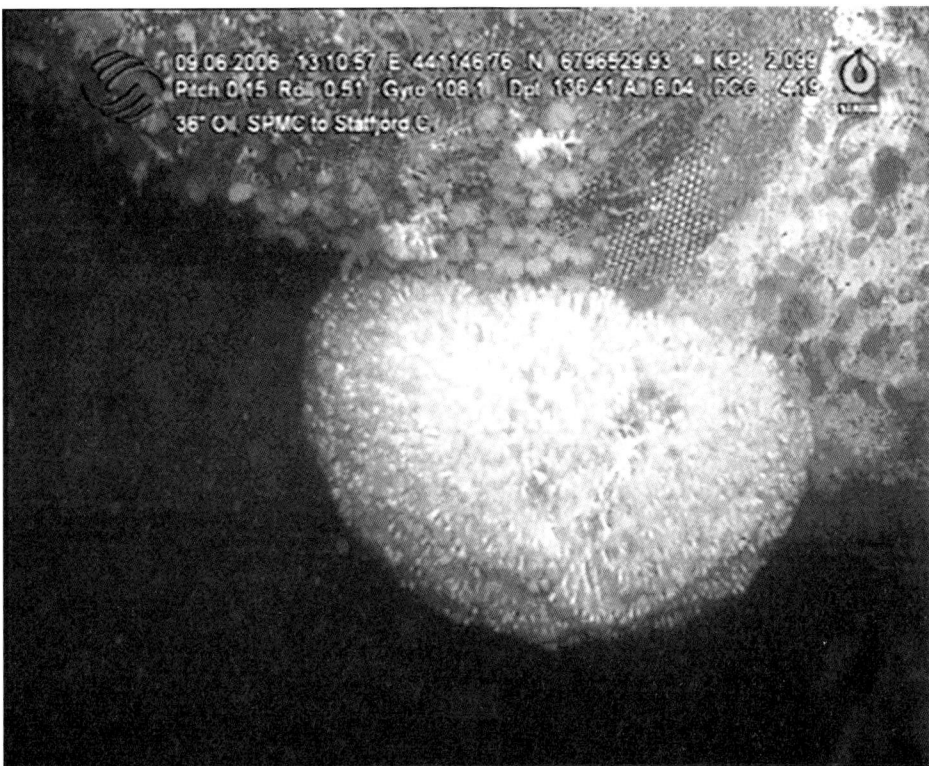

Figure 4.55. Same human-made structure as above, but a different colony. Here the whole colony, estimated to be 50 cm across, is seen.

Fredrikshavn, northeast Denmark. Shallow gas, gas seepage, and carbonate-cemented sandstones are quite common along the northeast coast of Denmark, around the island of Læsø and farther east in the Kattegat. The carbonates are lithified sandstones, which generally take the form of individual slabs and (more commonly) thin "pavements". Most spectacular are the pillars and mushroom-shaped bodies which stand up to 2 m above the seabed. Some pillars are almost 4 m tall and 1.5 m in diameter (Jensen et al., 1992). The pillars are not solid, but have multiple, poorly cemented, vertical, pipe-like structures inside them. It is probable that these features were originally formed beneath the seabed, and were subsequently exposed by the erosion of seabed sediments, probably during the post-glacial period of isostatic uplift (Jørgensen, 1992; Judd and Hovland, 2007).

Jørgensen (1992) reported that carbon isotope values of the cements are generally between −40‰ and −50‰, although values of between −33‰ to −61‰ were measured. These Danish bubbling reef carbonates, including the pillars, represent a specially enriched habitat for both microfauna and macrofauna (Jensen et al., 1992). Thus, these resemble the newly described methane reefs found in the South China Sea (Han et al., 2008).

4.11 REEFS OFF THE FAROE ISLANDS

As a last outpost against the 3,000 m deep Norwegian Greenland Sea lie the Faroe Islands. The shelf on which the islands reside is between 100 m and 300 m deep and stretches about 100 km out in all directions. During the period 1987–1990 the two fishery research vessels *Håkon Mosby* and *Magnus Heinason*, under the Marine Benthic Fauna of the Faroe Islands, BIOFAR programme, performed a systematic sampling and hydrographic study of the Faroe banks and also the adjacent banks (Lousy, Bill Bailey, and Faroe Bank). Major banks to the south, Hatton (500 km southwest of the Faroe Islands), Rosemary, and the deeper Wyville–Thomson Ridge, were also included.

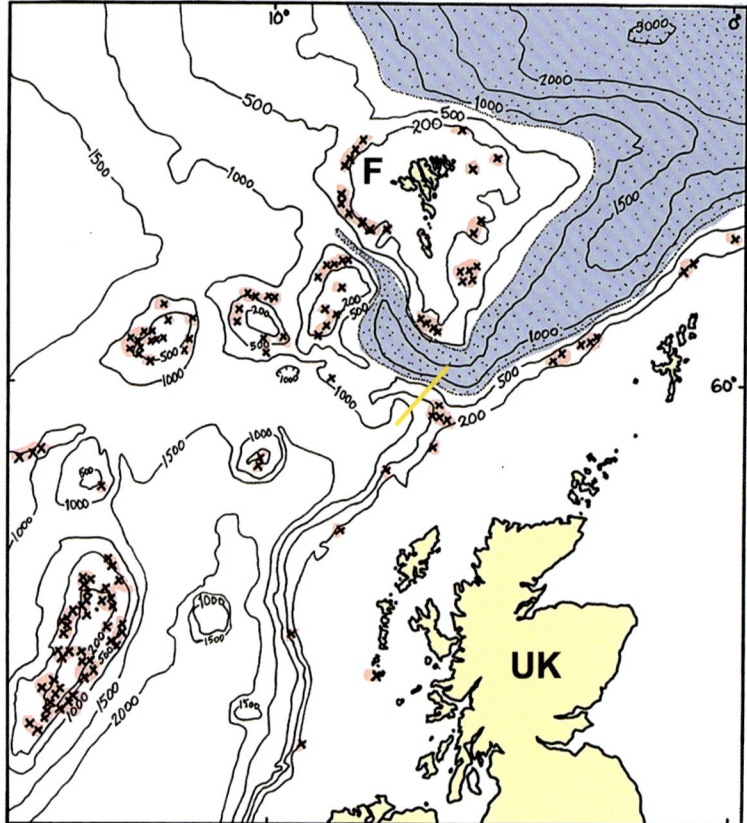

Figure 4.56. Map over parts of the northwest Atlantic margin showing the Faroe Islands, the Rockall bank (lower left), and adjacent areas. Red/black crosses: mapped *Lophelia* reefs. Light blue: deep, dense water from the Norwegian Sea Basin, which is normally colder than 0°C (i.e., North Atlantic Deep Water, NADW). The oil-fields Schiehallion and Foinaven (not shown) are located to the west of the Shetland Islands. The yellow line shows the location of Figure 5.16.

Figure 4.57. A rare photograph of a great fork-beard (*Phycis blennoides*) living inside the Troll pockmark, where clusters of *Acesta* bivalves, large octocorals, and methane carbonate (reefs) were found.

Lophelia reefs were most commonly found on the northwestern side of the Faroese shelf. The deepest occurrence was at 600 m, although fishermen say that they occur down to 800 m (Frederiksen *et al.*, 1992). However, there are no reefs towards the northeast, where deep, cool water from the Norwegian Sea Basin drifts over the shelf (Figure 4.56). The water is probably too cold at times for *Lophelia*. On Bill Bailey bank there is abundant *Lophelia pertusa* and *Madrepora oculata*. Here *Lophelia* occurs from 275 m to 1,020 m water depth. On Hatton bank, *Lophelia* and *Madrepora* are also abundant. This bank lies closer to Rockall bank than to the Faroe Islands. In one case it was noted on the echosounder that one of the *Lophelia* reefs was up to 30 m high (Frederiksen *et al.*, 1992).

The researchers concluded that *Lophelia* occurrences had a tendency to be concentrated on the shoulder of the continental shelf, and that it had to be special currents and other oceanographic conditions such as internal waves that determined their whereabouts. Their theory on internal waves is discussed in Chapter 8.

4.12 SUMMARY

Off Scandinavia, cold-water corals and coral reefs occur on top of morainic sills in some of the fjords, along parts of the upper portion of the continental slope, and at specific locations on the continental shelf. For each location described, the origin and source of a stable food source is discussed. My conclusions are as follows: fjord-dwelling reefs (i.e., in Stjernsund, at Agdenes, Tautra, and several other fjord locations) probably rely on a combination of nutrients concentrated by strong tidal currents and by nutrients brought up through the morainic sub-surface sediments together with groundwater and dissolved gases.

Coastal reefs, such as at Fugløy in the north and Fedje in the south are also located on ridges where pockmarks occur in adjacent sediment basins. Here, it is concluded that these also rely partly on nutrients produced locally by hydraulic pumping of porewater and gases through the seafloor. On the continental shelf there are two main reef types:

(1) those that grow on ridges and highs with dipping, often acoustically enhanced sediment bedding underneath (the Sula Ridge, HRC reefs, Træna Deep reefs, and the Malangsreef); and
(2) those that grow over hydrocarbon fields, such as at Husmus, where gas hydrates are suspected to form immediately below the sediment surface, and at Kristin, where they grow inside pockmarks, and at the Morvin field, where they grow in elongated patterns, indicating forcing by prevailing bottom currents.

At the Troll field, in the Norwegian Trench, *Paragorgia arborea* and *Acesta excavata*, two species commonly found together with *Lophelia*, occur inside a large, 8 m deep pockmark. This indicates that they have found good living conditions here, where gases migrate out of the seafloor at regular or irregular intervals. Why *Lophelia* is not found at this location is not known. The most recent discovery of deep-water *Lophelia* reefs off Norway was made off the small island Floholman in Vesterålen. They occur near the *Hola* (the hole), where gases are known (from echosounder images) to emit through the seafloor and where both bacterial mats and methane-derived carbonate crusts are found. So, here is a location where there is no doubt that a significant and stable amount of primary producers occur, independently of seasonal variations.

The largest reef occurrences are those that occur on the upper slope of the continental shelf, the Røst Reef Complex. No relationship with conditions in the sub-stratum has been found here, although a possibility is that they grow close to slope failures and fault scarps (similar to the Storegga scarp), where nutrients can easily escape from the sub-surface for long periods after slumping and slope failure events. Finally, there is a review of deep-water corals living on man-made structures, including *Lophelia* occurring on structural steel pipe members at the producing Statfjord field and also on steel wires from fishing activity near a gas blowout site in the UK sector of the North Sea.

5

North Atlantic coral reefs and giant carbonate mounds

The acceptance of a simple model of mound formation is unsatisfactory because of the observed variation in size, morphology and the prolific number of mounds with their sometimes very localized clustering.

Van Weering *et al.* (2001)

The key linking feature between a wide range of features, from small, seafloor build-ups, to isolated carbonate shoals such as Heywood Shoals, to large clusters of banks such as the Karmt Shoals, is that they are all associated with modern seeps. Where seeps are absent, so (generally) is the evidence for significant contemporary carbonate bank formation.

O'Brien *et al.* (2002)

5.1 INTRODUCTION

Since the pioneering visual documentation of deep-water coral reefs on the Porcupine and Rockall banks, performed by John B. Wilson, using a manned submarine (Wilson, 1979a, b), technology has come a long way. Today, the mapping of similar features along the northwest coast of Scotland, as performed by James Murray Roberts *et al.* (2003), is instead done using remote-sensing high-tech equipment, without having to put anyone in the water. This chapter mainly deals with the impressive and enigmatic giant carbonate mounds off Ireland. The story of their re-discovery in 1994 is told, when several survey vessels ventured there and found many different carbonate mounds and deep-water corals. In this chapter the Hovland, Magellan, Logachev, and Belgica Mounds are described and discussed. So are the strangest features of them all, the Darwin Mounds, located within the deep Rockall Trough, adjacent to a large field of pockmarks and adjacent to the

Figure 5.1. One of the "Kess-Kess" fossil reefs, Moroccan Sahara. The human figure at the top is Anneleen Foubert, now with Total (courtesy of J.-P. Henriet).

Wyville–Thomson Ridge. The findings of the first ever scientific drilling of a modern carbonate mound is also reviewed and discussed. The chapter includes the description and discussion of carbonate mounds and coral reefs found in the Gulf of Cadiz and no less important, near numerous mud volcanoes on the continental slope off Morocco. Finally, giant carbonate mounds forming long, linear structures off Mauritania are also reviewed.

5.2 THE ATLANTIC MARGIN

Along the northwest coast and elsewhere off the United Kingdom, there are many known "normal" deep-water coral reefs. However, after finding the impressive *Lophelia*-capped giant carbonate mounds off Ireland, these have more or less been overshadowed (Figure 5.2). The new types of reefs are so impressive and shrouded in such mystery that they naturally steal the limelight. However, for those who want

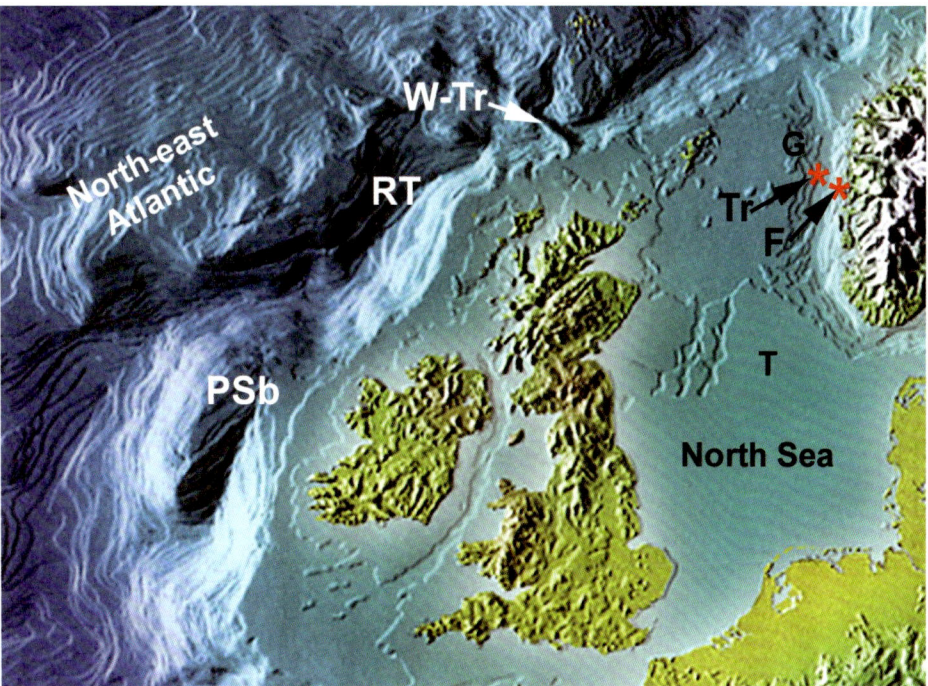

Figure 5.2. A general shaded relief map of the northeast Atlantic Ocean, Ireland, the British Isles, the North Sea, and parts of Norway and Denmark. It shows the locations of the Porcupine Seabight (PSb), Rockall Trough (RT), Wyville-Thomson Ridge (W-Tr), Gullfaks field (G), Troll field and pockmark coral location (Tr), Fedje *Lophelia* reefs (F), and Tommeliten gas seeps (T). See also map (Figure 5.8) for more details in the western region.

more information on, say, the Mingulay Reef Complex (deep-water coral reefs west of Scotland), the best is to either look up the original text, or to read the "bible" published after the great "Cold-water corals and ecosystems meeting" in Germany, 2002 (Freiwald and Roberts, 2005). There are currently many exciting developments in the field of marine sciences, auguring well for the discovery of many deep-water coral reefs:

> "The unexpected discovery of a cold-water coral reef of this scale in previously well-studied waters suggests that such reefs may be far more widespread than previously thought. Possibly, on continental shelves world-wide" (Roberts *et al.*, 2003).

The area to the west of the British Isles, extending from the Shetland Islands southwards beyond Ireland, has attracted significant interest as the oil industry has moved into deeper waters. As well as petroleum exploration, detailed environmental assessments have been undertaken. Although scientists became aware of deep-water coral reefs during the 1970s, the general public was mostly ignorant (Wilson, 1979a). But, this changed, especially after the following news story in the *Sunday Times*, June 22, 1997, which clearly portrays what is needed to achieve general public awareness ...:

> "**Oil team finds giant British coral reef**. A GIANT coral 'Barrier Reef' running parallel to Britain's northern coasts has been found deep under the Atlantic in areas previously considered among the coldest and most barren on earth. The reef, thought to be thousands of years old, was built by coral polyps ...
> ... it was discovered by researchers attempting to assess the environmental damage that may be caused by companies granted licences to drill for oil in the sea northwest of Scotland.
> ... Dr Sian Pullen, head of marine conservation at the World Wildlife Fund for Nature, said the government had issued the licences without any real idea of possible damage. 'There should have been a full assessment of just what is down there before the licences were handed out,' he said." (Leake, 1997).

Yes, indeed; but, it may be worth asking the following question: Who would have known about these fantastic occurrences without the oil industry doing the detailed mapping in the first place?

There is no doubt that the great carbonate mud mounds capped with coral off Ireland surpass "normal" *Lophelia* reefs, perhaps not in numbers, but in size. Thus, they are up to kilometres across at their base and up to 400 m in height! They could have been called "pinnacle reefs", a term used for ancient deposits in geology, but they are called "carbonate mounds". Mounds of various descriptions have been reported from both the Rockall Trough margin and the Porcupine Bight (Figure 5.5). Since 1994, work on them has been undertaken by the oil industry, the TTR (Training Through Research) cruises of the UNESCO-funded *Floating University*, and several large, multi-national/multi-disciplinary projects funded in part or wholly by the European Commission: ECOMOUND, GEOMOUND, ACES, IODP, and HERMES.[1]

A review and discussion of the Porcupine Bight Mounds; the strange, and much smaller Darwin Mounds, and the Rockall Trough Margin Mounds follows. They are all morphologically and biologically different from each other, and occur in differing geological settings.

5.2.1 Porcupine Seabight Mounds

A couple of months after publication of my *Terra Nova* article "Do carbonate reefs form due to fluid seepage?" (Hovland, 1990b), some puzzling mail arrived, two

---

[1] HERMES = Hotspot Ecosystem Research on Europe's Deep-Ocean Margins; ACES = Atlantic Coral Ecosystem Study; ECOMOUND = Environmental Controls on Mound Formation along the European Margin; IODP = Integrated Ocean Drilling Program; GEOMOUND = Geological Controls on Mound Formation.

brown envelopes containing remarkably similar 2-D seismic data. One had been sent by Peter Croker, of the Irish Petroleum Department. The other was from Mike Martin in Perth, Australia. They both wanted to know if the mound-shaped features seen on the enclosed 2-D seismic records could fit into my new model (Figures 5.3 and 5.4). One of the images was from 800 m water depth in the Porcupine Basin, off Ireland, the other was from the Vulcan Sub-Basin, off Western Australia (water depth 300 m).

After carefully checking numerous data concerning the areas in question, and also relying on previously published information, especially by Croker and Shannon (1987), a new paper was published: "Fault-associated seabed mounds (carbonate knolls?) off western Ireland and north-west Australia" (Hovland *et al.*, 1994a). However, as authors, we were ignorant of how fast this information would be utilised. It proved to become instantly used by several research teams, and it evidently spurred the Belgian research vessel *Belgica* and the Russian vessel *Professor Logachev* to cruise over the knolls off Ireland.

Hovland Mounds

Originally, the Porcupine Bight Mounds were thought to consist of only 31 large mounds which occur in a 15 km by 25 km area on the northern slope of the Porcupine Basin where the water is 600 m to 900 m deep. These "Hovland Mounds" are up to 150 m high and composed primarily of carbonate mud with large quantities of debris from *Lophelia pertusa* and *Madrepora oculata*. Today, there is an estimated 39 mounds in this province. Some of them are located inside sediment depressions, called "moats", possibly formed by currents sweeping away sediments. However, these depressions can also have formed as a consequence of gas seepage (Judd and Hovland, 2007) removing fine-grained sediments, just as in pockmarks. The oil-field Connemara is located immediately to the north, at water depths of 250 m to 400 m. In this area, 3-D seismic data show that there are pockets of shallow gas at various depths right up to near the seafloor surface. There are also numerous pockmarks, evidencing significant hydrocarbon migration through the seafloor (Games, 2001).

Belgica Mounds

Located southeast of the Hovland Mounds and the adjacent Magellan Mounds are the large and impressive Belgica Mounds (Henriet *et al.*, 2001; De Mol *et al.*, 2002; Huvenne *et al.*, 2002). A total of 66 conical mounds have been mapped with multibeam (swath) bathymetry and sidescan sonar. The discovery of the Magellan and Belgica Mounds in May 1997 was published in *Nature* as follows:

> "During a recent cruise in the Porcupine Basin, off southwest Ireland, we discovered two extensive and hitherto largely unsuspected deep-water reef provinces, including a giant cluster of hundreds of buried mounds. The ring shapes of many reefs suggest that they are caused by an axial fluid expulsion at the sea bed, a transient flow well confined in space and time. We are exploring various hypotheses, but a stimulating avenue for research is opened by a glacially

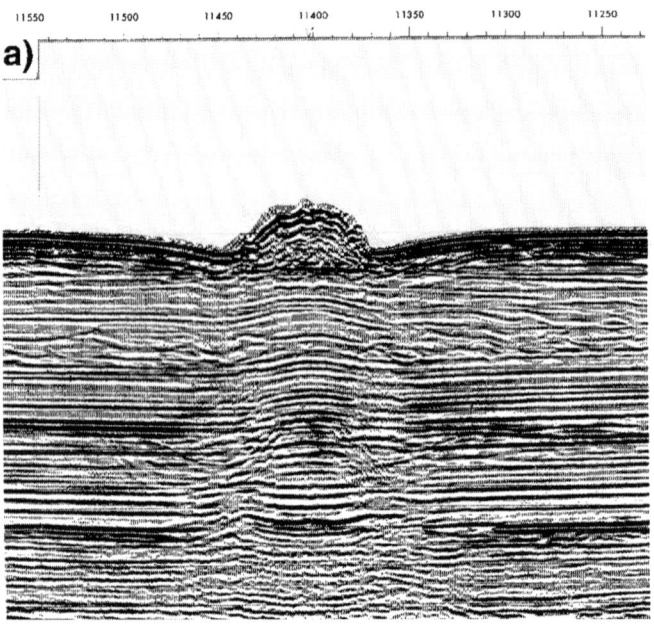

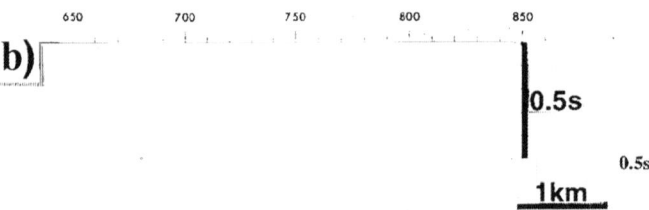

Figure 5.3. The original image that proved the existence of giant carbonate mounds off Ireland. This image was interpreted according to Hovland (1990b)'s ideas of seepage, and was combined with data from Australia (see Figure 5.4).

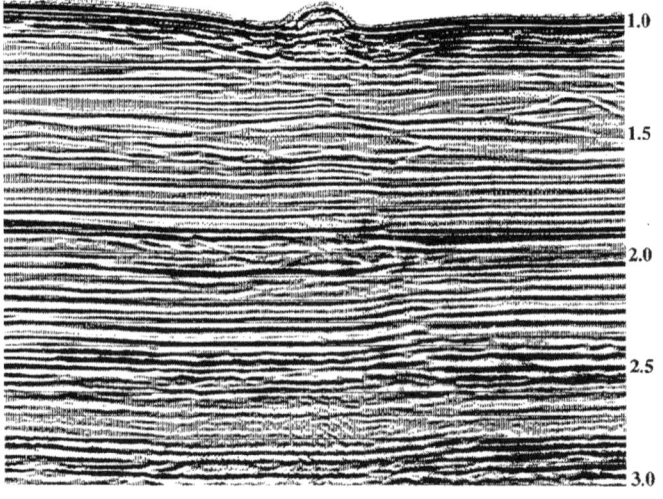

Figure 5.4. The remarkably similar 2-D seismic image of a live, giant carbonate mound from the Vulcan Sub-Basin, off northwest Australia (i.e., on the other side of the globe, Hovland et al., 1994a). This stirred our imagination and was the incentive to make us assemble enough comparable data to publish a new article in 1994 (Hovland et al., 1994a).

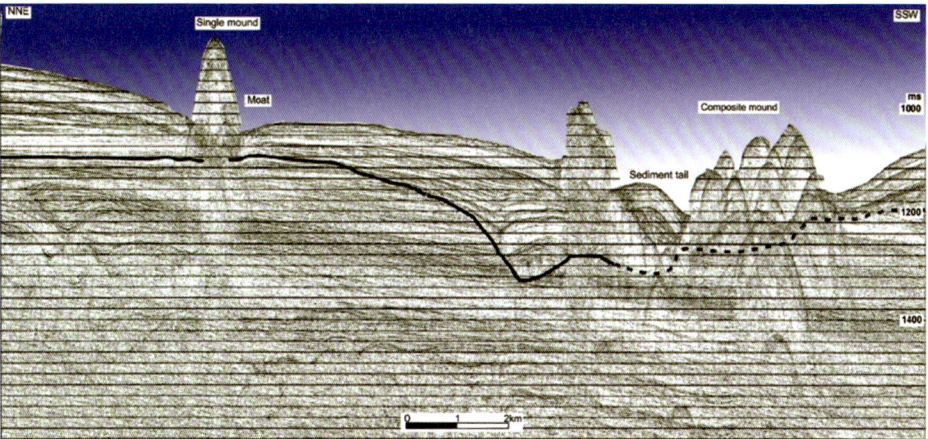

Figure 5.5. A high-resolution seismic image across the carbonate banks in the Porcupine Seabight. These are examples of Hovland Mounds and can be clearly seen sitting inside seabed depressions (courtesy of Ben De Mol, and modified from De Mol et al., 2002).

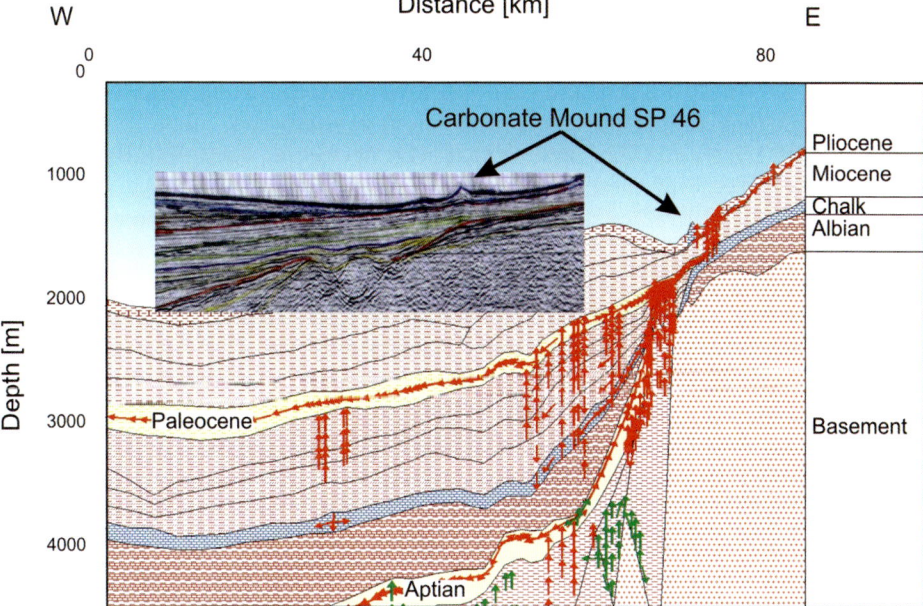

Figure 5.6. Results of a 2-D hydrocarbon basin modelling (migration pathway) study (Naeth et al., 2005). There is no doubt that hydrocarbons are able to form in the basin and that they will migrate up to the location where the carbonate banks are currently growing. The main reason for this is the large body of crystalline basement rock seen on the right. The arrows represent migrating hydrocarbons.

controlled growth pulse and subsequent decay of a shallow layer of gas hydrates as a methane buffer and probably indirectly as a ground for overlying biological communities" (Henriet et al., 1998).

This hypothesis is partly analogous to the interpretation of the Husmus reefs (Section 4.2.6), where dissociating gas hydrates are suggested as a source of local nutrients. However, I do not agree with the idea of "a transient flow well confined in space and time", as there is probably still some sort of active fluid flow today. However, which chemical is flowing (H_2, CO, CO_2, and/or CH_4) is unknown (see discussion in Section 5.1.6).

However, researchers searching for a proof of this "simple theory" could not find it:

"No increase in hydrocarbon concentrations was found in the mound and shallow sub-mound sediments although present at much deeper levels. Instead, cold-water corals occurred throughout the whole mound succession thus supporting an alternative hypothesis that suggests biological and environmental controls as driving factors for mound formation and development" (Dorschel et al., 2007).

The evidence used for this conclusion was reached after some current measurements performed over a period of only 17 days near Galway Mound, one of the "live" Belgica Mounds. These competing theories will be discussed in Chapter 8.

Later studies of Belgica Mounds have proven that they occur at water depths ranging from 550 m to 1,225 m and that they tower above a strongly eroded seafloor surface. Single and elongated clusters of the mounds partly root on an "enigmatic", deeply incised, faintly stratified seismic facies (De Mol et al., 2002; Van Rooij et al., 2003). Whereas some of these up to 170 m high giant structures have a variable coverage of corals, there are some that are more or less dead, and there are some that are very "lively":

"In the deeper part (>900 m water depth) of the Belgica mound province (Beyer et al., 2003), an extremely 'lively' mound was discovered in 1998 on the basis of a very diffuse surface acoustic reponse. This mound, known as the Thérèse Mound, was selected as a special target site to study processes involved in mound development for European Union (EU) Fifth Framework research projects. Video imaging revealed that Thérèse Mound, jointly with its closest neighbour, Galway Mound, might be one of the richest cold-water coral environments in Porcupine Seabight, remarkably in the middle of otherwise barren mounds" (Wheeler et al., 2007).

To me, this is exactly what I would expect if the living species within and on the mounds were fed by internal "hydraulic plumbing systems". The activity of such systems tends to vary in time and space; they turn on and off according to deep, complex fluid flow and pressure pulses (Hovland and Judd, 1988; Henriet et al., 1998; Judd and Hovland, 2007).

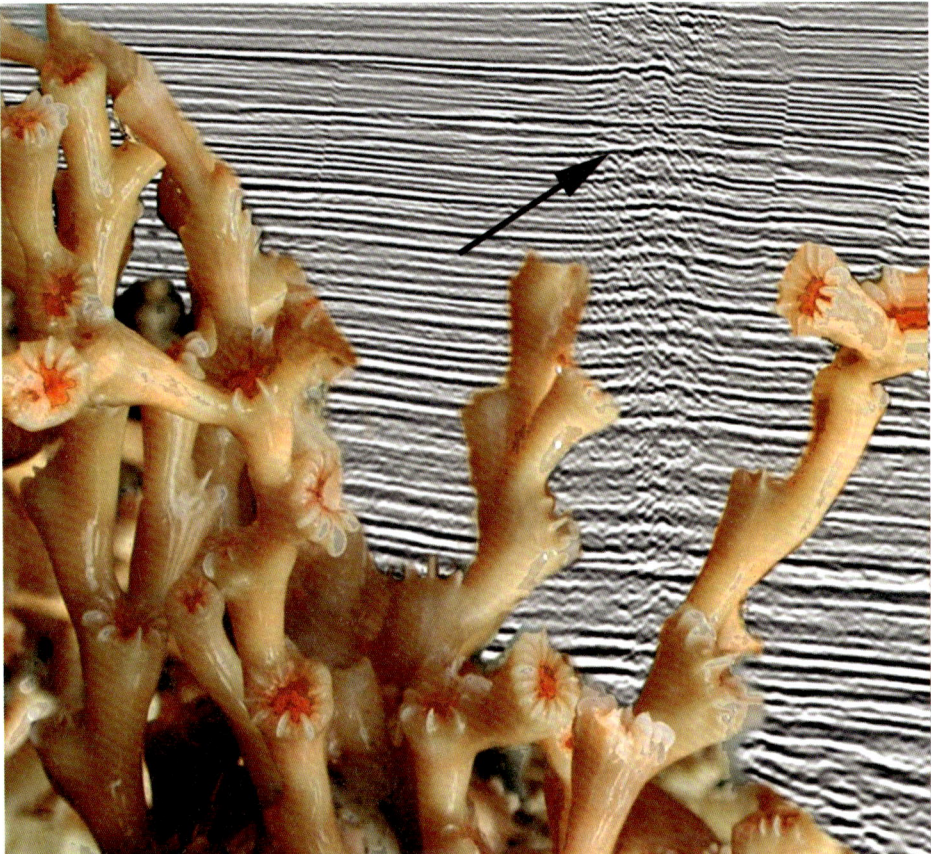

Figure 5.7. This is a photo-montage that figured on the front cover of the *Journal of Petroleum Geology*, where a paper by Naeth *et al.* (2005) suggests that hydrocarbon gases are migrating through the sediments at the Porcupine carbonate mounds. The arrow points at a feature on the 2-D seismic record suggesting a "gas chimney" structure. In the foreground is a newly picked (live) branch of pink *Lophelia pertusa* from one of the reefs in the area (slightly modified from Naeth *et al.*, 2005).

Van Rooij *et al.* (2005) provided a good description of how Belgica Mounds are associated with sediments moving along the seafloor in the form of sandwaves. This description vividly proves the plentiful occurrence of mobile silty sand in parts of the Porcupine Bight. The available mobile sand must be the main reason some of the mounds become buried and die:

"Side-scan imagery of [Belgica] mounds reveals that they are largely covered in sediment waves. The topography of the mounds (often up to several 100 m high) provides an obstacle to basal currents causing acceleration in current speed and facilitating sediment transport in the form of migratory waves. A grading of the

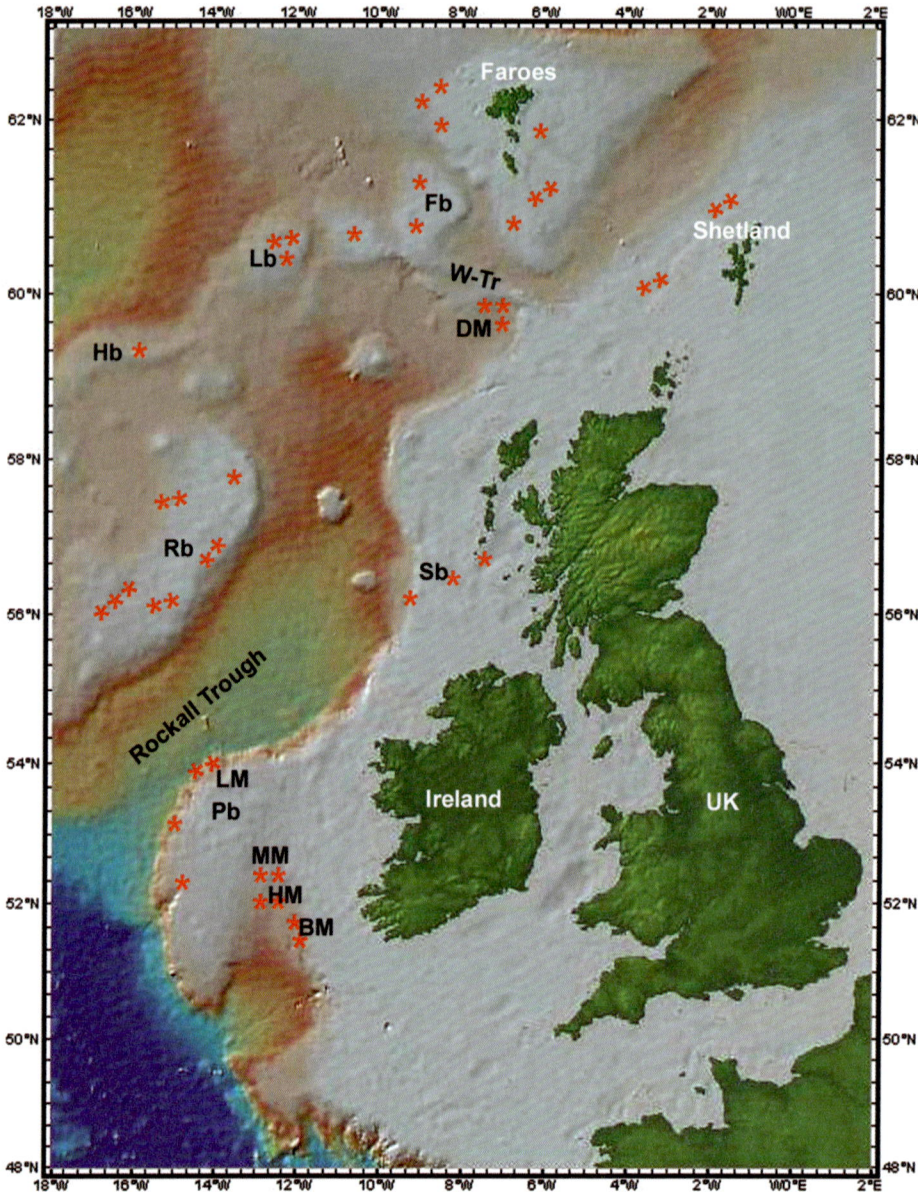

Figure 5.8. Map from the Faroe Islands, in the north, to the Porcupine Basin, in the south, showing various types of coral reefs and carbonate mounds. The red asterisks indicate documented *Lophelia* corals (according to Roberts et al., 2003 and others, see text). BM = Belgica Mounds, HM = Hovland Mounds, MM = Magellan Mounds, Pb = Porcupine bank, LM = Logachev Mounds, Rb = Rockall bank, Sb = Shanton bank, Hb = Hatton bank, DM = Darwin Mounds, W-Tr = Wyville–Thomson Ridge, Lb = Lousy bank, Fb = Faroe bank.

sediment wave styles exists in relation to mound proximity and position up the flank of the mounds. Coral accumulations occur on all parts of the sediment wave although a tendency for preferential growth on crests and upper flanks has been observed. Further away from the foot of the mound, coral accumulation density decreases and they occur on the wave crests and upper flanks and not in the troughs. Other mounds in the Belgica Mound area showed no evidence of coral accumulations and may therefore be relic mounds or possess limited coral activity and are becoming buried by the sediment waves.

On the flanks of the easternmost Belgica mound ridges (where some of the mounds are partly buried below drift sediments), current velocities were sufficient to produce a succession of barchan dunes, each supplied with sediment from the tails of its predecessor. These were often associated with inferred gravel sediment waves. Inter-ridge areas are characterised by a homogenous seafloor with isolated sediment wave fields. On the peripheries of some of these sediment wave fields and in isolated areas between the carbonate mounds, small coral mounds 50 m in diameter infrequently occur (named Moira Mounds). Trawl marks were also common in this area" (Van Rooij et al., 2005).

To me, this description suggests that corals survive even in a moving sand field. And, it clearly demonstrates a high degree of robustness (i.e., that they grow despite a "hostile", corrosive bottom-layer environment, which is contrary to the bio-expectations mentioned in Section 4.3.2).

Magellan Mounds

Immediately upslope of the Hovland Mounds, there is a province of mostly buried features called Magellan Mounds (Figure 5.8). The area is crescent-shaped, about 90 km long by 8 km to 20 km wide, and it is apparently depth-controlled. The mounds are smaller than the Hovland Mounds, being 50 m to 100 m (exceptionally 130 m) in height, but they are all buried beneath a few tens of metres of sediment, and therefore probably pre-date Hovland Mounds. There are a total of 641 Magellan Mounds recorded on 2-D and 3-D seismic data (Huvenne et al., 2003). But this is an underestimate of the true number, as they must also occur between the seismic lines. Thus, it is estimated that there are at least 1,600 such mounds (Huvenne et al., 2005; Foubert et al., 2005; De Cock, 2005).

Huvenne et al. (2005) mapped the Magellan Mound Province with the TOBI (Towed Ocean Bottom Instrument) sidescan sonar and inspected some of the mounds using an ROV. They only found four "active mounds" in the northwestern sector of the Magellan Mound area. One of these mounds is called Mound Perseverance (Figure 8.5). The distribution of live corals and coral facies is described as follows:

"The facies map shows that the largest part of the area is covered with bioturbated fine sediments. Only on the mounds themselves, coral debris, dead and live coral are encountered, causing part of the high back-scatter response on the TOBI imagery. Live corals are only found on the highest mound tops, or on a shoulder of Mound Perseverance. *Lophelia pertusa* and *Desmophyllum cristagalli* are the most abundant species, occurring in bushy colonies of some 1 to 1.5 m across and 0.75 to 1 m high. These are built of a framework of older (dead) material with live polyps at the outer shell, often facing the downslope direction (especially *Lophelia*). The colonies occur in groups of 5 to 10, generally situated in areas of coral debris with patches of plain sediment" (Huvenne *et al.*, 2005).

In my view, the reason the extinct Magellan Mounds occur as ring-shaped structures is most probably the self-sealing nature of seeps, mentioned in Section 4.6.2, for the pockmarks at Troll (Hovland, 2002). This means that when seepage of methane and other substances (CO_2, H_2, etc.) occurs through the seafloor, the seepage first leads to the development of bacterial mats, then to the precipitation of carbonate crusts, which seal off the main seep conduit. Subsequent seepage has to find its way around this central seal, and the ring-shaped (doughnut-shaped) seeps occur:

> "The large surface reefs [Hovland and Belgica mounds] seem to be associated with fault-controlled methane seeps from deeper hydrocarbon reservoirs, but the numerous buried reefs [Magellan], rising from, or rooted to, a common and virtually undisturbed geological horizon, do not show any obvious correlation with deeper faults. Reflection seismograms of the Magellan reefs reveal the widespread twin association of symmetric lung- or butterfly-shaped structures, suggesting central cross-sections through ring structures" (Henriet *et al.*, 1998).

The lack of "obvious correlation with deeper faults" is a well-known problem when dealing with the detection of minute conduits (on the centimetre scale), such as those existing beneath pockmarks, on geophysical data (seismograms). The geophysical data only have a resolution in the metric scale (see Judd and Hovland, 2007), and are therefore normally "blind" when it comes to detecting vertical seep conduits.

5.2.2 Seep or non-seep, could drilling resolve the issue?

Although the theory of seepage fuelling the growth of Porcupine Bight mounds seems to be supported by geophysical and geological data (Naeth *et al.* 2005; Henriet *et al.*, 1998; Hovland *et al.*, 1994a; Figure 5.6), to date, however, there are no surface-geochemical, or other "hard" supportive evidence. So, a European group managed to arrange for the scientific deep-sea drilling vessel *Joides Resolution*, in 2005, to drill into and through the core of one of the Belgica Mounds. Their hopes and expectations were of achieving a clearer insight into the formation and development of these structures (Ferdelman *et al.*, 2006).

Drilling a "dead" reef

The mound targeted for scientific drilling was the Challenger Mound alleged to be "... a prominent mound structure covered with dead cold-water coral rubble on the southwest Irish continental margin ..." (Ferdelman et al., 2006). The reason for selecting this particular mound was that it is apparently dead. Thus, any severe damage to a living mound structure, by "puncturing" a hydraulic system, would be avoided. The scientific objectives of this drilling were to

1. Establish whether the mound roots on a carbonate hardground of microbial origin and whether past geofluid migration events acted as a prime trigger for mound genesis.
2. Define the relationship between mound initiation, mound growth phases, and global oceanographic events.
3. Analyze geochemical and microbiological profiles that define the sequence of microbial communities and geomicrobial reactions throughout the drilled sections.
4. Obtain high-resolution paleoclimatic records from the mound section using a wide range of geochemical and isotopic proxies.
5. Describe the stratigraphic, lithologic, and diagenetic characteristics, including timing of key mound-building phases, for establishing a depositional model of cold-water carbonate mounds and for investigating how they resemble ancient mud mounds.

In short, the immediate results of the drilling, showed that " ... the mound rests on a sharp erosional boundary", underlain by "drift sediments" consisting of glauconitic and silty sandstone of Miocene[2] age, which is capped by a "firmground" (Ferdelman et al., 2006). Thus, the root of the Pliocene[3]–Pleistocene[4] aged Challenger Mound proved to be a several-million-years-older sandstone. In my view, such an old (dipping) sandstone layer may perfectly well act as an efficient conduit for migrating geofluids, including methane gas. The reason is that a sandstone layer is normally much more permeable and porous than the other (surrounding) clay-dominated sediments. As such, this geological setting is more or less analogous to the interpreted structure of the Norwegian Sula Ridge reefs.

Some of the conclusions gained from the 2005 IODP 307 Expedition were also reviewed by Williams et al. (2006):

"By drilling Challenger Mound, project scientists were able to demonstrate for the first time that such mounds are built and structurally supported by cold-water corals. The branching skeletons of the corals buffered fine-grained sediments and constructed the remarkable 155-meter-high elongate conical structure of the mound. Though cold-water corals are organisms sensitive to environmental change (e.g., temperature), they were able to maintain a deep-water mound community under glacial/interglacial climatic change, waxing and waning with the prevailing oceanographic regime. Light hydrocarbon seeps do not appear to

---

[2] The Miocene lasted from 23.8 million YBP until 5.3 million YBP.
[3] The Pliocene lasted from 5.3 million YBP until 1.8 YBP.
[4] The Pleistocene (the current geological period) began 1.8 million YBP.

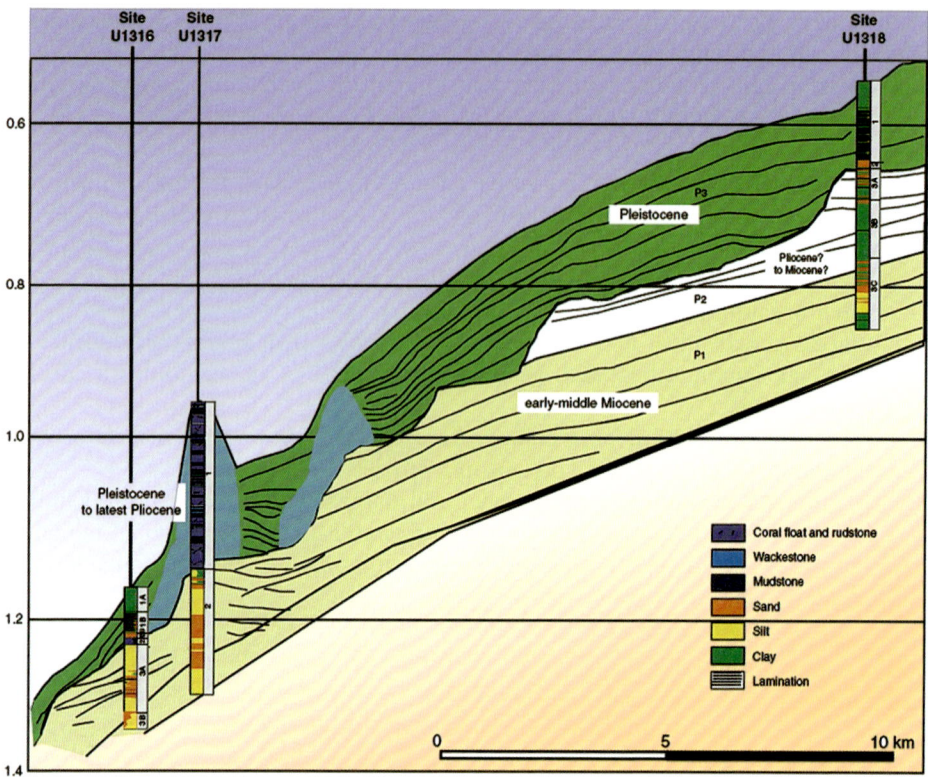

Figure 5.9. The resulting geo-profile over the Challenger Mound constructed after the preliminary data from IODP Drilling Expedition 307. Lithostratigraphy of the three sites projected on the seismic profile of Challenger Mound along a north–northwest to south–southeast transect. P1–P3 refers to seismic units as defined by Van Rooij et al. (2003). Lithostratigraphic units are numbered next to the lithologic column. Based on the lithostratigraphic data, seismic Unit P3 is younger (<0.26 MYBP) than the upper mound succession (Ferdelman et al., 2006).

have played a significant role in mound growth, although the role of prokaryotes in the lithification of the underlying Miocene sediments remains to be clarified. Further, it was demonstrated that the corals do not require a lithified hard ground to root on, as had previously been suggested."

In the first official IODP 307 Expedition Summary it is stated that "A role for hydrocarbon fluid flow in the initial growth phase of Challenger Mound is not obvious either from lithology or geochemistry and microbiology results", even though the following irrefutable geochemical and microbial results were found during the drilling campaign:

"Prokaryotes are present in all samples counted, but abundances appear to be low throughout much of Site U1316; a zone nearly 30 m thick between 56 and 85 mbsf

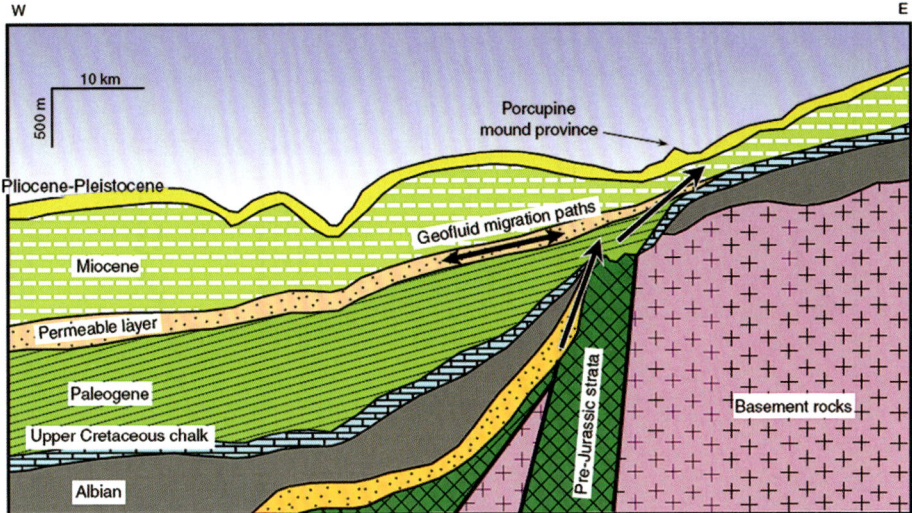

Figure 5.10. This figure, included in the IODP 307 Expedition preliminary report, shows clearly how unique the location of Belgica Mounds is in relation to the underlying geology: the location is right at the "upstream" edge of a prominent underground structure of impermeable "basement rocks". Any geofluids migrating from the Porcupine sediment Basin will most probably emit into the water column above this major vertical discontinuity, as also indicated with arrows in the sketch (after Naeth et al., 2005; Ferdelman et al., 2006).

[metres below seafloor] appears to represent a 'dead zone' based on the absence of evidence for cell division. However, prokaryote abundances increase in the zone of apparent methane oxidation coupled with sulfate reduction. Enhanced ethane/methane ratios suggest preferential removal of methane over ethane through the methane–sulfate transition (82–130 mbsf). Generally, methane concentrations are low within Subunit 3A sediments and only increase to concentrations of 2 mM at 130 mbsf" (Ferdelman et al., 2006).

According to the IODP 307 Expedition Weekly Report (accessed at *www.iodp.org*, 2005): in hole U1317 (on mound) "Methane is first detected at 130 mbsf (2.1 ppm) and gradually increases concentration with depth to 16,000 ppm at 220 mbsf." In hole U1316 (off mound), methane was first detected at 80 mbsf, but here "Low concentrations of ethane were also detected at 80 mbsf and increased in depth to the bottom of the hole."

To me, these results are very significant, as they clearly show that there is not only methane in the porewater system increasing rapidly with depth over 90 metres (from 130 mbsf to 220 mbsf), but also higher hydrocarbons, such as ethane! Furthermore, bacteria utilize these substrates as is clearly shown by the cell counts, just like they do at gas hydrate sites (i.e., ODP Leg 146, Cascadia Accretionary Prism, Westbrook et al., 1994). The gases also increase where I would expect to see an increase, near the base of the reef, adjacent to the Miocene sandstone mentioned previously.

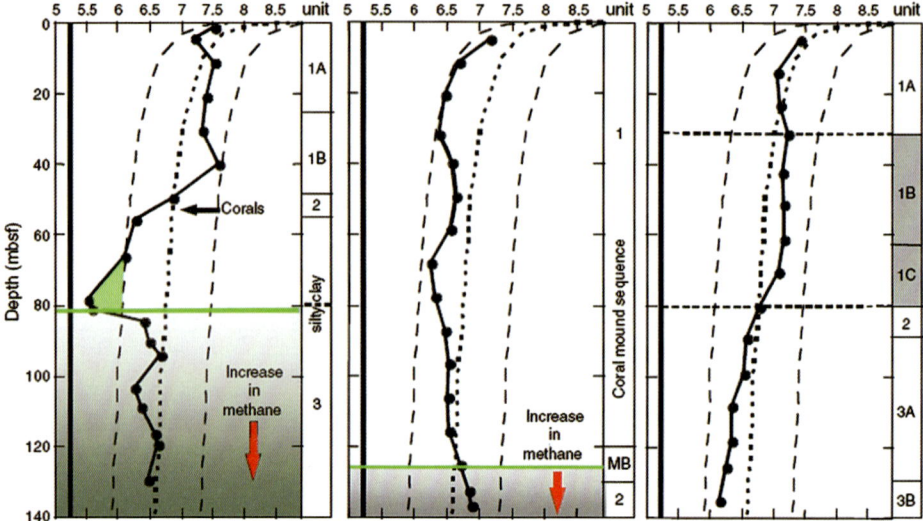

Figure 5.11. Preliminary data from IODP 307 Expedition, showing the variation in the number of live bacterial cells per cubic square centimetre at the three sites, U1316 (left), U1317 (middle), and U1318 (right, see Figure 5.5 for location). The increasing methane content in the sampled sediments is also shown. Note that methane kicks in at MB, which is the mound base. Note also that the number of bacterial cells is highest just below the level where methane starts entering the upper sediments (site U1316) from below, just as expected in the hydraulic theory. Dotted line = global prokaryote profile (Parkes *et al.*, 2005), dashed lines = upper and lower prediction limits for this profile. MB = mound base (Ferdelman *et al.*, 2006). The prokaryote (bacterial) cell counts are given in $\log_{10}$ numbers/cm^3.

In addition, "a distinct increase in chloride concentrations between 80 mbsf and 150 mbsf of as much as 580 mM (at 140 mbsf)" was observed (Ferdelman *et al.*, 2006, p. 12). But, instead of linking such an excursion in salinity (chlorinity twice that of normal seawater) with the formation and dissociation of gas hydrates (as observed at Hydrate Ridge, off Oregon, see Milkov, 2004), the Expedition 307 scientists conclude: "This excursion may be correlated to a major oceanographic change in seawater salinity." Such a "major oceanographic change" is unknown from other northwest Atlantic sedimentary cores and is therefore not confirmed.

But, perhaps the most significant conclusion gained from this scientific drilling, so far, is the following:

"... it is clear that the Challenger Mound is not a present-day analog for microbially formed Paleozoic–Mesozoic mounds. Rather, Challenger Mound is in many ways more reminiscent of the Cenozoic bryozoan mounds located at the shelf edge margin of Great Australian Bight (James *et al.*, 2000). A significant difference from Great Australian Bight mounds, however, is the preservation of carbonate mound or reef structures in an essentially siliciclastic environment" (Ferdelman *et al.*, 2006, p. 13).

This link has some very interesting consequences, not pointed out by Ferdelman *et al.* (2006). First, although the mounds in the Great Australian Bight started out as Cenozoic, they lived right up to the present! Second, nobody really knows how the Great Australian Bight bryozoan mounds formed and were maintained through millions of years. Thus, they are amongst the most enigmatic mounds ever to be found. Furthermore, these mounds seem to be clearly seep (fluid flow) related, as they are associated, not only with very high methane concentrations (up to 12,000 ppm), but a rare combination of high CO_2, high H_2S (>150,000 ppm), and high alkalinity (up to 10.6%, i.e., 137 mM) levels (Feary *et al.*, 2004). Therefore, these mounds seem to have relied on seeping fluids originating from a rare occurrence of underground H_2S–CH_4 hydrates (Feary *et al.*, 2004). And, the corollary must logically be that, if the Great Australian Bight bryozoan mounds are seep-related, by inference the Belgica Mounds are too.

But, even though a lot of factors point in the direction of a hydraulic association with the Porcupine Bight mounds, there exist some competing theories as to why they are located there. One was presented by Freiwald and Roberts (2005). It states that the mounds in the Porcupine Seabight are located at depths near the transition between eastern North Atlantic Water and the underlying denser Mediterranean Outflow Water, MOW (originating from the Straits of Gibraltar). The idea is as

Figure 5.12. From the Thérèse Reef, Porcupine Slope, Belgica Mounds.

follows: when organic matter produced in the photic zone falls from the sea surface, it slows down its descent and concentrates at this density boundary, which intersects at the tops of the tallest of the Belgica Mounds (Williams *et al.*, 2006). This interesting theory will be discussed further in Chapter 8.

Drilling "juvenile" reefs

Whereas Challenger Mound is considered a fossil mound (although it resides amongst "lively" mounds), a new IODP drilling campaign is under planning. Jean-Pierre Henriet of Ghent University, Belgium and others mainly want to find out more about how "microbial mediation may play a primordial role" in carbonate mound development:

> "The success of Expedition 307 paves the way for the further exploration of the diverse world of Atlantic carbonate mounds. A comparative study of Challenger mound with a juvenile mound, featuring active fluid flow processes, is the logical second step. IODP Proposal 673-Pre2 focuses on young mounds on the newly discovered Pen Duick Escarpment (Van Rensbergen *et al.*, 2005), off Morocco. These mounds occur amidst giant mud volcanoes and various emanations of methane seepage. Within the targeted mounds, horizons with fresh corals alternate with layers featuring coral dissolution (Foubert *et al.*, 2005)" (Henriet *et al.*, 2006).

This is of course a very good idea, although drilling a live reef (or mound) may cause lethal damage to it. But, as with most living organisms, it may be necessary to kill and dissect in order of find the truth. This is a difficult moral and ethical issue that has to be discussed before drilling commences. One reason we probably could dispose of a couple of reefs or mounds is that there seems to be many of them, while only a few years ago we did not even know they existed. However, their main future threat is perhaps not mechanical disturbance induced by trawling/drilling, but, rather, changes in the seawater chemistry caused by anthropogenic CO_2, to be discussed in Chapter 9.

5.2.3 The Rockall Trough margin mounds

Rockall Bank, the Logachev Mounds

About 400 km northwest of Porcupine Bank, there is another broad bank, the Rockall Bank. The two banks are divided by the Rockall Trough, an up to 3,000 m deep and 300 km wide trough running from southwest to northeast (Figure 5.8). Rockall Bank is best known for the small island Rockall, which is of volcanic origin. The broad Rockall Bank is made up of sedimentary layered strata, capped by volcanic rock (i.e., intrusive sills). These are near-horizontal strata produced by lava flows. Large portions of Rockall Bank lie at 200 m–500 m water depth.

John B. Wilson documented some of the *Lophelia* structures on the Rockall Bank by using manned submersibles in the late 1970s (Wilson, 1979a, b). But, the most impressive structures on Rockall Bank resemble the giant Belgica Mounds. These are located on the southeastern limb of the bank, where they are perched above the Rockall Trough. They were first mapped from the Russian vessel *Professor Logachev* in 1997 (Figure 5.13). The tallest are up to 100 m high and 5 km at their base. These are capped with *Lophelia pertusa* and *Madrepora oculata*. The mounds occur in a 15 km wide zone, which is almost parallel to the slope, located between 500 m and 1,200 m water depth (Akhmetzanov et al., 1998; van Weering et al., 2003a):

> "The mounds here are of strongly variable dimensions and shape, they may occur as steep pinnacles rising up to 300 m above the adjacent seafloor, with a diameter of 2–3 km, or they occur as large, slightly less high, irregularly shaped clustered mound complexes. The large mound clusters appear to form irregularly shaped ridges of up to 15 km length, separated from each other by valleys and gullies, and aligned with their longest axis almost perpendicular to or at a high angle with the bathymetric contours and the general trend of the slope. Up-slope, the mounds become less high, occur more isolated and locally are completely buried under a relatively thin sediment cover" (van Weering et al., 2003a).

Because van Weering and co-workers performed their mapping by means of seismic data, they could also say something about the sub-stratum on which the

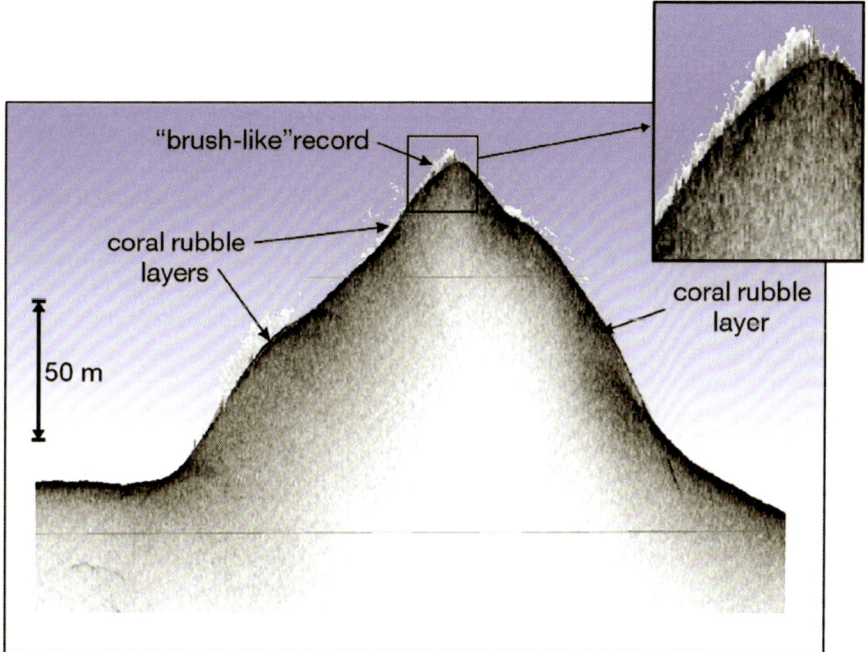

Figure 5.13. One of the giant carbonate mounds on the Rockall Bank, capped with *Lophelia* (Kenyon et al., 2003). No vertical scale exaggeration.

mounds are constructed:

> "The mounds form the upper part of a complex seismic sequence, formed by three seismic units above a well-defined and distinct, locally irregular and faulted, probably basaltic, acoustic basement (AB) reflector. The most recent mounds are underlain by an erosional unconformity, as in the SE Rockall Trough area R1 [i.e., the NW Porcupine Bank mounds, see below]. We have tentatively correlated this unconformity with the youngest regionally recognized unconformity in the Rockall Trough. The stratigraphic position of the mounds above this unconformity indicates that these mounds have a relatively young, early Pliocene to Holocene age" (van Weering et al., 2003).

The biology of these mounds was also studied:

> "The tops of the mounds are covered with an extremely dense and thriving, rich benthic fauna of ahermatypic cold-water corals (mostly *Lophelia pertusa* and *Madrepora oculata*, *Stylaster* sp.) and associated rich benthic fauna such as foraminifera, gastropods, brachiopods, bivalves and echinoids. However, in addition to the cold-water corals, crinoids are extremely abundant on the SW Rockall margin mounds" (van Weering et al., 2003).

Their conclusions are partly based on seismic interpretations and are very interesting, as they relate development of these carbonate mounds to the substratum:

> "If our initial interpretation of two stages of mound development is correct (at least for the SW Rockall Trough margin study area), escaping pore-water fluid of unknown origin, but possibly related to cooling of volcanic sediments extruded after initial formation of the Rockall Trough, might explain that the first phase of mound formation tends to occur concentrated in a restricted part of the slope. The dating of the onset of the mound formation is uncertain; however, development has started directly upon the acoustic basement (AB). It is inferred that the first sequences above the AB containing the mounds are at least of early-Tertiary age" (van Weering et al., 2003).

This surprising conclusion means that the mounds have been actively growing for several millions of years, a conclusion that has a very important corollary (not mentioned by van Weering et al., 2003), namely that they have lived through numerous cycles of glacials and interglacial periods, with very variable currents and water masses, sea-level and temperature changes. Perhaps, therefore, these mounds (i.e., the cold-water corals) witness less climate-dependency than previously believed possible according to bio-expectations (see, e.g., Roberts et al., 2006).

Northwest Porcupine Bank Mounds

One of the largest deep-water carbonate mound populations is located at 700 m–900 m water depth on the northwest limb of Porcupine Bank (Figure 5.8); that is, on the southeast Rockall Trough margin (Croker and O'Loughlin, 1998; van Weering et al., 2003b). O'Reilly et al. (2003) mapped these mounds using the long-range sidescan sonar system TOBI in 1998, during the TRIM (TOBI Rockall Irish Margin) survey. From this survey, individual mounds were found to be circular to elliptical in plan view, and that they ranged in width from 50 m to 850 m and were up to 200 m high. The sedimentary structures that became apparent on the TOBI records indicated strong northeast-flowing currents at 800 m water depth. The large-scale sedimentary bedforms included slope-parallel escarpments produced by mass wasting. However, around some of the largest mounds these sharp escarpments seemed to have been smoothed by vigorous contour currents (O'Reilly et al., 2003). The researchers found evidence that current streamlining effects control the shape of the mounds as they grow larger. The largest mounds are more elliptical in plan, which probably means that they minimise the hydraulic drag force. They therefore developed a theoretical model based on the statistical properties of the mound population and the shape of the individual mounds. Their results are very interesting:

> "The frequency distribution of mound size follows a general power law, which is determined by the growth rate of the framework-building coral species and the rate at which they colonised the substrate. Initially, bottom currents support mound growth until the mounds become so large that hydraulic drag forces retard their growth. A model for the evolution of the population predicts that increased hydraulic drag forces on the larger mounds cause a sharp decrease in their number, in agreement with the observations. The model also allows an age structure for the population to be determined and correlations between the growth of the mound population and palaeoclimatic variations in the NE Atlantic to be attempted" (O'Reilly et al., 2003).

Even though the currents have a significant effect on the growth and development of these mounds, there may also be other factors that determine where they occur.

5.2.4 *Lophelia* within the deep Rockall Trough, the Darwin Mounds

The Porcupine Bight mounds may occur as enigmatic and mysterious structures. But, the Darwin Mounds are even more so. Although they are far from impressively high (up to 5 m only!), they occupy an area of about 1,200 km^2 (Figures 5.14 and 5.15). They are up to 75 m in diameter and were first identified as areas of high backscatter on sidescan sonar records. There are hundreds of them and they are all sub-circular. They are composed of sand with some bioclastic (carbonate) material, "blocky rubble", which may be cemented carbonate sediments and/or coral debris, and sands (Bett, 2001). Their most impressive trait, however, is their enigmatic dwelling location:

"The Darwin mounds occur in the northernmost Rockall Trough between 900 and 1,060 m water depth. The Rockall Trough is a steep-sided, deep-water basin, closed to the north by the Wyville–Thomson Ridge [W-TR]. This ridge, with a sill depth of 400–500 m, separates the Rockall Trough from the Faroe–Shetland Channel and Faroe Bank Channel to the north, and forms an important oceanographic barrier that largely prevents the cold bottom water of the Norwegian Sea from entering the trough" (Masson et al., 2003).

The highly reflective (black) spots on the sidescan sonar records of these mounds represent individual *Lophelia pertusa* colonies (Figure 5.15). The colonies are denser on the mounds to the north, at the foot of the W-TR, and generally, their heights decrease southwards. Some have little relief, or even a *negative* relief! Those in the north have distinctive tails; elongate to oval patches of moderate to high backscatter (Figure 5.14). These are up to 500 m long, aligned southwest–northeast, and they all lie to the southwest of the mounds, suggesting that they are oriented in the direction of the prevailing bottom current. These tails have no topographic signature, and their sediments appear from cores and photographs to be the same as the surrounding seabed (Bett et al., 1999). However, they have an unusual fauna characterized by xenophyophores, a giant protozoa.

South of the mounds, where the seabed sediments are muddier and the water depth is 1,000 m to 1,200 m, there is a very large area (>3,000 km^2) covered by

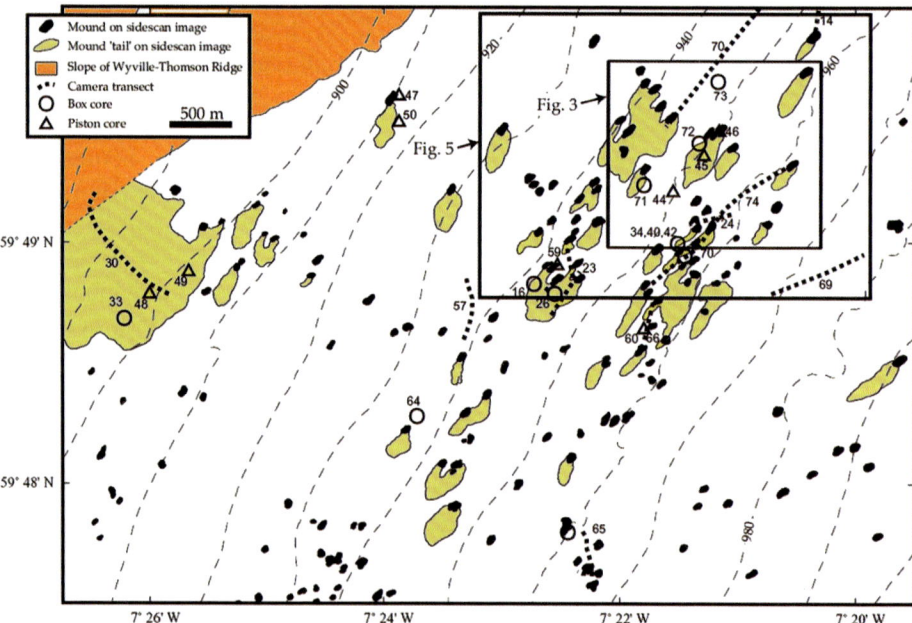

Figure 5.14. Detailed interpretation of the area of Darwin Mounds (black spots) with tails (grey zones). Locations of sediment cores and camera stations are indicated (from Masson et al., 2003).

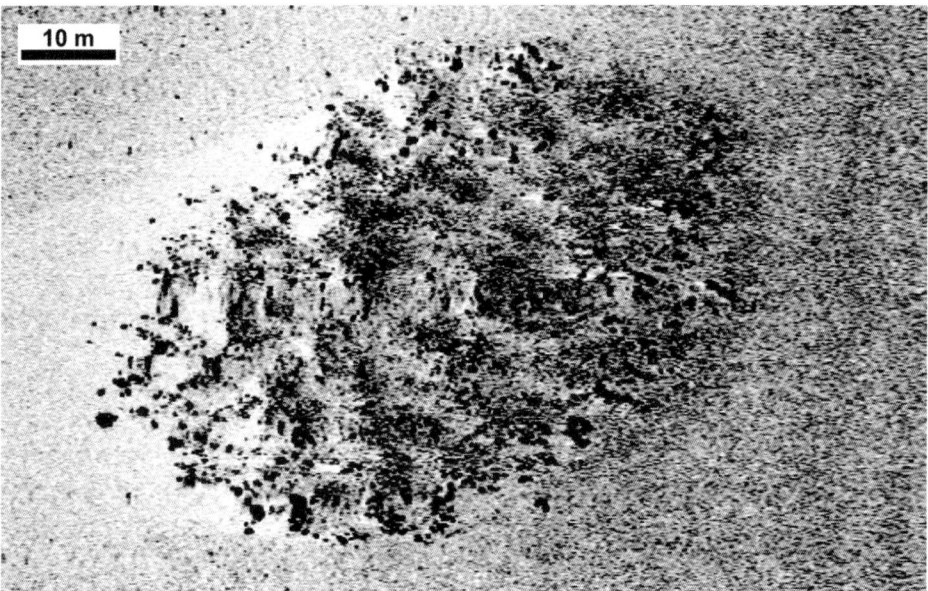

Figure 5.15. A sidescan sonar record of a typical Darwin Mound with parts of its "tail", seen as the light (low-reflective) sediment on the left. This mound is 61 m across and the high backscatter spots correspond to "individual biological colonies of coral and other organisms" (Masson *et al.*, 2003).

pockmarks. These are typically circular, 50 m in diameter, and have a low relief. Unlike the mounds, the pockmarks appear to have the same fauna as the "normal" seabed (Masson *et al.*, 2003). Somehow, the Darwin mounds and the pockmarks share some trait (factor) in common. Thus, it seems there is a north–south, depth-related trend in the nature of the seabed features associated with the fining of the sediments. The features gradually change from having a positive relief in the north, to a negative relief in the south. According to Masson *et al.* (2003), this suggests variations in a single geological process:

> "The variation of the 'mound-like' sonar targets between positive relief mounds and negative relief pockmarks is probably the key to understanding their origin, and leads us to suggest that all of these features are related to fluid escape from the seabed."

Thus, the consensus view is that all these features ranging from the positive relief Darwin Mounds, in the north, to the negative relief true pockmark depressions, in the south, actually represent a type of sand volcano formed by the expulsion of fluids (i.e., porewater from the sediments, Judd and Hovland, 2007).

Where the sediments are coarse, sub-surface sands are brought to the seabed where they accumulate and form mounds (positive features). The escaping fluids are generally thought to be sourced by de-watering or slumping on the southwest side of the W-TR. However, Bett (2001) asked why *Lophelia* grows on the mounds. He noted that they may benefit from the elevated position, but also recognised that they need a hard substrate. He suggested that this might indicate the presence of cemented sediments, in which case, authigenic methane-derived carbonates may have formed.

But, according to Judd and Hovland (2007), there is another possibility:

"We will not be convinced by either interpretation until either seeps are documented, or MDAC [methane-derived authigenic carbonate] is found. Instead, we favour a less conventional idea. Consider that the W-TR acts as a massive dam that prevents the very cold water (about $-1°C$ at 1,000 m depth) in the Norwegian Sea Basin from entering the Rockall Trough. On the south side of the W-TR the temperature is about $7°C$. The density contrast between the two sides of the ridge is therefore large, and should be more than enough to drive water through inhomogeneities (cracks and fissures) of the ridge. We therefore suggest that the cause of both the pockmarks and the mounds on the south side of the ridge is pycnoclinal, that is, a hydraulic head caused by a horizontal density contrast."

The possibility of subsurface seepage of seawater through a geological barrier, like the W-TR, has recently been assessed theoretically by Nunn and Harris (2007) and found to be a viable process. If this is true, then it is not "only" porewater which comes to the surface at the Darwin mounds, but also exotic seawater of a North Atlantic Deep-Water (NADW) origin (Figure 5.16). In any case, there is certainly a hydraulic component to the Darwin Mounds and their associated *Lophelia* colonies.

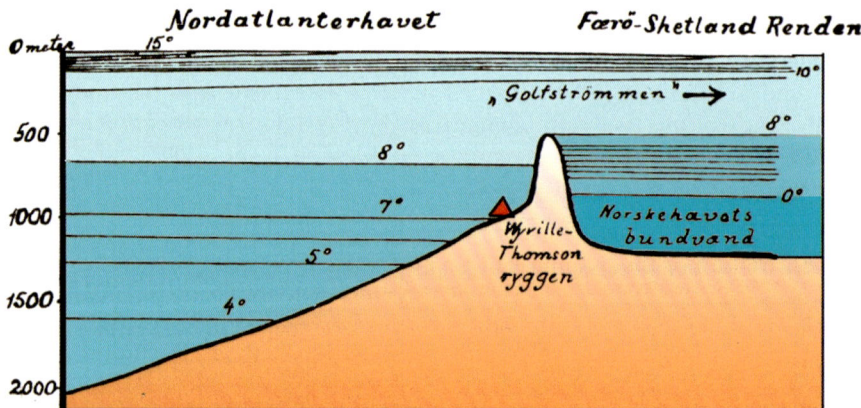

Figure 5.16. An early drawing by Bjørn Helland-Hansen (1916), showing the W-TR (Wyville–Thomson Ridge) acting as a dam between two very different water masses in the northern Atlantic, with cold, dense water to the north (NADW), and warmer water to the south (see also Figure 4.56). The Gulf Stream (*Golfströmmen*) crossing over the W-TR, as it flows through the Faroe–Shetland channel. Vertical water temperatures (in °C) and depths (in metres, left) are indicated. The location of the Darwin Mounds is shown by a red triangle.

This component will, undoubtedly, spike the production of subsurface and seawater bacteria and archaea, and thus "fertilise" the surrounding seawater, in full accordance with hydraulic theory.

5.2.5 Observed acoustic plumes associated with mounds

Acoustic detection of sub-surface (shallow) gas has been located above and near the Connemara field, north of the Magellan Mounds (Games, 2001). But, according to Henriet (1999), "acoustic plumes" (flares) (see Figure 4.34) were observed, but never published in association with some of the mounds off Ireland:

> "In July, 1997, the Prof. Logachev survey has detected torch-like geysers on the seabed, 2 in association with the Logachev mounds, 3 fainter examples on the flanks of a Hovland mound. Though the nature of these acoustic anomalies is still a matter of debate (gas seep or fish shoals?), their shape and the fact that they are systematically rooted to the seabed, in areas characterized by peculiar morphological anomalies, suggest a possible nature as gas seeps or gas-bearing fluid "geysers", in such case impressive in size (up to 40 m high). Anyhow, re-visting such sites, with acoustic surveys, side-scan sonar imaging, video and water and sediment sampling is imperative" (Henriet, 1999).

Yes, indeed, as these acoustic observations probably represent the most concrete evidence of ongoing seepage in this vast carbonate mound area to date, it is imperative to perform a systematic mapping of them. However, as known by most who have done acoustic surveys, they take time and careful planning to document. Some of the clear acoustic plumes we have seen defy visual documentation and can only be imaged acoustically and documented by water sampling and proper chemical analysis (Judd and Hovland, 2007).

5.2.6 Possible sources of seeping fluids off Ireland

Although thermogenic hydrocarbons undoubtedly occur in abundance in and around the Porcupine and Rockall Banks, there is evidence of even more "exotic" deeply sourced fluids. According to Reston *et al.* (2001), there is a deep, crustal transform fault which crosses the region. Here, the authors suggest that there has been and probably still is mantle serpentinisation and deep serpentinite mud volcanism occurring. If this is the case, then any of the serpentinisation-associated fluids, such as H_2, CO, CO_2, and CH_4, can be produced in abundance over millions of years. As pointed out by Früh-Green *et al.* (2004), the fluids associated with serpentinisation of mantle peridotite feeds directly into the biological cycles of the ocean. This suggests that the carbonate mounds off Ireland are manifesting these fluids in a geobiological manner. Serpentinisation produces vast amounts of fluids that permeate both regionally and locally through the entire sediment pile and exit through distinct focused locations on the surface. The plethora of giant carbonate mounds in the area would therefore form more or less independently of climate changes, as they are constructed over perhaps millions of years and are still being "fuelled" from below. This indicates a long-lived stable energy source, exactly what would be the case if long-lasting serpentinisation was active.

5.2.7 Serpentinisation, the most important geobiological process?

Because about 60% of our planet is surfaced by the oceanic crust and the oceanic crust is hidden by seawater, at a mean depth of over 3 km, the fluid flow processes occurring in this crust may be very important. Most of us know about the spectacular black smokers and other such hydrothermal hot vents occurring on the oceanic crust, but only few realize that there is much more fluid flow occurring beneath and through that crust than elsewhere on the planet's surface. Generally, we can say that the oceanic crust consists of fractured and layered basalt and volcanic material. The high density of this material makes it reside at great depths (it sinks into the mantle more than the continental crust, which is of a lower density, being mainly composed of granites and gneisses). We can perhaps say that the oceanic crust is sandwiched between two components that chemically "hate" each other (i.e., thus they react dramatically when they come into contact with each other). Above the crust is seawater, and below it is peridotite, which is a ferric magnesium silicate (see the equation below). From simple physical experiments, we can see what happens when water comes into contact with the elemental metal magnesium (Mg): there is a strong reaction, which also produces hydrogen (H_2), which is a very strong reducing agent.

Because the oceanic crust has many cracks that cut through it, even though it is up to 10 km thick, seawater comes into contact with peridotite. Furthermore, the regional porosity of the oceanic crust can be up to 25%, and therefore it acts as a large reservoir of seawater. At the boundary between the oceanic crust and the underlying mantle (where the peridotite is), the temperature may be up to 1,200°C. The chemical reaction whereby water reacts with the olivine in the peridotite of the mantle is called "serpentinisation", as the end product is the rock serpentinite. The reaction may start at temperatures down to 200°C, but normally occurs at around 400°C. As the equation shows, a lot of hydrogen is released in this reaction. Because of the special composition and layering of the Earth, this process has been ubiquitous all over our globe, and throughout most of geological history (i.e., for several billions of years). According to Konhauser (2007) and others, this may be one of the processes responsible for the origin of life on Earth:

> "Hydrothermal vents additionally represent localized environments where a continuous source of free energy and nutrients are released to sustain a microbial community. Some hydrothermal gases (H_2, H_2S) and solids (FeS, S^0) were prerequisites to prebiotic organic synthesis, and later they provided the redox potential for primitive metabolic pathways (e.g., Russell *et al.*, 2005)."

Off-ridge seepages can also be extremely reduced and alkaline through the serpentinisation of the ultramafic crust and the subsequent dissolution of a fraction of the metal hydroxide and serpentine:

$$2Mg_{1.8}Fe_{0.2}SiO_4(\text{olivine}) + 2.933H_2O \longrightarrow$$
$$Mg_{2.7}Fe_{0.3}Si_2O_5(OH)_4(\text{serpentine}) + 0.9Mg(OH)_2 + 0.033Fe_3O_4 + 0.033H_2$$

This leads to vent fluids enriched in a number of major ions and transition metals. Ammonium was probably released at hydrothermal vents through its extraction from basalts and high temperature N_2 fixation (Lilley *et al.*, 1993).

> "Iron and phosphate, though absent from alkaline hydrothermal fluids, would have been present in the Hadean ocean as a result of discharge from the more acidic black smoker vents (Russell and Hall, 1997)" (Konhauser, 2007).

Carbonates at Lost City vents from serpentinites

Tall, slender carbonate chimneys towering 60 m above the seabed at the Lost City site near the Mid-Atlantic Ridge were found more or less by accident in December 2000 (Kelley *et al.*, 2001). These and other less spectacular structures on seamounts in the southeastern Mariana forearc are formed by the precipitation of minerals from cool fluids derived from the serpentinisation of mantle peridotites. The Lost City structures are dominated by aragonite ($CaCO_3$) and brucite $Mg(OH)_2$. The venting fluids are warm, 40°C to 75°C, and have a high pH of 9.0 to 9.8, compared with 8.1 of normal seawater (Kelley *et al.*, 2001). The fluids were found to be depleted in Mg, enriched in Ca, and to contain high concentrations of dissolved methane, hydrogen, and higher hydrocarbons (Holm and Charlou, 2001; Früh-Green *et al.*, 2004). Cann and Morgan (2002) found it odd that calcium-rich fluids were associated with peridotite, which is calcium-poor. They suggested that this is because serpentinite contains even less calcium, so the left-over calcium is removed in the water.

Chimneys found on the Mariana forearc seamounts were composed of aragonite and calcite in varying proportions. This is quite remarkable because the site, at a depth of 3,150 m, lies below the carbonate compensation depth (CCD), below which carbonates are soluble in seawater, which accounts for the evidence of chimney corrosion described by Fryer *et al.* (1990).

5.3 GULF OF CADIZ, WEST OF GIBRALTAR

The Gulf of Cadiz is the large embayment located between Portugal, Spain, and Morocco, adjacent to the Straits of Gibraltar. Here, Mediterranean Outflow Water (MOW) may dominate sedimentation. However, the tectonic framework of the Gulf of Cadiz reflects the complex boundary between the Iberian, African, and Eurasian Plates. This area has experienced several phases of rifting, convergence, and strike–slip faulting as the North Atlantic opened, Tethys closed, and then the basins of the western Mediterranean were formed (Maldonado *et al.*, 1999). There is wide consensus that subduction is occurring in this region and probably also widespread shear faulting. Over the last 20 years, numerous pockmarks, mud volcanoes, and carbonate chimneys have been documented in the Gulf of Cadiz, even down to water depths of several thousand metres (Pinheiro *et al.*, 2003). This bears witness to an area subjected to widespread fluid flow through the seabed, both focused and diffuse. The main cause of such fluid flow must be the deep tectonic forces which are in action here, probably combined with ongoing serpentinisation at plate boundary zones, between the oceanic crust in the west and the continental crusts in the east. Leg 210 of the ODP proved that there has been widespread active serpentinisation in this region (Tucholke *et al.*, 2004; Judd and Hovland, 2007).

5.3.1 The lower slope, mud volcanoes and gas hydrates

West of Gibraltar, there have been several periods of extension and compression since the Triassic (Gardner, 2001) associated with plate boundary movements. According to Baraza *et al.* (1999), more than 1,000 km^2 of the seabed is affected by features associated with seabed fluid flow resulting from a combination of tectonic compression and along-slope gravitational sliding (Somoza, 2001; Somoza *et al.*, 2003).

The very first discovery of deep-water corals in this region was made by Cunha *et al.* (2001) on samples from the top 30 cm of gravity cores obtained from eight different mud volcanoes. In the core material, they identified pogonophorans and thyasirid bivalves (which are known to be seep-associated), and also debris of the corals *Lophelia* and *Madrepora*, amongst other macrofaunal species. Pogonophorans were particularly common in the samples from the Bonjardim mud volcano (Deep Portuguese Margin field) and mud volcanoes of the Spanish Moroccan field.

Since this discovery, corals have also been found elsewhere in the Gulf of Cadiz. However, so far, no typical *Lophelia* reefs have been found there (Van Rensbergen *et al.*, 2005). It seems that researchers expect to find either haystack reefs or carpets of deep-water corals, similar to those found on the Porcupine carbonate mound structures. During several TTR cruises to the Gulf of Cadiz, many coral colonies were observed. Most of them seemed to be dead cold-water corals. But core samples and underwater TV data often showed twigs of living corals on the edge of dead ones (Foubert *et al.*, 2008). But, several TV lines taken near the shelf part of the Gulf of Cadiz (Vernadsky ridge) and in some absolutely levelled surface areas (approaching the Mekenes mud volcano) have also shown the presence of corals.

This is certainly a paradox, as expressed by Foubert *et al.* (2008):

> "The paradox is that nearly no live corals are presently being observed at the surface of the mounds, while the mound cores display throughout a high number of reef-forming cold-water coral fragments (scleractinians) in association with numerous associated fauna formerly inhabiting the econiches provided by the coral framework. Environmental and oceanographic conditions during the recent past (glacials/stadials) were probably more favourable for cold-water coral growth."

The researchers then discuss why there seems to have been an extinction of corals in the Gulf of Cadiz, a discussion which may be rooted in erroneous bio-expectations.

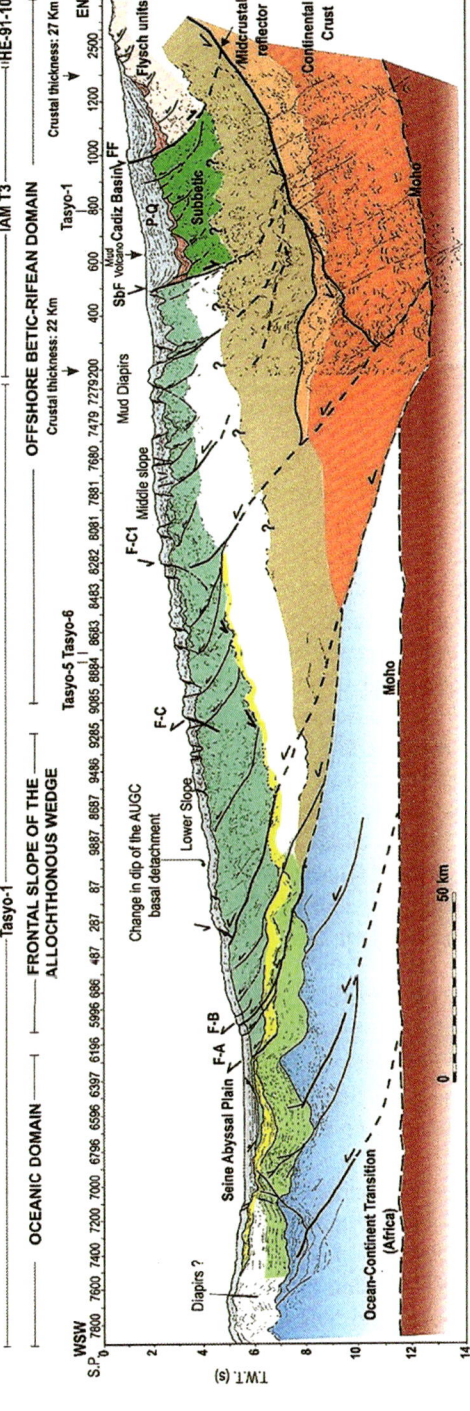

Figure 5.17. This interpreted regional geological transect across the Gulf of Cadiz, from Medialdea *et al.* (2004) and De Haas *et al.* (2006), clearly shows that there has been and probably still is an active accretion of sediments. From comparison with similar sediment piles and structures, off Japan and off Oregon, the situation resembles one of active subduction. The implication of this is that accretionary wedges of this type are associated with fluid flow (Westbrook *et al.*, 1994).

Why should the corals not be able to live in sparseness, rather than in coppices and haystack reefs? Also, off Norway we have examples (i.e., the Husmus structures) of more dead than live twigs, as if they are really stressed and only barely maintain a living, but even so they are able to maintain a few live colonies which accumulate as debris over time. Furthermore, on mud volcanoes, the clay has a very low mechanical bearing capacity, even much lower than that found on the Norwegian outer shelf, discussed earlier. Perhaps the live twigs mentioned above are rooted on other twigs or dead coral colonies that have sunk into the soft mud. I think this may be an important issue for future research to look into.

A geochemical study of cores from several of the Gulf of Cadiz mud volcanoes allowed Blinova and Stadnitskaia (2001) to conclude that two mud volcanoes, Ginsburg and Bonjardim, are "very active" because "their deposits are characterised by extremely high concentrations of hydrocarbon gases and the presence of gas hydrate". Gas hydrates also occur on Yuma, Carlos Ribeiro, and Olenin (Mazurenko et al., 2001). In contrast, Rabat, Jesus Baraza, Tasyo, and Carlos Ribeiro were found to be less active, although hydrocarbon concentrations were still above background. In the Gulf of Cadiz, there is no reason to believe that corals do not benefit from upward migration and associated fertilisation of near-bottom seawater by hydrocarbons (i.e., from the hydraulic process).

5.3.2 The Pen Duick occurrence

Off Morocco, there are also mud volcanoes and associated corals. From coring and geochemical analysis, widespread fluid flow has been documented and the origin of deep-water corals has recently been discussed by Maignien et al. (2006), thus:

> "Our work hypothesis is that carbonates, produced by oxidation of light hydrocarbons including methane, may contribute to the sediment cementation and stabilization of the mound. Recently, on top of Pen Duick escarpment (Gulf of Cadiz), carbonate mounds were discovered in the close vicinity of structures such as pockmarks, carbonate crusts and mud volcanoes, bearing witness of fluid migration. This provides a new opportunity to test our work hypothesis. Hence, pore water profiles of methane, sulphate, sulphide and carbonate, were measured within the sediments of a mound and compared to background concentrations."

In the context of this book, their most interesting observations and conclusions were stated as follows:

> "Interestingly, *Lophelia* coral rubbles were present all along the sediment column suggesting that this mound is a potential babitat for cold water corals and associated communities.
>   ... anoxic oxidation of hydrocarbons, and subsequent carbonate production, may play a key role in the mound formation and/or stabilization" (Maignien et al., 2006).

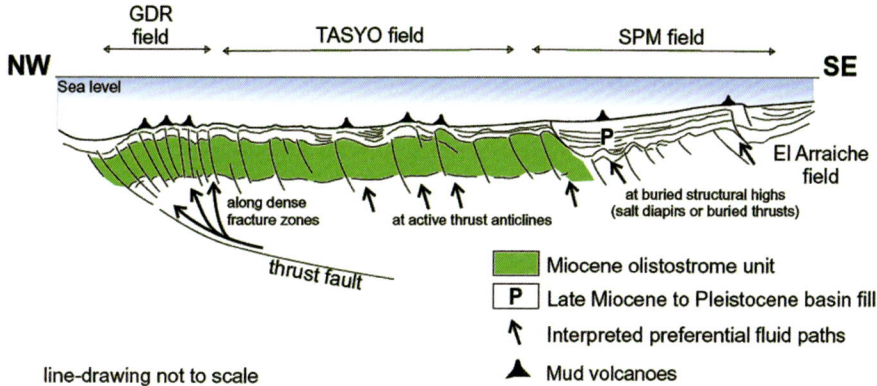

Figure 5.18. El Arriche mud volcano geological setting interpreted by Van Rensbergen *et al.* (2005). Note the many locations where the geologists expect and interpret upward-migrating fluids from depth. The mere existence of numerous mud volcanoes in this area also suggests that there is a large dewatering and degassing of sediments, just as expected in an accretionary subduction setting (see also Figure 5.17).

Thus, it seems seafloor hydraulics is more or less obvious at this mud volcano (Pen Duick) escarpment location (Figure 5.18), and could be a highly interesting candidate for investigating how such a linkage works physically, chemically, and biologically.

But, for me, the most interesting result is provided in the statement: "... *coral rubbles were present all along the sediment column.*" Which confirms my suspicion, mentioned previously, that when deep-water corals grow on a very weak substratum, such as on mud volcano breccia (soft clay), they sink into the sub-stratum before they manage to build large colonies. On the surface, this would look like a carpet of live twigs, rather than lush, live colonies. However, when there is simultaneous sub-surface formation of methane-derived carbonates, larger colonies and even reefs may build up as the carbonates stabilise the soft clays.

In a proposal to drill into a juvenile reef structure (located at $35°17.48'$N, $06°47.05'$W), more details about the Pen Duick occurrences are given:

> "A 9 m-long CASQ core was recovered in 2004 on top of one of the largest buildups—60 m high. It contained cold-water corals over its whole length, a striking similarity with Porcupine mounds, but a very strong smell of hydrogen sulfide was reported. Pore water analyses gave evidence of a sharp sulphate–methane transition zone at 4 mbsf.
>
> Off-mound cores brought possible evidence of decomposing gas hydrate crystals."

And furthermore,

> "the systematic observation of mound nucleation on an escarpment suggests a primary hydraulic control. Drilling should clarify the controls and expression of

the overall fluid migration pattern in and below such a mound, and verify the potential role of hydraulic factors on the specific site of mound nucleation" (Henriet and Dullo, 2005).

I fully agree, as the location described here could be founded on a "self-sealing seep" methane carbonate (Hovland, 2002).

A cruise onboard the German research vessel *Maria S. Merian* (her maiden voyage) disclosed new information on corals off Morocco:

"The presence of numerous carbonate mounds, in the close vicinity of mud volcanoes (Gemini MV, Fiuza MV, Don Quixote MV, etc.), and ovelying large faults (Pen Duick escarpment, Renard Ridge), suggests a possible relationship between carbonate mound distribution and fluid migration through the seafloor. A previous coring of a carbonate mound in this area (Privilege cruise—R/V *Marion Dufresne*) showed that mounds were indeed the place of an enhanced flux of methane-bearing sulphate-to-methane transition zone of 3.5 m below the mound surface. At this depth ^{13}C depleted carbonate, as well as sulphide are released to the sediment column, suggesting an active zone of anoxic methane oxidation. Hence a possible relationship between focused fluid flow and the carbonate mound distribution" (Pfannkuche *et al.*, 2006).

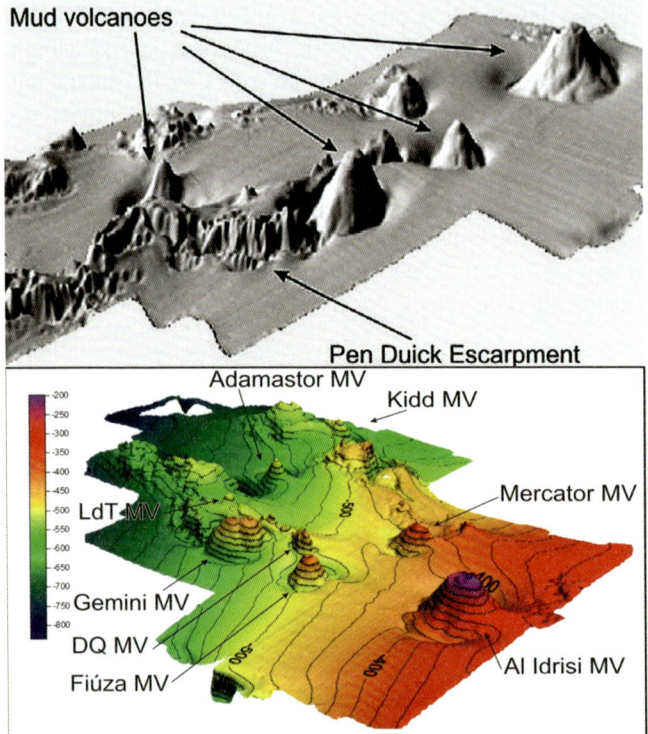

Figure 5.19. These two images show perspective views of the topography of the Pen Duick mud volcano and carbonate mound province, off Morocco, where deep-water coral reefs occupy some of the summits of escarpments and mud volcanoes. The upper image is viewed from the south, looking northwards, and the lower, coloured image is viewed from the east. See text and Figures 5.20 and 5.21 for more information (courtesy Jean-Pierre Henriet, Tjeerd van Weering and the MOUNDFORCE and MICROSYSTEMS project partners).

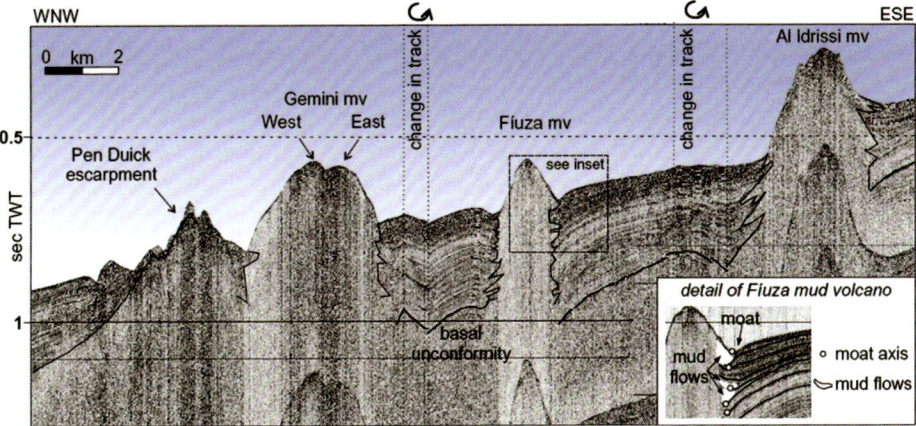

Figure 5.20. A 2-D high-resolution seismic section through the Pen Duick mud volcanoes and carbonate mounds (here only visible on top of the Pen Duick escarpment). Note how the sub-surface sediments of the mud volcanoes penetrate through and inter-finger with the upper, layered sediments (from Van Rensbergen et al., 2005). TWT = two-way time.

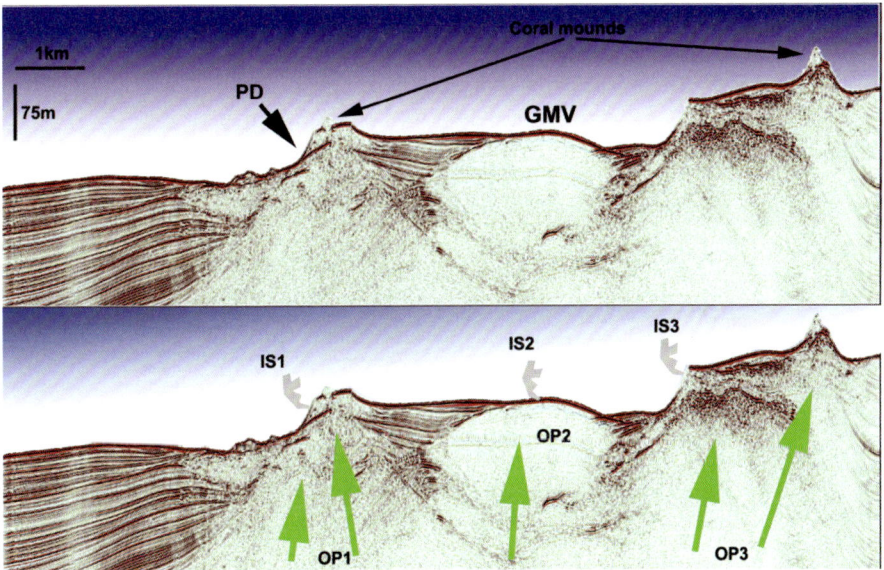

Figure 5.21. A 2-D seismic record through the Gemini mud volcano (GMV), with an interpretation shown beneath (the seismic record is courtesy of Tjeerd van Weering, NIOZ, the Netherlands). Because mud volcanoes have conduits that are in direct (fluid) contact with deeper over-pressured zones, they represent shallow zones of over-pressured sediments and porewater, indicated in the lower figure as three distinct OP zones (over-pressured zones). Natural leakage of fluids (gas and porewater) is indicated from these three zones, as three IS (inferred seep) zones. Note the location of giant carbonate mounds ("coral mounds", upper image) relative to these inferred seep zones. See text for further discussion. PD = Pen Duick escarpment.

The Pen Duick mud volcano and giant carbonate mound province is shown in Figures 5.19–5.21. The association between the mud volcanoes and the occurrence of the carbonate mounds is seen especially well on the seismic record through the Gemini mud volcano (Figure 5.21). From previous studies of over-pressured marine sediments, it is known that over-pressured zones can be mapped with such data (Hovland and Curzi, 1989; Hovland et al., 2002). Because mud volcanoes have conduits that are in direct (fluid) contact with deep, over-pressured zones, they represent shallow zones of over-pressured sediments and porewater. Furthermore, such zones also have topographical associations (i.e., they are often either dome-shaped or conical, Yusifov and Rabinowitz, 2004). Because they are made out of low-permeable material (clay), seepage of the over-pressured fluids can only be bled off where conduits (channels) occur at the seafloor. In Figure 5.21 three such inferred seep locations have been indicated. From the same figure it can be seen that the two giant carbonate mounds are located relatively close by the inferred seep locations.

Our last example, the Mekenes mud volcano (off Morocco) is located centrally in an area where scattered deep-water coral reefs are found (Figure 5.22). According to descriptions by Michael Ivanov (pers. commun., 2008), live *Lophelia* have been found in many cores near and on this mud volcano. The coral locations are seen as some of the highly reflective patches in the sidescan sonar image (Figure 5.22). It is inferred

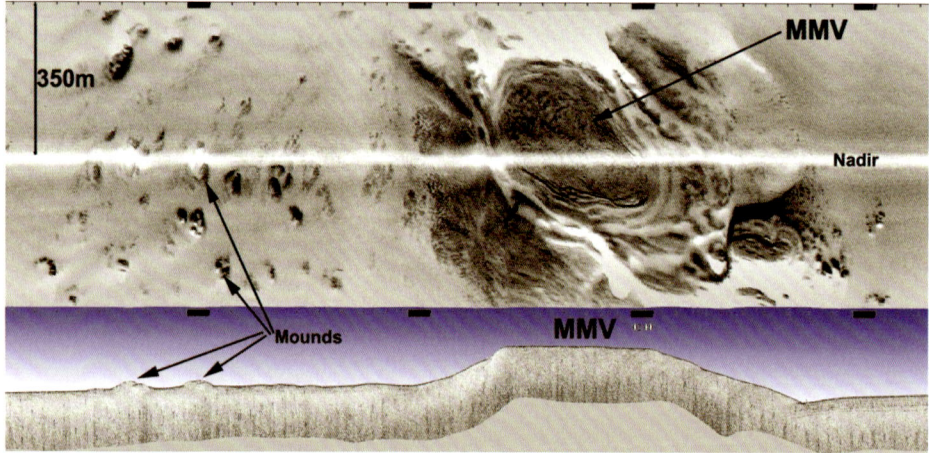

Figure 5.22. Mekenes mud volcano (MMV), off Morocco, shown on a sidescan sonar record (upper image) and on a simultaneous high-resolution seismic record (lower image). The centreline of the sidescan sonar image (nadir) shows where the seismic section is acquired. This pie-shaped mud volcano is about 700 m long and 400 m wide and has a small "parasite" volcano showing at the lower right on the sidescan sonar image. Deep-water corals are found on and adjacent to this mud volcano, and can be seen as some of the highly reflective (dark) zones on the sidescan sonar image. Note the smaller mounds scattered on the seafloor, especially to the left of the volcano. These may represent gas hydrate pingoes (Hovland and Svensen, 2006) (the records are courtesy of Michael Ivanov, Moscow State University, Russia and the UNESCO-funded, TTR-16 cruise onboard the *Professor Logachev*).

that they cannot build large mounds because of the weak bearing capacity of the soft clay, as discussed previously.

5.3.3 Carbonate mounds off Mauritania

Colman *et al.* (2005) reported the discovery of numerous large, up to 100 m high carbonate mounds at 450 m to 550 m water depth off Mauritania, Northwest Africa. These structures were found during various kinds of seismic mapping, sea-floor sampling, and visual inspection for hydrocarbon exploration. Some of them occur within 10 km (upslope) of the oil-field Chinguetti and the Banda gas-field (downslope).

According to the researchers, these may represent the world's largest currently living deep-water carbonate mound systems. They cover an area of at least 95 km^2 and they occur in a linear extent of at least 190 km, with an average width across their base of 500 m. Even though a geochemical exploration survey, where numerous

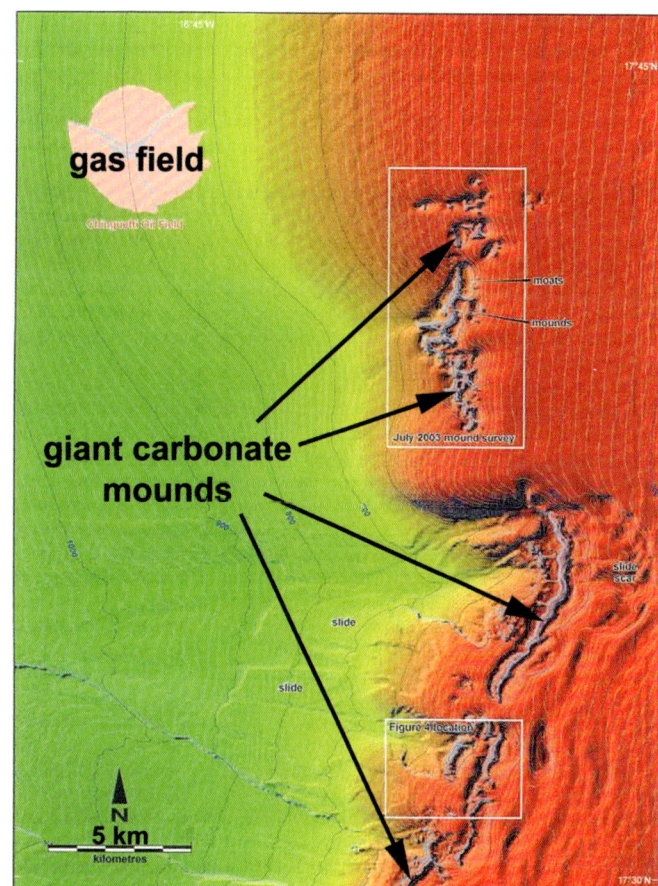

Figure 5.23. A regional shaded topography map from off Mauritania. The large, linear carbonate mounds can be clearly seen, together with their associated linear depressions (moats), which occur on either sides (upslope and downslope) of the mounds (from Colman *et al.*, 2005).

seabed soil samples showed hydrocarbons from methane to pentane, was run in the area (results unpublished), the researchers conclude that they do not "appear to be seepage related", and that they may be related to "oceanogaphic conditions" instead.

Perhaps the most surprising find is that the mounds grow in linear trends, where they more or less follow the shelf-break trend, at 500 m. Another significant find is that they are all associated with adjacent moats (i.e., relatively deep depressions), resembling those surrounding some of the Porcupine Bight mounds.

Compared with its immense total size, very little visual documentation has been obtained from this complex. Thus, very few of the inspected sites have shown live *Lophelia pertusa*, and the researchers conclude (perhaps wrongly) that:

> "Live coral cover on the mounds appears to be much less than was previously present ..." (Colman *et al.*, 2005).

This is where expectation may surpass reality. Even if abundant skeletal debris occur within the core samples obtained from the mounds, this may be a factor of time (accumulation) and sorting (erosion of fines). Their final concluding point is therefore perhaps premature at this stage of exploration:

> "It is unclear whether the decline in coral communities has occurred as a result of demersal trawling activity, or from past changes in oceanographic conditions, or from a combination of these two factors" (Colman *et al.*, 2005).

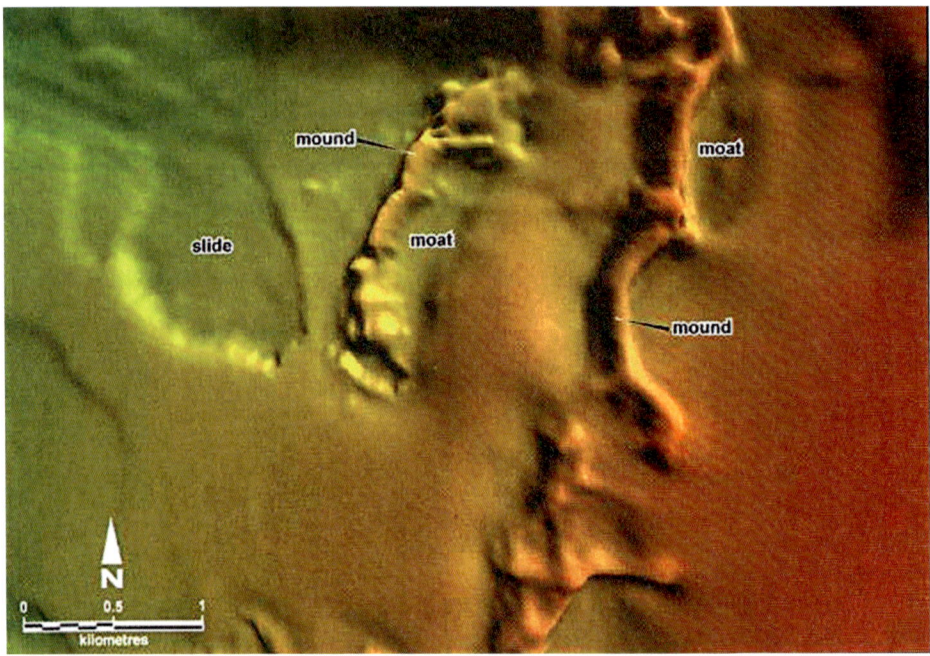

Figure 5.24. A detailed shaded map showing the strange Mauritanian giant carbonate mounds (from Colman *et al.*, 2005).

However, this very interesting study clearly proves that we urgently need information about the world's inventory of deep-water carbonate mounds, so that bottom trawling can be stopped at such locations. It also proves that carbonate mounds may crop up at the least expected locations in water depths greater than 500 m (see also Zibrowius and Gili, 1990).

5.4 SUMMARY

This chapter concentrates on deep-water coral reefs that build up giant carbonate mound structures along specific portions of the continental margin (i.e., off Ireland at Porcupine Bight and Rockall Bank, and off Portugal/Spain and Morocco, in the Gulf of Cadiz), and also some large, linear structures off Mauritania. One of the giant carbonate mounds belonging to the Belgica Mound group, the Challenger mound was targeted for scientific drilling. It proved that the mound had been growing there continuously for several hundred thousand years. Despite major climate change episodes (i.e., periods of glaciations and non-glaciations, or interglacials), the deep-water corals continued to survive. This indicates that the deep-water corals such as *Lophelia pertusa* may be more robust than previously thought. The mound is rooted on a permeable, silty sedimentary layer which had increasing methane concentrations with increasing sediment depth. Thus, the results indicate a flux of light hydrocarbons up towards the seafloor below and adjacent to the giant carbonate structures. A very special case of *Lophelia*-associated deep-water coral reefs are the Darwin Mounds, located on top of low sediment mounds with their telltale "comet-trails" (or tails), indicating focused fluid and sediment influx from the sub-stratum.

In the Gulf of Cadiz and off Morocco, no giant carbonate mounds have so far been studied, although they have been recorded by remote-sensing instruments (on seismic records). However, *Lophelia pertusa* occurs especially near and partly on mud volcanoes in the area. The most spectacular, but still poorly studied giant carbonate mounds are the linear structures mapped along parts of the continental slope off Mauritania. They are flanked by deep linear moat depressions, suggesting either strong currents flowing along their bases, or a flux of gas and porewater emitting through the seafloor at their bases.

6

Other deep-water coral reefs, worldwide

Deep water hydrothermal, groundwater, and hydrocarbon seep communities were all reported for the first time within the last 30 years (by Lonsdale, 1977; Paull et al., 1984; and Kennicut et al., 1985, respectively). Seeps in shallow water were perhaps discovered earlier, but realisation of the processes involved has really taken place over a similar time scale.

Judd and Hovland (2007)

In high methane flux areas, the benthic biomass produced through chemosynthetic processes can be 1,000 to 50,000 times greater than the deep-sea biomass resulting indirectly from photosynthetic production.

Weaver et al. (2004)

6.1 INTRODUCTION

Off eastern USA and Brazil, in the western part of the North and South Atlantic Ocean, in the Mediterranean Ocean and in the vast Pacific Ocean, there are yet more occurrences of deep-water coral reefs and carbonate mounds. Whereas I have had direct personal involvement and knowledge of such structures in the northeast Atlantic, I have had little involvement with those from the rest of the world, which are reviewed in this chapter. The ones selected herein are mainly those we included in our original book (Hovland and Mortensen, 1999). For a more complete and perhaps comprehensive review of reef occurrences across the globe, I recommend the book by Freiwald and Roberts (2005), *Cold-water Corals and Ecosystems*.

142   Other deep-water coral reefs, worldwide   [Ch. 6

Figure 6.0. Deep-water coral reefs, or lithoherms, in the Florida Straits (from Neumann *et al.*, 1977). The length of the manned submersible *Alvin*, illustrated here, is about 5 m.

6.1.1 The Campos and Santos Basin reefs, off Brazil

An area in the Campos Basin off Brazil is covered by *Lophelia* reefs. It occurs more than 120 km from the coastline and covers about 600 km^2. It stretches along the continental slope between known large oil-fields and gas-fields, including parts of the Marlim and Bubira fields, from water depths of 570 m to more than 850 m (Viana *et al.*, 1998). The Bubira reefs (see Figure 6.2) cover about 15% of the seafloor and occur in thickets and mounds that are about 70 m wide and 10 m high. In contrast to the living *Lophelia pertusa*, the coral *Solenosmilia variabilis* has only been found as fossils and skeletal remains in core samples.

So far, the most southerly deep-water coral reefs published for the Atlantic Ocean are found in the Santos Basin off Brazil, the Sumida reefs. The Sumida reefs bathe in water of the north-flowing Antarctic Intermediate Current, which has a temperature range of 4°C to 9°C and a salinity of 34.5‰. The current speed is about 1 knot. In contrast to the corals occurring on the other side of the Atlantic Ocean, which grow on top of carbonate mounds, the Campos reefs occur within and around the edges of large depressions, pockmarks, most probably caused by fluid flow through the seabed (Sumida *et al.*, 2004):

Sec. 6.1] Introduction 143

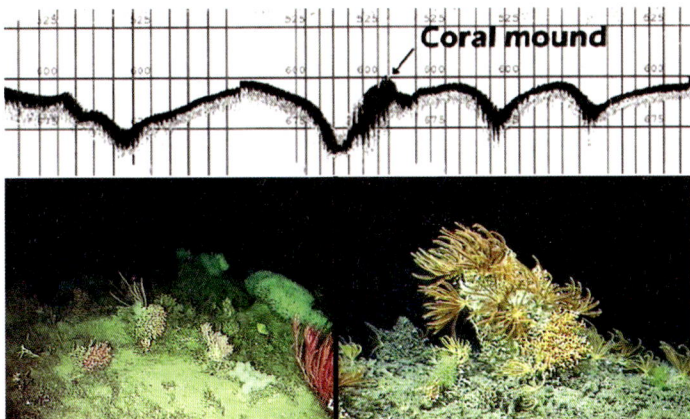

Figure 6.1. The upper image is an echosounder record showing pockmark-associated deep-water coral reefs (shown as "coral mounds"), the "Sumida reefs", off Brazil (area "Sr" in Figure 6.2; Sumida et al., 2004). The two lower colour images were acquired recently from the same general area (area "Sr" in Figure 6.2). The image on the left is from a water depth of 851 m, and the other one, with numerous crinoids, is from a water depth of 747 m. These images were acquired using an ROV, and proved that within the surveyed areas (Sr and Br in Figure 6.2) about 9% of the seafloor is covered by coral banks. These images are courtesy of Maria Patricia Cubelo Fernandez, obtained during 2004 during the "Campos Basin Deep-Sea Coral Assessment" project (organized by the R&D Center, Petrobras).

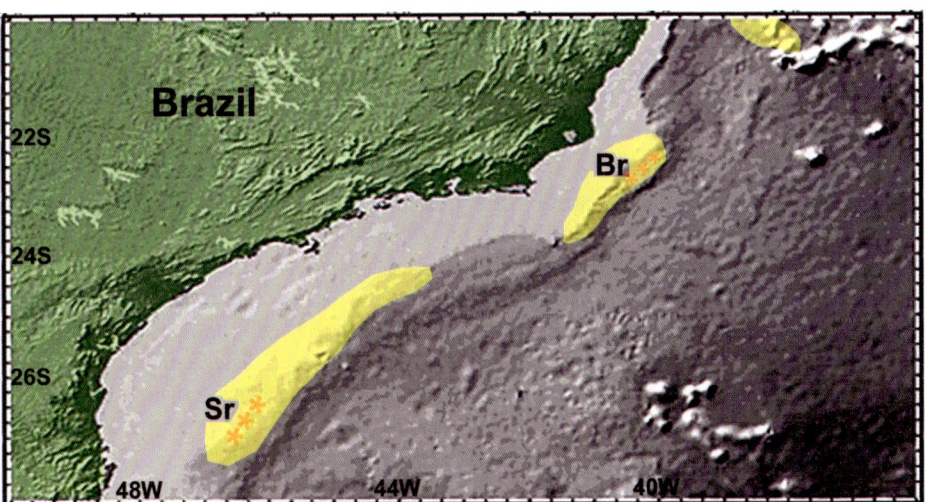

Figure 6.2. A shaded relief map of the Campos and Santos Basins off Brazil. Locations of the Bubira reefs (Br, Viana et al., 1998) and the Sumida reefs (Sr, Sumida et al., 2004) are shown as regional (yellow) areas, recently re-surveyed by a group of scientists from Petrobras, Brazil. The Sr area is dominated by *Lophelia* reefs, whereas the Br area is dominated by *Solesmilia* reefs (see Pires, 2007). There are also deep-water coral reefs farther north, as indicated by the northen yellow area. They occur at least down to water depths of 1,000 m, according to recent results from the "Campos Basin Deep-Sea Coral Assessment" project (R&D Center, Petrobras, Cubelo Fernandez, pers. commun., 2008).

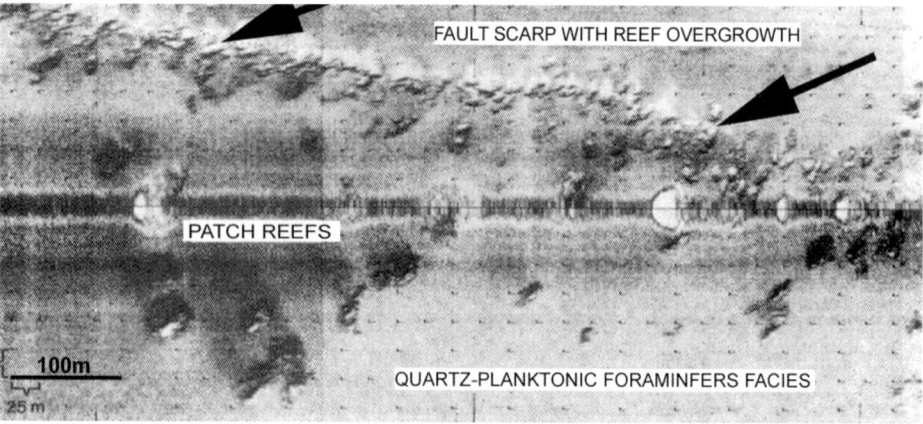

Figure 6.3. A sidescan sonar record showing a surfacing fault scarp with coral reef overgrowth in the northern part of the West Flower Garden Bank (Lease Block A-384). Water depth is about 105 m (from Rezak et al., 1985). Why should corals colonize a fault scarp if it was not for material (fluids) transported up along the fault and out into the water column at the surface? It should be noted that the fault scarp does not have any topographical signature.

"The echo-sounder records also showed the presence of big deepwater mounds associated with the edge of the pockmarks. These features are ~20 m in height and from 180 to 360 m wide. Records of demersal fishes were also registered right above coral mounds. The presence of deepwater coral mounds at the edge of pockmarks may be related to the flow of the cold, oxygen-rich Antarctic Intermediate Water over the bottom, or else may benefit from the fertilization of the water by the microseepage of hydrocarbons from the sediment" (Sumida et al., 2004).

Perhaps the most interesting find here is the association with pockmark depressions. According to new results from off northeast USA, it seems that most of the gas that migrates from pockmarks actually comes from the rims and shoulders, and not, as we have previously expected, from their centres (Newman et al., 2008). This result may not be generalised, but indicates that the seepage process associated with pockmarks is more complex than previously thought. These new results stem from continuous transect measurements of methane in the water column inside and across large pockmarks. This also agrees well with previous sediment-sampling results, like that reported from the HRC (Hovland and Mortensen, 1999), where the highest gas concentrations were found at the inside rim of the pockmark.

6.1.2 The Gulf of Mexico (GoM)

During the 1950s American geologists became interested in deep-water coral reefs (Teichert, 1958; Squires, 1964; Cairns and Chapman, 2001). *Lophelia* reefs living inside seafloor depressions also occur in the Gulf of Mexico (GoM). The depressions are suspected pockmarks (Moore and Bullis, 1960). These deep-water coral reefs were first discovered in 1955 as strange reflections on echosounder records. They occur on relatively soft, sandy seafloor sediments and make up a linear, 60 m high coral reef group of 1.2 km length. They occur at water depths of 400 m to 600 m and all grow inside depressions. Sampling (bottom scraping) showed that the reefs were mainly made up of *Lophelia pertusa*; there were also corals (*Cariophyllia* sp.) (Moore and Bullis, 1960).

Different kinds of corals occur farther north in the GoM, in shallower water. These are on carbonate banks which partly penetrate up into the photic zone. They are reckoned to be the world's most northerly occurring tropical reefs. They are not visible from the air. The most astounding common aspect of these reefs, however, is that they are all associated with sub-surface salt diapirs or salt stocks. Although these banks have been studied over a long period (Rezak et al., 1985), there is no consensus on how they were formed in the first place. Even though there is seepage of gases and brines on banks like the East and West Flower Garden banks and also numerous others (Rezak et al., 1985; Hovland, 1990b), none of the researchers has proposed seepage to be of any influence. According to a new geological model, buried salt stocks actually represent giant fluid conduits, which channel hydrothermally associated liquids and gases from great depths (>5 km) up to the surface (Hovland et al., 2006). The saline fluids are also charged with carbonate and may cause carbonate cap-rocks to form above the diapirs, as observed on the fossilized Damon Mound by Roger Sassen et al. (1994). They estimated that there is 32.7×10^6 t of $CaCO_3$ over the Damon Mound salt diapir in Texas. Their detailed study demonstrated that these cap-rocks, which have $\delta^{13}C$ values in the range $-24‰$ to $-32‰$, were derived by the microbial oxidation of migrating hydrocarbons. Furthermore, the reef fossils include colonial corals and other photic zone organisms, showing that these carbonates were formed in warm shallow water at a time when the sea-level was higher than it is today.

In the GoM, there are also numerous bioherms which are chemosynthetically associated at hydrocarbon seep sites. These have been conserved by American law, which means that the oil and gas extraction industry has to document that no such occurrences will be damaged during construction and extraction (Fisher et al., 1998). Research into the nature of these complex physical/chemical and biological organic/ inorganic structures has been performed nearly continuously since they were first discovered by Jim Brooks and his team at GERG (Geochemical Exploration and Research Group, College Station, Texas) during the mid-1980s, when bottom scraping and trawling over "anomalous" sections of the seafloor turned up pieces of bitumen, carbonates, and large tubeworms (Brooks et al., 1984, 1986). One of the sites discovered in this way was Bush Hill, in Green Canyon Concession Blocks 184 and 185 (210 km southwest of Grand Isle, Louisiana, at 27°47′N, 91°30.4′W). This is where Ian R. MacDonald (1990), then at GERG, performed his pioneering thesis work and documented the link between hydrocarbon seepage, gas hydrates, and biology/ecology:

"The site, known as Bush Hill, occurs over a salt diapir that rises about 40 m above the surrounding sea floor to a minimum water depth of 540 m.
   However, much of the sedimentary stratification of Bush Hill itself has been eliminated by rising gas and liquid by in situ formation of authigenic carbonate and sulfides, thereby creating a seismic wipe-out zone" (MacDonald, 1990).

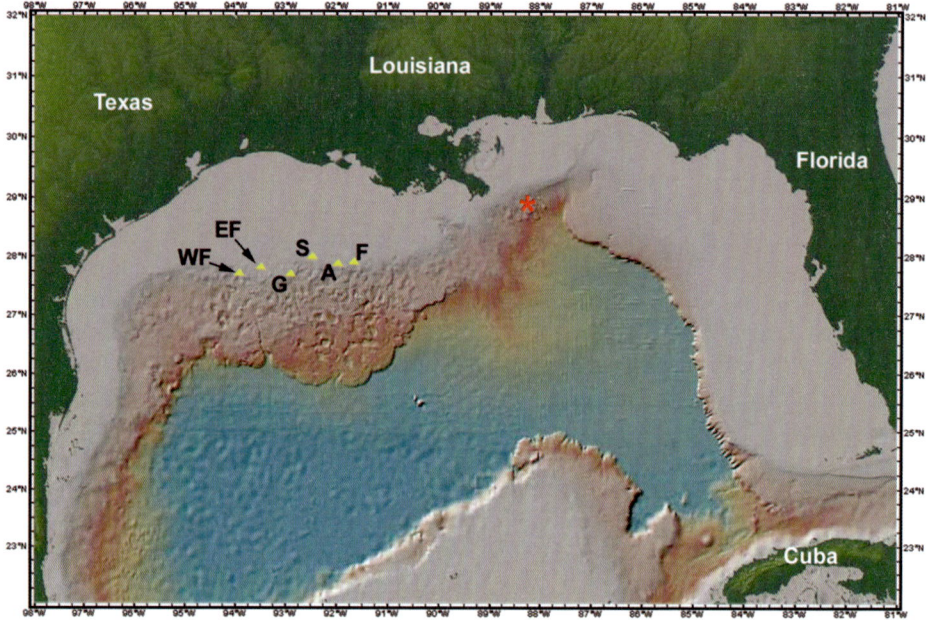

Figure 6.4. Map of the northern Gulf of Mexico. It shows the tropical coral reefs called the East and West Flower Garden Banks (WF and EF). The yellow triangles indicate known carbonate banks which have gas seeps (according to Rezak et al., 1985 and others). One of the very few known deep-water (ahermatypic) reef occurs at 400 m–500 m depth, about 75 km east of the Mississippi delta, approximately where the red asterisk is shown (the map has been modified from Hovland, 1990b and Hovland and Mortensen, 1999).

A relatively modern review of deep-water *Lophelia pertusa* and *Madrepora oculata* occurrences in the GoM seems to neglect much of this knowledge. Thus, Schroeder et al. (2005) provides a summaric (statistical) review of locations where these species have been found. The reason for not saying much about the sub-stratum is perhaps that very few of the locations were documented visually. Most of the information stems from reports of bottom trawling. A couple of interesting sites have turned up, however: for example, their Site 3, at 520 m water depth on the upper De Soto slope, where there is a low-relief knoll, which rises up to 90 m above the surrounding seafloor to a minimum depth of about 430 m:

> "Authigenic carbonate occurs on the crest and sides of the knoll. The most extensive and highly developed *L. pertusa* assemblages noted in the GoM to date have been observed at this site. Colonies display the typical bushy morphology and range in size up to 2 m in diameter, while aggregations of closely associated colonies with linear orientations have been observed to attain 1.5–2 m in width and 3–4 m in length. Many of the colonies are beginning to coalesce and appear to be in the first phase of the 'thicket' building stage described by Squires (1964)" (Schroeder et al., 2005).

Video documentation of this site was performed using a US Navy atomic research NR-1 submarine, the same vehicle used to document some of the *Lophelia* occurrences and bacterial mats on the continental slope off mid-Norway, mentioned on p. 90.

At their Site 7, there is a further interesting description, which links *Lophelia* to a probable seep site. Here, *Lophelia* occurs on

> "a low-relief mound located on the northern edge of an exposed carbonate rock complex which extends southward for over 2 km to the eastern rim of a submarine canyon" (Schroeder *et al.*, 2005).

There can be little doubt that the "exposed carbonate rock complex" has a methane-derived authigenic origin. Their Site 13 is at Bush Hill:

> "Site 13; *Lophelia* sp. reported in MacDonald *et al.* (1989). This site, known as Bush Hill, is a low-relief knoll located at approximately 580 m of water. It rises up to 40 m above the surrounding seafloor to a minimum depth of approximately 540 m. In 1986, during four dives with the research submersible Johnson-Sea-Link-I, colonies were observed and photographed attached to the larger, exposed boulders along an escarpment on the western side of the site" (Schroeder *et al.*, 2005).

When it comes to discussing which factors may be responsible for the occurrence of deep-water corals in the GoM, the researchers state:

> "... coral distribution is potentially regulated by a number of biological (e.g., recruitment, food availability) and physical (e.g., current regime, temperature) processes and mechanisms. Since our understanding of deep-water coral distribution, abundance, and ecology in the GoM is minimal, it is not yet known which of these processes or mechanisms are most influential" (Schroeder *et al.*, 2005).

> "The salt dome underlying the bank [East Flower Garden Bank] has furnished the framework for coral reef development, as well as provided a reservoir for oil and gas resources. Within a four-mile radius of the Flower Garden Banks, there are currently 10 production platforms and there is one gas production platform within the East Sanctuary boundary" (NOAA, 2002).

Seeping oil and gas is ubiquitous in the GoM, a fact that has been known for over a century, as can be seen from the map published by Captain Soley in 1919 (Figure 6.7). Therefore, future scientific work in the GoM should be able to work out the impact, if any, on benthic fauna, including deep-water coral reefs.

6.1.3 Deep-water corals off Belize

Belize is located on the southeastern part of the Yucatan Peninsula, in the eastern GoM. The beaches and tropical coral reefs of Belize have over millennia been plagued by oil spills and tar. There are historical accounts going back to 1775 of such

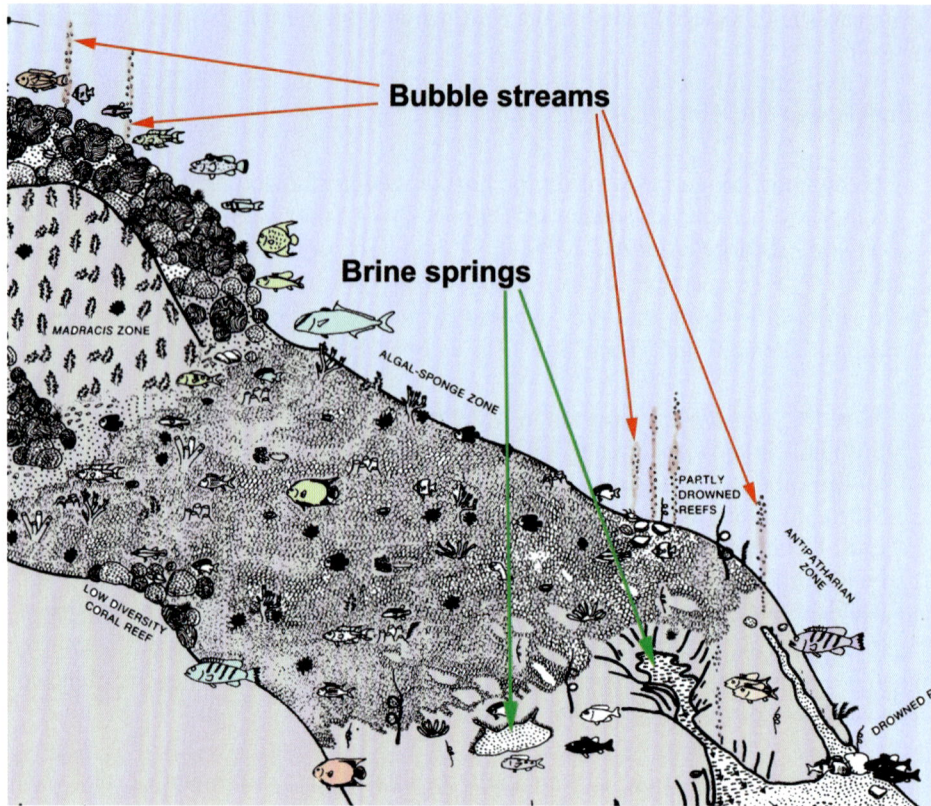

Figure 6.5. A detailed drawing of the southeastern part of East Flower Garden Bank, from Rezak *et al.* (1985). It illustrates zonation, but also shows evidence of seepage, which is not discussed by the authors. There are two types of seepage on both East and West Flower Garden Banks: brine seeps (green arrows) and gas bubble seeps (red arrows). Both these banks and many others in the depth range 70 m to 100 m in the GoM are located on top of buried salt domes (see also Figure 6.2) (from Hovland and Mortensen, 1999).

occurrences. In 1920 a very large oil patch (slick) was reported off Lighthouse Reef Island. Because these accounts were made during a period pre-dating oil tankers and offshore oil exploration and production, it means that natural oil and tar pollution is the norm for this area, most probably because of Belize's tectonic setting (Hovland and Mortensen, 1999).

Placenames along the coast of Belize reflect this polluted situation. The largest island off Belize is called Ambergris Caye, and is suspectedly named after tarballs floating ashore here. Cay-Caulker island means the island where the wooden boats can be caulked (with tar). Amber Head and Ambergris Creek are other examples. Along the beaches of Reef Point of Ambergris Caye, the old coral reefs are coloured brown and black by tar (Jean Cornec, pers. commun.). There are still tar lumps floating around and stranding on the beaches.

Figure 6.6. An interpreted high-resolution shallow seismic section shot across the East Flower Garden Bank, published in Berryhill (1987). The suspected seep-related carbonate reefs (hatching) occur at locations on the seafloor where gas-charged porewater emits through the seafloor. Gas-charged sediments are shown partly in pink (left) and with dots (interpreted from the seismic image). Salt bodies are shown in dark pink, and gas plumes in the water column are shown in red (from Hovland, 1990b).

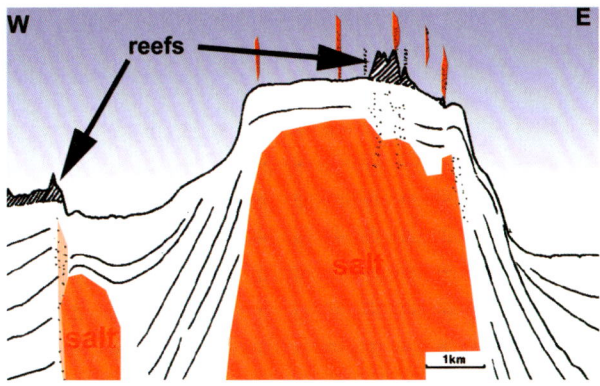

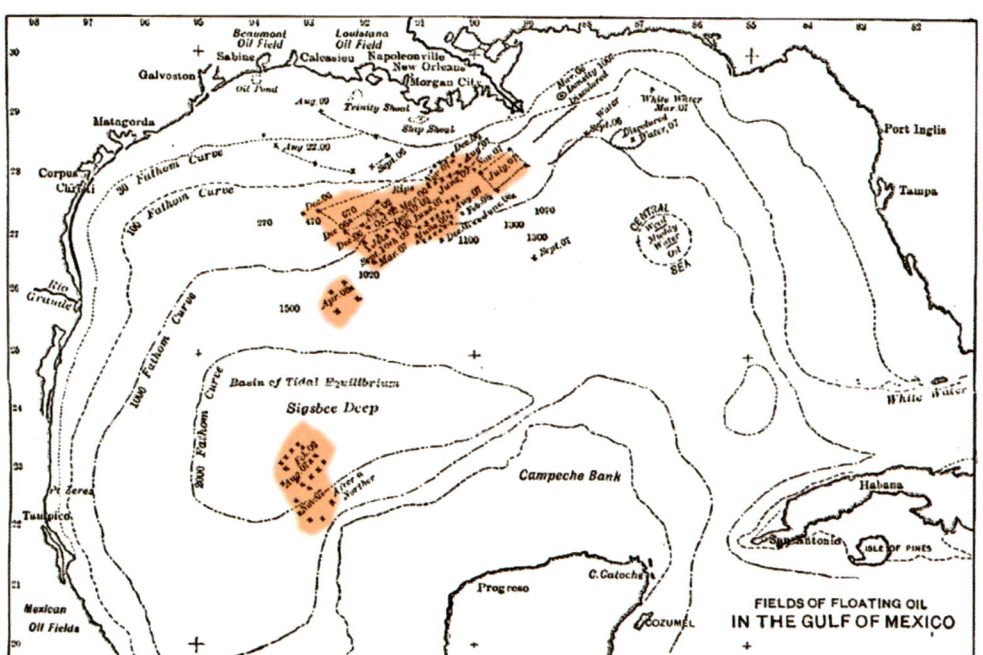

Figure 6.7. "Fields of Floating Oil", GoM, a map constructed by Captain Soley, 1910 (from Rezak *et al.*, 1985). It clearly shows that pelagic oil is a natural occurrence in the GoM. The only place it can come from is natural seeps through the seafloor. Such seeps have now been documented in abundance (see, e.g., MacDonald *et al.*, 1989; MacDonald and Leifer, 2002).

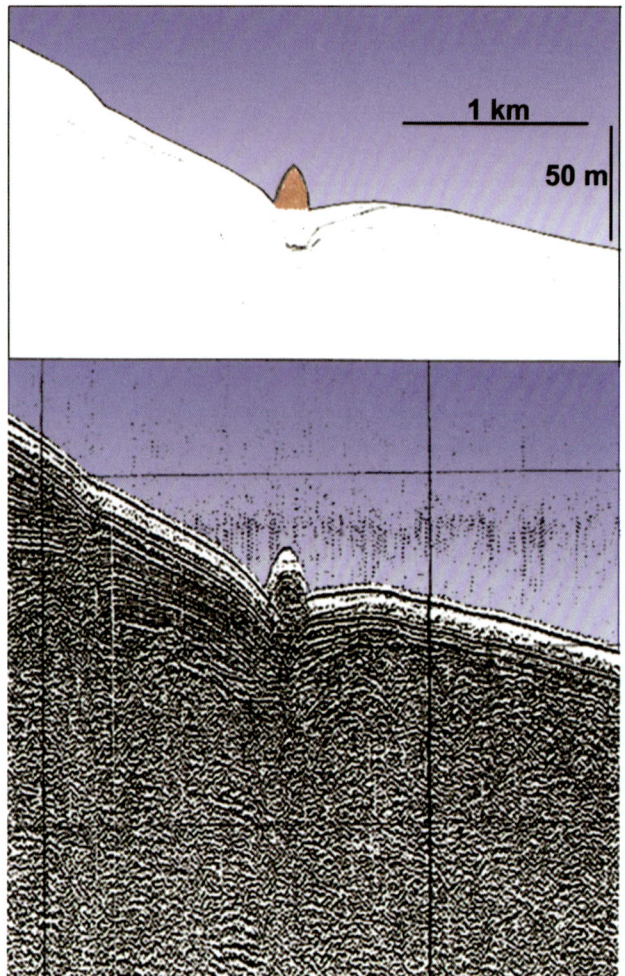

Figure 6.8. A 25 m high mound located inside a depression on a seismic image from off Belize. It is interpreted to represent a deep-water coral reef (seismic record courtesy of Jean Cornec and BNR Ltd).

6.1.4 The Florida Straits

On its way from the Gulf of Mexico, the Gulf Stream passes through the straits between the carbonate platform of Florida to the west and the Bahama banks to the east (Figure 6.10). These are the Florida Straits, which have a width varying between 90 km and 110 km. They are deeper than 1,000 m just south of the Florida Keys (north of Cuba), and have their shallowest portions, of just more than 600 m, west of Little Bahama Bank. The northward-flowing current has a speed of up to 3 knots and a temperature of up to 26°C near the surface. Both the Florida peninsula and the Little Bahama and Great Bahama Banks consist of layered carbonate rock, which is normally relatively smooth although subject to karst development, vertical erosion by groundwater. Such banks, which are found in numerous places worldwide, often host large accumulations of both fresh and saline groundwater (aquifers).

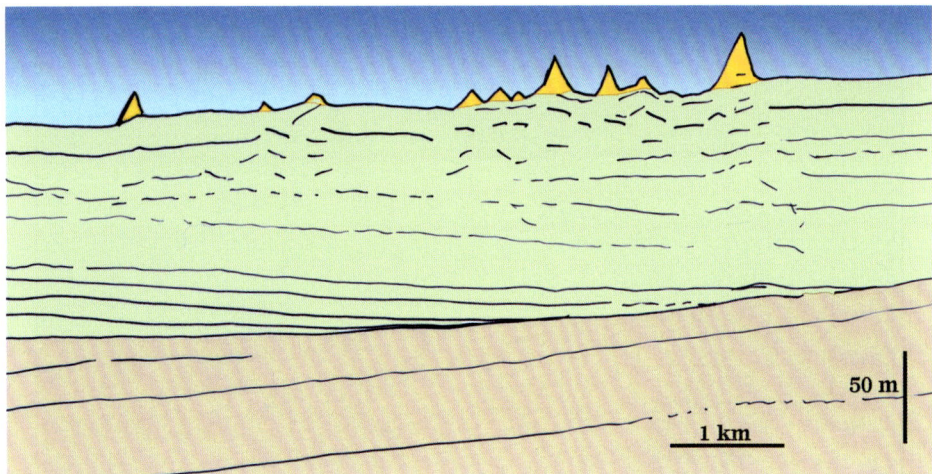

Figure 6.9. Interpretation of a seismic record across reefs at the bottom of the Florida Straits, between the Florida peninsula and Little Bahama Bank (see Figure 6.10 for location). Reefs (or lithoherms) are shown in yellow. They occur over distinct discontinuities in the sub-stratum layers (the seismic record is from Eberli et al., 1996).

In 1971 the American manned submersible *Alvin* was used to ascertain why the seafloor on the eastern side of the Florida Straits was so rough and had such a high relief. The structures that were previously interpreted as large sandwaves or slide deposits proved to represent innumerable lithified biological accumulations capped by living coral species (Neumann et al., 1977). The mounds resembled streamlined haystacks, between 30 m and 50 m high. Because their base turned out to be carbonate-cemented sediments, they were termed "lithoherms". The most common organisms are crinoids, sponges, and ahermatypic corals that grow with their polyps facing into the in-coming current (upstream). Between these haystacks the seabed is smooth and flat (Figures 6.0 and 6.9).

Alvin was also used for diving into a 500 m deep canyon on the eastern side of the Florida Straits, only 6 km from the edge of the Little Bahama Bank (Figure 6.10) and close to the lithoherms. However, only sand and some limestone blocks were found there. No theory for this "headless (blind) canyon" is offered by Neumann et al. (1977). However, Chanton et al. (1991) have investigated how they form. They dived into headless canyons along the Florida Escarpment, in the Gulf of Mexico, farther to the west (and also in Monterey Canyon off California). These canyons were found to have formed by seepage of brines and groundwater. Thus, even on Little Bahama Bank it may be possible that brines are leaking out and forming the canyons by a process called "spring sapping" (Orange et al., 1999; Paull et al., 1984).

Neumann et al. (1977) also wanted to ascertain why the lithoherms only formed on the eastern side of the straits. They reached the following conclusion: the water being forced upwards on the eastern side of the straits (by the Coriolis force) attains the saturation point for CO_2. This is because the rising water is warmed up and

152 Other deep-water coral reefs, worldwide [Ch. 6

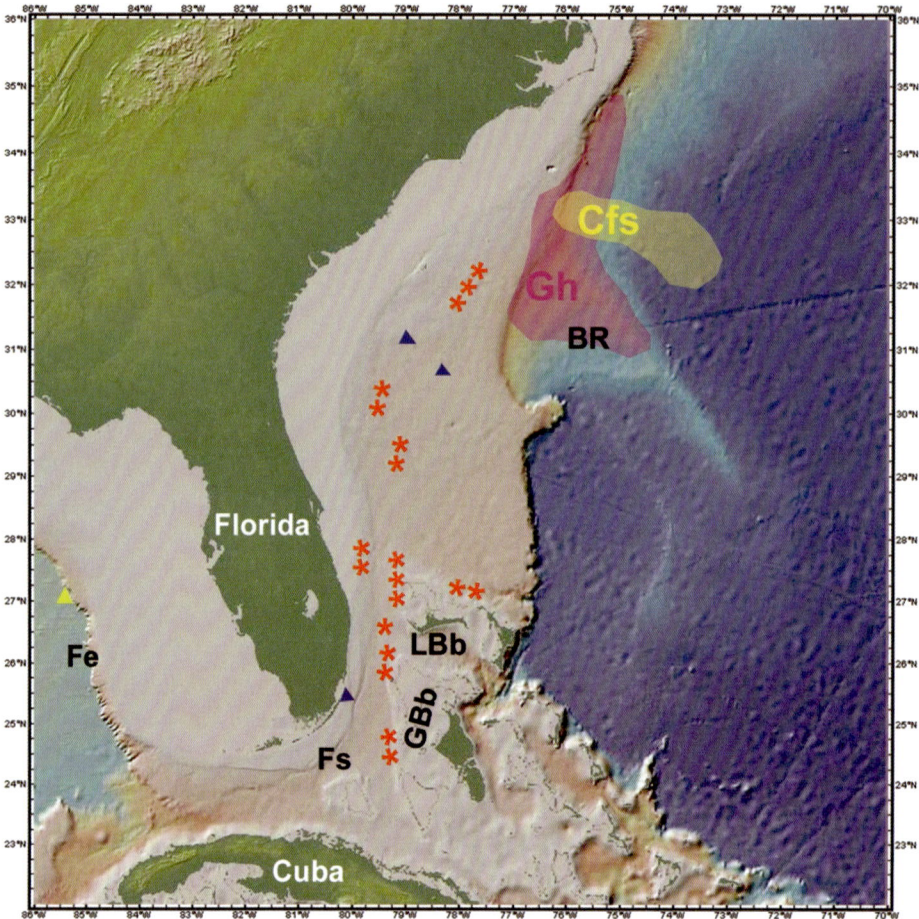

Figure 6.10. A shaded relief map of the northwestern part of the Atlantic Ocean and the southeastern USA, including Florida. Red asterisks are coral reefs and lithoherms (Neumann *et al.*, 1977). Fe = Florida Escarpment, Fs = Florida Straits, GBb = Great Bahamas bank, LBb = Little Bahamas bank, BR = Blake Ridge, Gh = occurrence of gas hydrates on Blake ridge (shown in pink), Cfs = the Cape Fear slide (shown in yellow). The three blue triangles are known freshwater seeps. The yellow triangle is the brine seep at the foot of the Florida Escarpment (Paull *et al.*, 1984; see also Judd and Hovland, 2007) (adapted from Hovland and Mortensen, 1999).

pressure-released. Calcite and aragonite will then supersaturate and small crystals will grow (i.e., auto-precipitate). However, this model raises several questions, leading me to doubt its validity. First, the bioherms do not occur high up on the slope towards the Little Bahama Bank (where the rising water becomes supersaturated with respect to CO_2); they always occur near the bottom of the straits. Second, the auto-precipitation of the crystals should not occur within the sediments, but only in the water column, in the form of "whitings" (Judd and Hovland, 2007). Furthermore, lithoherms have subsequently also been found in places where upwelling does not occur; for example, on the western side of the Florida Straits (Reed, 1980, 1992) and north of the Little Bahama Banks (Mullins et al., 1981). It is therefore tempting to find another explanation.

The ODP performed a drilling campaign (Leg 166) in the Florida Straits, near Great Bahama Bank, not far from some of the lithoherms (Eberli et al., 1996). The main objective was paleoclimatic (i.e., to find out more about climate changes using geological coring and analysis of the layered sedimentary rocks). A secondary objective was hydrologic (i.e., to find evidence of any movement of groundwater and pore fluids). On the basis of heat flux measurements, Kohout (1967) had previously found extremely high temperature gradients both on the Florida platform and on the Bahama banks. Therefore, he postulated the existence of areas where seawater is sucked through the seabed and into the groundwater system. There would also be areas where this water exits through the seafloor and into the seawater column, just like the springs, cold vents, found at the base of the Florida Escarpment, where they fuel a special chemosynthetically based fauna (Paull et al., 1984).

Results from ODP Leg 166 proved Kohout right. Anomalous chemical layering was found in the porewaters of the cores acquired (Eberli et al., 1996). Furthermore, it was found that the groundwater moves through the sedimentary rocks and that there are great variations in the methane and H_2S concentrations. Therefore, a logical conclusion is that there must be exiting porewater (groundwater) at or near the locations where the lithoherms occur. This conclusion is also supported by the observation of groundwater flow out of the Florida platform near the Florida Keys, with groundwater fluxes of up to 8 litres per square metre being measured (Manheim, 1967; Shinn, 1993; Hovland and Judd, 1988).

6.1.5 The Blake Ridge occurrences

By following the Gulf Stream another 500 km northwards, we arrive at the Blake Ridge, which is a massive, finely layered sediment drift deposit (Paull and Dillon, 1980). Here, at a water depth of 700 m, about 230 km from Cape Fear, North Carolina, researchers using an echosounder in 1956 discovered some irregular mounds on the seafloor. Subsequent investigations revealed that these mounds indeed represented deep-water coral reefs. There were at least 200 reefs within an area of 450 km^2 (Stetson, 1962; Paull et al., 2000). Most of the reefs are up to 160 m high and 800 m in diameter and follow north–south running lines, parallel with the general bathymetric trends. Similar reefs have also been found farther south, east of northern Florida (Newton et al., 1987). Near both of these regions, especially the Blake Ridge, gas hydrates, pockmarks and mud volcanoes are found (Judd and Hovland, 2007).

6.1.6 Off Nova Scotia

It has long been known that there exist deep-water corals off Nova Scotia, Canada. They are especially abundant in abrupt deep depressions, types of gorges or canyons

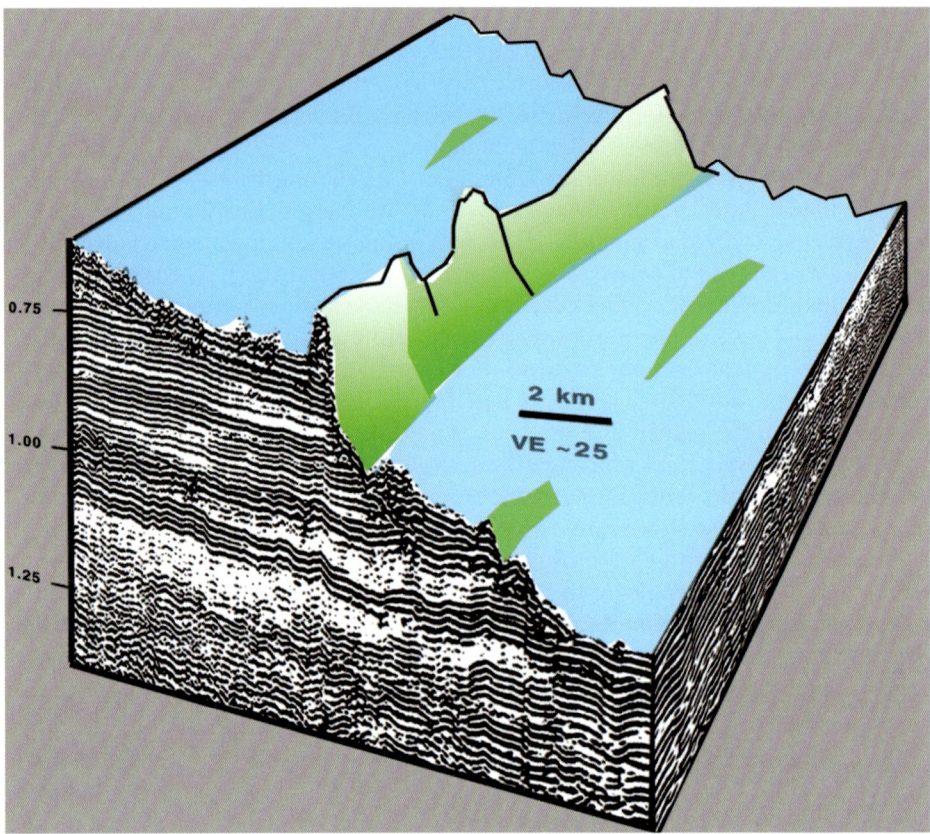

Figure 6.11. This 3-D perspective sketch has been made on the basis of a reflection profile crossing from Florida's Hatteras Slope onto the Blake Plateau, provided by Paull *et al.* (2000, their fig. 2). The green structures represent "... huge, slope-parallel, coral capped lithoherm complexes". According to Paull *et al.* (2000): "Rough appearing seafloor elsewhere along the profile is also associated with other 'lithoherms', coral reefs or carbonate mounds. Tracing of seismic units indicate that Eocene strata sub-crop along this profile."

on the continental shelf, according to local fishermen and Buhl-Mortensen and Mortensen (2005) and Mortensen *et al.* (2006). The octocoral *Paragorgia arborea* is the most common, although there are also finds of *Lophelia pertusa* (Breeze *et al.*, 1997).

"The coral 'trees' grow out from cliff walls, at depths of 170 to 300 fathoms [300–550 m] in the deep gulleys" (Breeze *et al.*, 1997).

6.1.7 Orphan Knoll, off Newfoundland

Orphan Knoll is a relatively small seamount at 2,000 m water depth, located 250 km east of Newfoundland, Canada. There are 200 m high and 2 km wide mounds

Sec. 6.1] Introduction 155

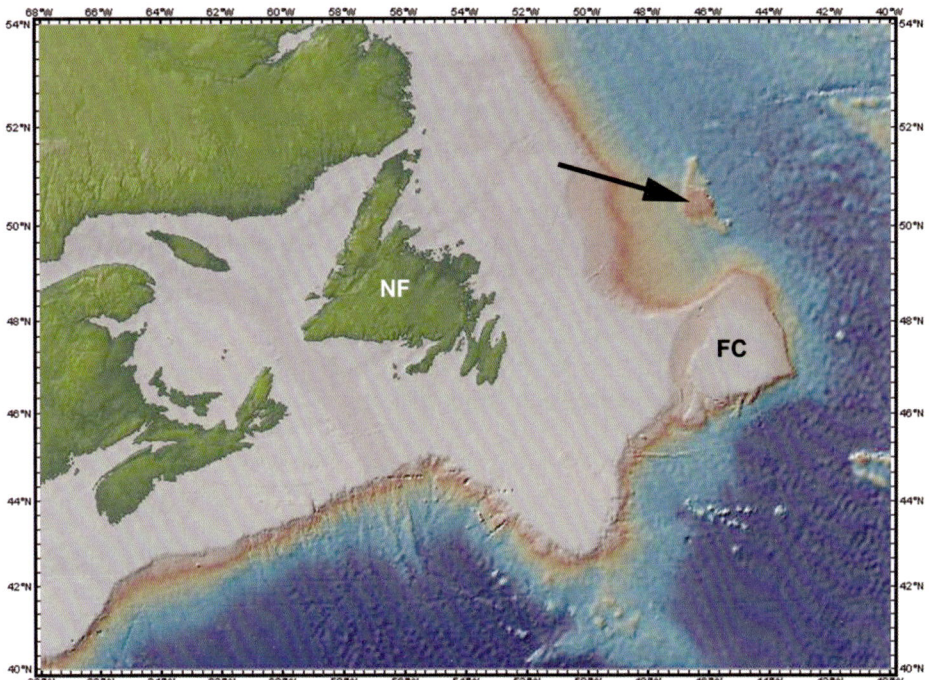

Figure 6.12. A shaded relief map showing Newfoundland (NF), the fishing ground Flemish Cap (FC), and, immediately to the north, Orphan Knoll (arrowed). Orphan Knoll represents a volcanic seamount, which has some large pinnacles, suspected to represent carbonate mounds (based on Grant and Jackson, 1994). Numerous pockmarks occur in the area (see King and MacLean, 1970; Judd and Hovland, 2007).

occurring on the summit of this seamount. So far, they have only been mapped with echosounders and other remote-sensing equipment. However, the mounds are expected to represent giant carbonate mounds of the type seen in the Porcupine Basin off Ireland (Grant and Jackson, 1994). There are at least 40 such mounds within an area of 70 km by 30 km, spanning a water depth of 1,900 m to 2,100 m.

6.1.8 The Mid-Atlantic Ridge

During a combined study of the fauna and ecology along the Mid-Atlantic Ridge (MAR) with scientific bottom trawling and ROV inspection, deep-water corals including octocorals and *Lophelia pertusa* were found (Mortensen *et al.*, 2008). Along the section of the MAR stretching from the southern part of the Reykjanes Ridge (off Iceland) and the Azores archipelago, eight sites were surveyed using ROVs. Corals were found at all sites, at depths ranging between 800 m and 2,400 m! Although octocorals dominated the coral fauna, *Lophelia pertusa* was one of the most frequently observed corals, present at five of the eight sites:

"It occurred on basaltic outcrops on the seamounts but always as relatively small colonies (<0.5 m in diameter).

The deepest record of *Lophelia* was 1.340 m, south of the Charlie Gibbs Fracture Zone. Accumulations of dead debris of coral skeletons could indicate a presence of former large *Lophelia* reefs at several locations. The number of megafaunal taxa was 1.6 times higher in areas where corals were present compared to areas without corals. Typical taxa that co-occurred with *Lophelia* were crinoids, certain sponges, the bivalve *Acesta excavata*, and squat lobsters" (Mortensen *et al.*, 2008).

These observations are very interesting, especially the facts

(1) that there are large accumulations of dead Lophelia, suggesting previous reefs; and
(2) the increased number of megafaunal taxa (including *Acesta excavata*) with the occurrence of corals.

To me, this suggests that where there could be a local food source (i.e., a nearby seep (vent) location, there is more life on the seafloor, including corals. From the work by Charlou and Donval (1993), Charlou *et al.* (2000), Kelley *et al.* (2001), Holm and Charlou (2001), and others, it is now well known that there exist numerous warm-water seeps all along the MAR, and especially along some of the fracture zones (transform faults), and that these sites are centres of primary (bacteria and archaea) production.

6.2 THE EASTERN ATLANTIC

6.2.1 Small mounds off Congo

Relatively small seabed mounds occur amongst numerous pockmark craters within an area in the Lower Congo Basin. They have been described by Gay *et al.* (2007). In the northeastern part of their study area, at about 400 m water depth, two circular structures, 200 m to 400 m wide and about 30 m high, were visible on the bathymetric map of the seafloor as dome-like structures. On the multibeam reflectivity map, they appear as black (highly reflective) patches (Figure 6.13). A sediment core collected from within the main dome-like structure shows carbonate-rich sediments and debris of cold-water corals interbedded with hemipelagic muds.

"Despite the less permeable nature of the Quaternary[1]–Pliocene sequence, this interval is affected by tight polygonal faults that have possibly channeled fluids from the west-dipping normal fault and fed the fluid seep site. In this area the gravity-driven normal fault does not affect the seabed. However, fluids coming

---

[1] The Quaternary began 2 MYBP and continues to the present day.

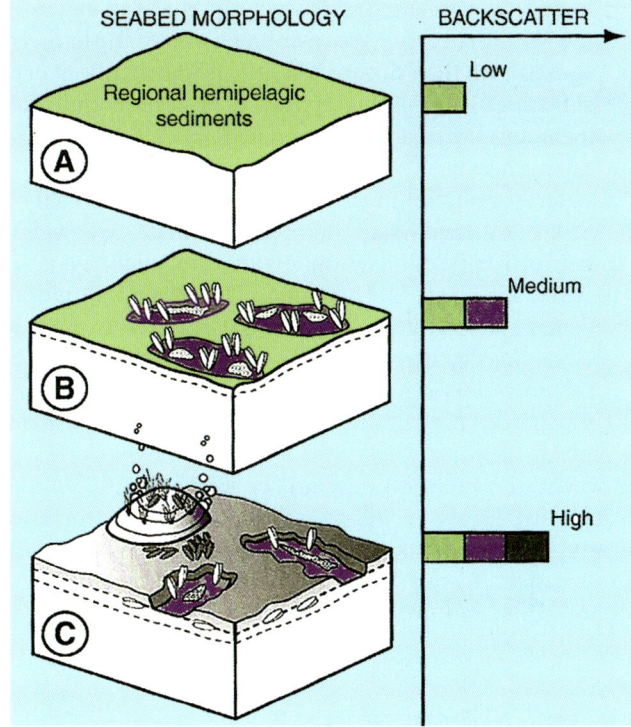

Figure 6.13. This drawing is from Gay *et al.* (2007). It shows how they envision the 30 m high carbonate mound in (C) has formed, starting with an undisturbed seafloor in (A), to a pockmarked floor with bivalves and other seep-related organisms in (B), to mounds, pockmarks, and gas seeps in (C). High acoustic seabed reflectivity (backscatter) is shown to be a consequence of seepage and organic biodiversity (adapted from Gay *et al.*, 2007).

from the upper turbiditic Miocene channels reach the seabed to form seafloor fluid seeps. In this case, polygonal faults act as a shortcut, as they represent preferential pathways for fluid migration with a high drainage potential at the node of three contiguous polygonal cells" (Gay *et al.*, 2007).

The researchers have made a conceptual model (see Figure 6.11) based on the hydraulic seafloor processes that cause the mounds, which have deep-water corals growing on them.

6.3 THE MEDITERRANEAN SEA

The situation for deep-water coral reefs in the Mediterranean is best explained by Taviani *et al.* (2005):

"The Mediterranean basin represents an excellent biological archive of past and modern deep coral growth whose study may help to understand taxonomic, biogeographic, ecological, and evolutionary patterns of modern deep coral bioconstructions, best embodied by the *Lophelia* reefs and mounds of the Atlantic Ocean. In fact, the occurrence of extant deep coral genera in the Mediterranean basin is documented, although not continuously, since the Miocene. Following the Messinian crisis, the re-colonization of the basin by deep coral is likely to have started with the Pliocene but little is known about deep coral biota linked to hard substrates during this epoch. It is certain that Atlantic-type deep-sea corals including the scleractinian triad *Lophelia–Madrepora–Desmophyllum* have been established in the basin since the Late Pliocene–Early Pleistocene, as proven by outcrop evidence in southern Italy, especially Sicily and Calabria, and Rhodes. Still-submerged dead coral assemblages are widespread in the entire basin between ca. 250–2,500 m depth; the majority is aged at the last glacial by AMS, C^{14}, and U/Th dating. The present situation (post-glacial) is a general decline of such deep corals in the Mediterranean, and this is especially true for *Lophelia*, which appears to be more severely affected by local extinctions. To date, the only exception to this general rule is represented by the recent discovery of prosperous *Lophelia* populations in the eastern Ionian Sea" (Taviani et al., 2005).

However, the situation may not be as depressing as stated here, because there are many other places, including off Norway, where there seems to be more dead coral around than live and lush coral. Again, this has to do with our bio-expectations. Why do we expect to see a lot of live coral compared with dead coral? Perhaps this is the natural state, and not a sign of impending extinction.

6.4 THE PACIFIC OCEAN

6.4.1 Coral gardens off the Aleutian Islands

According to new data from off the Aleutian Islands, there is a dense, diverse, and prolific cold-water coral region in these waters. The coral gardens observed during a recent seabed survey around the Aleutian Islands consist of a highly diverse community, including

> "stolon corals, true soft corals, sea whips, and sea pens, gorgonian corals, stony cup corals, hydrocorals, and black corals. To date, a total of 86 different taxa (genera, species, or subspecies) of cold-water corals have been collected, with 25 potentially endemic to the islands (found nowhere else)" (*www.lophelia.org/images/jpeg/Aleutian_Garden.jpg*, Lindner, 2007).

So far, however, it has not been possible to acquire more details of these exciting occurrences.

Because there are clear signs of heavy impact by bottom fishing (trawling), the American government has banned bottom contact fishing in a 370,000 km² area around the Aleutian Islands, according to *www.lophelia.org* (2007). This, therefore, represents the largest area of the ocean floor protected against trawling activity, to date. Many more countries should follow the American example, including Norway, before deep-water corals are totally obliterated by destructive trawling practices.

6.4.2 Davison Seamount off California

Measures to protect resources on a large seamount, the Davison Seamount, off California, are currently being made. The reason for this move is the discovery of a prolific fauna there. According to De Vogelaere *et al.* (2005),

> "Davison Seamount ... is one of the largest known seamounts along the western United States. It is 42 km long and 13.5 km wide. From base to crest, Davison Seamount is 2,400 m tall; yet, it is still 1,250 m below the surf. Davison Seamount has an atypical seamount shape, having northeast-trending ridges created by a type of volcanism only recently described (Davis *et al.*, 2002); it last erupted about 12 million years ago.

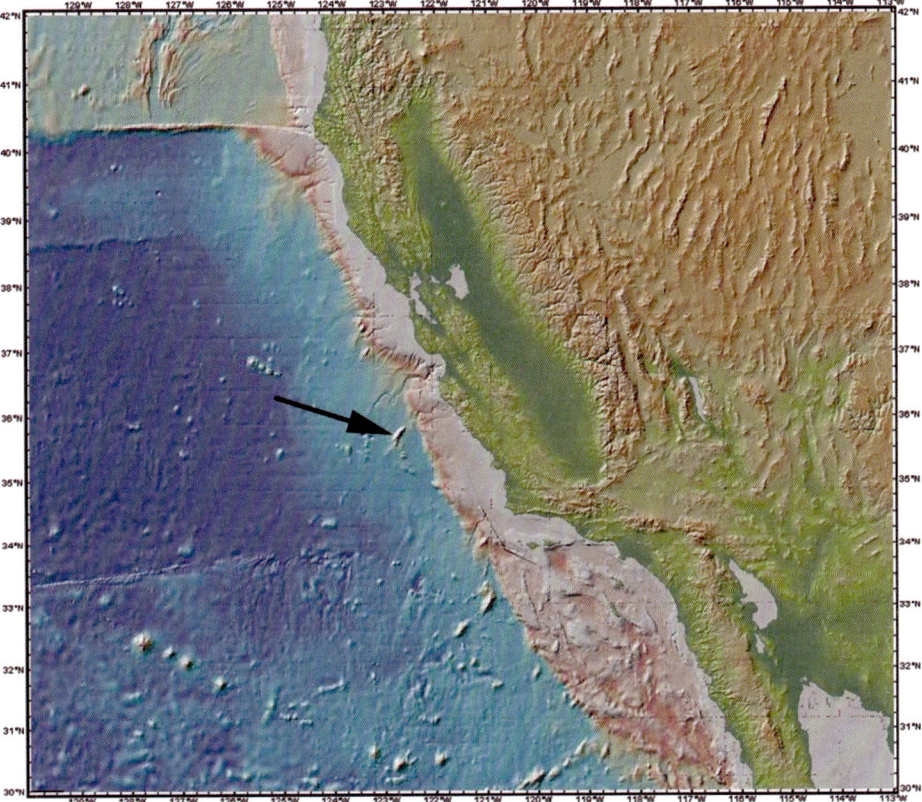

Figure 6.14. A relief map showing California and the ocean floor west of California. The arrow points at Davison Seamount, south of Monterey Canyon, where 20 different deep-water coral species were found. This is one of the few that have been visually inspected by an ROV and scientifically trawled and sampled. Obviously, there must be many more such seamounts, judging from this image.

Preliminary analyses (of video) indicate that 20 coral taxa were found, and they were almost exclusively located on high relief, ridge areas." (De Vogelaere et al., 2005).

More about seamounts and their tendency to produce fluids, long after their active volcanic period, can be found in Section 7.4.

6.4.3 Australian carbonate knolls

Although information on deep-water carbonate mounds (knolls) was published by Hovland et al. (1994), they have not been sampled or studied visually since. Therefore, little is known about their true nature. In a recent hydrocarbon seepage study performed in the Arafura Sea, offshore northern Australia, by Nadège Rollet and her team (2007), several large mounds and carbonate knolls have been documented. They are 3 m to 30 m high and up to 3 km across:

> "The sea bed mounds have a strong back-scatter side-scan sonar signal and cause underlying signal attenuation in sub-bottom profiles with high amplitude at the sea bed horizon. They are mostly rooted in the U7 erosional surface. The buried mounds observed in sub-bottom profile data are also rooted on surface U7, and cause signal attenuation within the underlying section. Both these sea bed and buried mounds may be formed from either carbonate or cemented sand ridges formed during a small transgression event, after a eustatic low-stand (unconformity U7). Although the sea bed and buried mounds appear to be developed over areas of deeper faulting, observed in the sub-bottom profile data, there is no clear link between them and evidence of any fluid flow. Compared to other mounds on the North West Shelf of Australia, they are smaller in scale than those described in the Vulcan Sub-basin, and larger than the mounds on the Yampi Shelf" (Rollet et al., 2007).

In the Timor Sea, north of Australia, spectacular carbonate mounds and reefs are important for two reasons: because of their association with petroleum, and because they are important marine biological habitats; Ashmore Reef, for example, is a Marine Protected Area because of its rich biodiversity and range of endemic species (Glenn, 2002). Integrated surveys, similar to that described by Rollet et al. (2007), combining sea surface seep detection, seabed mapping, and detailed seismics, have here (in contrast to the other Australian reefs mentioned) demonstrated a strong link between the banks and petroleum seeps:

> "The key linking feature between a wide range of features, from small, seafloor build-ups, to isolated carbonate shoals such as Heywood Shoals, to large clusters of banks such as the Karmt Shoals, is that they are all associated with modern seeps. Where seeps are absent, so (generally) is the evidence for significant contemporary carbonate bank formation" (O'Brien et al., 2002).

These Australian occurrences proves how difficult it is to document obvious or clear links between coral mounds and seeps, even with integrated methods.

6.5 SUMMARY

This chapter has presented a brief world tour to visit coral reefs located on the rims of pockmarks off Brazil, similar reefs in the Gulf of Mexico, where they also occur aligned along faults and on the tops of salt dome cap-rocks (such as at West and East Flower Garden Banks). In such settings, it is natural to suspect a seepage relationship. In the Florida Straits, they occur as lithoherms in strong current settings. However, there is also reason to suspect some seepage of groundwater and perhaps gases through the seafloor. On the Outer Blake Plateau, off Cape Fear, North Carolina, large lineations of deep-water coral reefs and suspected giant carbonate mounds occur at specific locations, which are correlated with out-cropping (exposed) sedimentary bedding planes.

Deep-water corals also occur off Nova Scotia and on Orphan Knoll, a seamount located east of Newfoundland, Canada. In the southeastern Atlantic Ocean, small carbonate mounds hosting deep-water corals are associated with pockmarks. And, on the strange Davison Seamount, at 1,250 m depth, off California, 20 coral taxa were found. Although this volcanic seafloor structure formed up to 12 million years ago, it may still be actively producing volcanic gases, as is known from many other extinct volcano features both on land and in the ocean. Finally, many carbonate knolls off Australia are not found to be associated with seepage, although others definitely are.

7

Ancient and modern analogues

This study provides an example of the formation of such mounds that represents an important hydrocarbon reservoir in Alberta.

Al-Aasm and Vernon (2007)

These remarkable organisms are chemosynthetic, rather than photosynthetic: they spin poison into food by living in a symbiotic relationship with bacteria that break down the methane or hydrogen sulphide for their own energy.

Gleick (1997)

7.1 INTRODUCTION

Analogies to our modern deep-water coral reefs and giant carbonate mounds have long been known to geologists and palaeontologists. They were the first to find them and describe them from exposures on the ground and also from drilling into them. In this chapter, I review some of the features, like the famous Waulsortian Mounds, that clearly resemble the Porcupine Bight giant carbonate mounds. This review describes some petrol-filled carbonate reefs that act as reservoirs for hydrocarbons. It continues with a description of fossil seep carbonates, the Devonian[1] fossil reefs in Algerian Sahara, the Kess-Kess Mounds in Morocco, seamounts capped by carbonates, and a review and discussion of modern-living stromatolites and microbialites.

[1] The Devonian lasted from 417 MYBP until 354 MYBP.

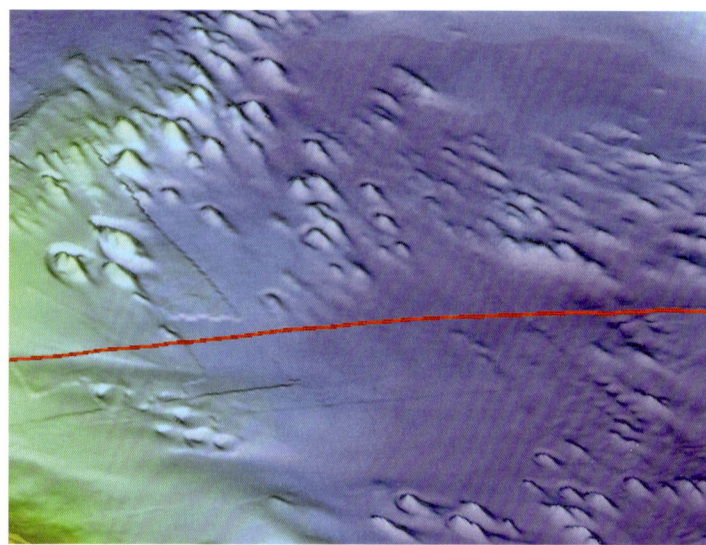

Figure 7.1. An llustration showing the typical "canopy" appearance of deep-water coral reefs, as they were found during multibeam mapping at Breisunddjupet, off mid-Norway. These are between 2 m (smallest) and 5 m high. This perspective view also shows the route of an intended pipeline that was moved farther west because of these reefs.

7.2 WAULSORTIAN-LIKE MOUNDS FROM THE CARBONIFEROUS

Waulsortian mounds are fossil mounds found in ancient sedimentary strata, as described by Ariel A. Roth (1995):

> "The nature and origin of Waulsortian-like mounds have been topics of interest for more than 125 years. A compelling reason for the interest is their uniqueness in the geological reef record owing to their total lack of frame-building organisms."

There are three main theories to explain the origins of Waulsortian-like mounds:

— the lithoherm theory, suggesting the material was cemented into place, by penecontemporaneous[2] processes, diagenesis (Neumann *et al.*, 1977);
— a microbial origin, calling for organic films and filaments to stabilize and support the carbonate mud during mound development (Miller, 1986);
— baffling (trapping) of carbonate fines within stands of crinoids and bryozoa in mounds which contain relatively high skeletal fossil content (King, 1986);
— a combination of all three of these mechanisms, which acted in concert to produce some of the mounds (Pratt, 1982; Brown and Dodd, 1990).

---

[2] A penecontemporaneous process is one that originates soon after the rocks in which it is displayed are formed.

When drilling into these structures or studying them in the open, it is hard to understand how they could form, even at very steep angles, if the mud was not cemented:

> "The massive reef core consists mainly of fine, calcium-carbonate mud. The robust wave-resistant reef frame builders of our present reefs are missing. There are some sponges but sponges are not known to produce great reefs; and, there is insufficient algae to bind the sediments" (Roth, 1995).

This is exactly the same as found in the modern giant carbonate mounds of the Porcupine Seabight off Ireland. According to Dorschel *et al.* (2007), the carbonate mounds off Ireland manage to build their tall structures despite the lack of "frame-building organisms" because of the deep-water corals they are capped by:

> "Due to the sediment buffering capacity of the corals and lower energetic environments within the thickets, silts and clay deposit during slack water are protected against re-suspension."

Thus, the fine-grained and un-cemented sediments keep in place and accumulate into tall structures because they are "armoured" by deep-water coral skeletons.

7.2.1 A key to the present?

Because skeletons and fragments of calcifying organisms are also found inside some Waulsortian-like structures, they resemble modern carbonate mounds. This was also the conclusion after the IODP Leg 307 when drilling through Challenger Mound. The accumulation of fine-grained carbonate mud is kept in place, not by cementation or by organic filaments, but rather by a meshwork of durable dead (*Lophelia*) skeletons and fragments (Dorschel *et al.*, 2007). In geology, there is a saying: *The Present is the key to the Past*; however, in this case it is not only the key to the past, but also to the present! So, the new, modern catchphrase (according to Hans Konrad Johnsen, Statoil) should instead be: *The Present is the key to the Present*. Which, in effect, means that one needs to be aware that what is at first interpreted as unusual, anomalous, or outright curious, may in fact represent the norm for that particular situation, feature, or process.

Although we can use present knowledge on how carbonate build-ups occur in deep and low-dynamic environments, we may be able to utilize information from the ancient structures to learn more about what to expect internally, inside the giant modern carbonate mounds. Roth (1995) provides us, further, with one interesting example, the Muleshoe Mound (Carboniferous) of southern New Mexico. This mound is about 100 m high and represents one of the Waulsortian mounds formed in fine lime mud. There are many other such mounds worldwide:

> "These mounds are characterized by a core composed mainly (50–80%) of calcium carbonate mud. Some are spectacularly conical with relatively steep

166 Ancient and modern analogues [Ch. 7

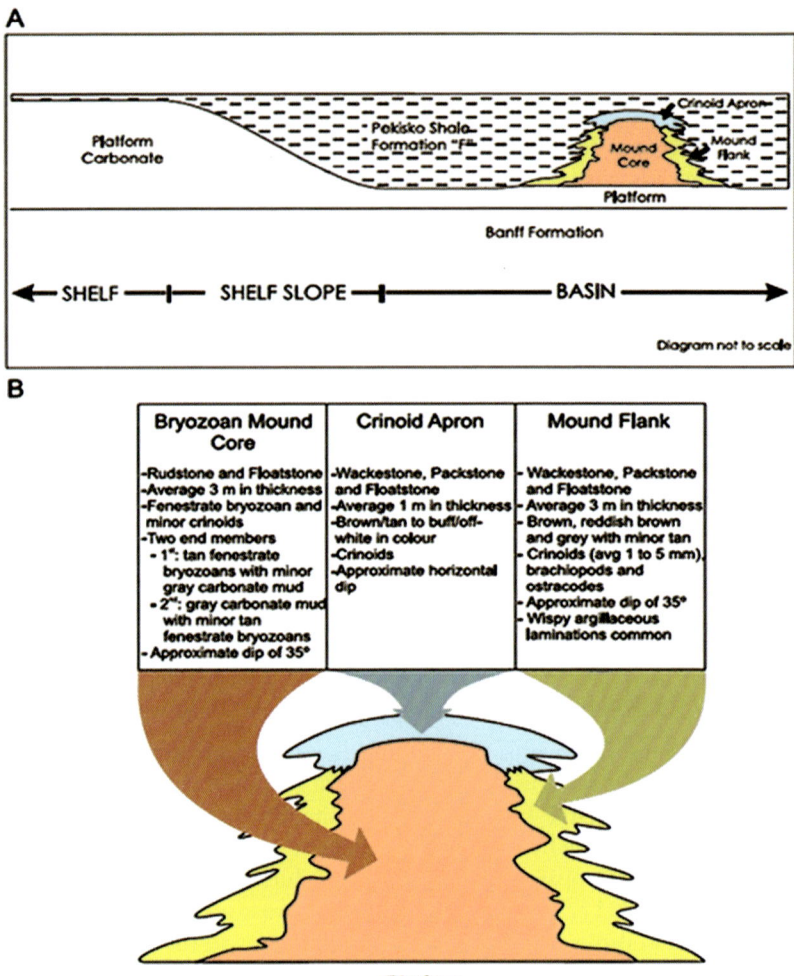

Figure 7.2. (A) Schematic sketch of a (fossil) Waulsortian mound in the Normandville field, eastern Canada. (B) Sketch of a Waulsortian mound (see A), showing the different facies encountered during drilling into the structure. Such mounds seem to be true fossil analogues to the living giant carbonate Belgica Mounds off Ireland (adapted from Al-Aasm and Vernon, 2007).

sides. In some mounds, the mud core gives evidence of bedded layers which can suggest transport of sediments. Pray (1965) has described the intrusion of dikes into these mounds coming from soft sediments below them. This indicates that the layers below were still soft when the intrusion took place" (Roth, 1995).

Such details (intrusion dikes) will be difficult to find by drilling, should they exist inside living modern carbonate mounds.

7.2.2 Petroleum-filled mounds

Alberta occurrences

Unique, deep-water carbonate, mud-rich build-ups, which contain a significant percent of skeletal components (crinoids and bryozoans) occur in Tournasian age (Carboniferous) carbonate rocks of the Normandville oil-field in northwestern Alberta, Canada (Al-Aasm and Vernon, 2007). But, although these structures today represent prolific hydrocarbon traps (reservoirs), their origin during the Carboniferous period is unknown:

> "The main control on localization of mound growth was either structural or topographic highs. The physical characteristics, factors controlling localization and the growth sequence of these mounds have been difficult to establish and remain enigmatic" (Al-Aasm and Vernon, 2007).

When they lived, these structures were equivalent, and perhaps even more impressive in size than the current carbonate mounds (Belgica Mounds) off Ireland. There are, however, two remarkable aspects of this Waulsortian-like mound, which I interpret as being telltale for concluding that the structures were fuelled by seepage of deep-sourced fluids, including light hydrocarbons. These are

(1) the fact that the structures are filled with hydrocarbons today;
(2) the fact that portions of the calcite of the mounds have been transformed into dolomite.

Added together, these two aspects suggest that seepage of light hydrocarbons (methane, ethane, propane, butane, pentane, etc.) and Mg-rich fluids (porewater) fuelled the mound build-ups while they were still forming on the seafloor. Then, when they became buried and extinct, their porous structure first filled up with seawater, and as they were buried deeper under more sediment, the seepage fluids (light hydrocarbons) replaced the water and preserved the porosity (as previously suggested by Hovland, 1990b). Simultaneously, Mg-rich fluids (possibly brines and hydrocarbons) aided in transforming calcite into dolomite (seep-induced diagenesis).

Williston, Canada

Deeply buried, oil-filled Waulsortian-like mounds occur in the lower Dickinson Lodgepole sedimentary unit located at 3,300 m depth. The build-ups are composed predominantly of bryozoan and crinoid grainstone and packstone. They are up to 100 m high and 1.6 km in diameter. According to Randolph B. Burke of the North Dakota Geological Survey, they are distributed at between 3 km and 4 km depth around the deepest part of a deep (>15 km) sedimentary basin. Site-specific mound types are located on and around intrabasinal blocks, and they are definitely associated with fault zones. However, according to these researchers, "the nature of the mound/fault association is unclear". This supports the hydraulic theory. We shall now look more closely at how to identify seep-related structures in fossil reefs of this kind.

7.3 FOSSIL SEEP CARBONATES

In recent years several fossil seeps have been identified. Early descriptions of fossil carbonate mounds came from Monferrato, northwest Italy (Clari *et al.*, 1988; Cavagna *et al.*, 1999). Judd and Hovland (2007) have listed specific aspects common to the identification of fossilised marine and lacustrine (lake-associated) seep carbonates:

> "Following the suggestions of Cavagna *et al.* (1999), and others (Belenkaia, 2000; Thiel *et al.*, 2001), we consider the following features to be diagnostic of fossil MDAC [methane-derived authigenic carbonates]:
>
> 1. Patchy distribution, which is clearly evidence of a phenomenon occurring only in some parts of a sedimentary basin.
> 2. Restricted lateral extent of carbonate cements, usually within siliciclastic sediments. Boundaries may be transitional or sharp: 'in a few clear cases the percentage of cement and the abundance of chemosynthetic taxa remains gradually decrease away from the paleoseep; more commonly, however, lateral relationships of limestone masses and blocks are sharp and more enigmatic' (Cavagna *et al.*, 1999).
> 3. Evidence of fluid flow pathways, in the form of networks of fractures and cavities with 'complex and polyphase infillings by carbonate sediments and cements' (Cavagna *et al.*, 1999).
> 4. Diagenetic products and fabrics anomalous in clastic sediments and contrasting with those shown by the surrounding sediments. Cavagna *et al.* (1999) identified the following: 'abundant intergranular cement, in "loosely" packed, uncompacted, sandstone or mudstones; cavity-filling cements showing different mineralogies (aragonite, calcite), morphologies (botryoidal, fibrous, sparry) and commonly displaying bandings due to a variable amount of inclusions'.
> 5. Associated macro- and/or micro-fossils.
> 6. Biomarkers.
> 7. Carbon isotope ratios in the range -60 to $-20‰$."

Fossil seeps are important, not only because their recognition helps to explain the formation of related rocks, but also we can take advantage of geological exposures to examine the whole system, including the sub-seabed plumbing system. Some fossil cold-seep communities have been identified because chemosymbiotic-related fauna, such as tubeworms and bivalves, have been recognised. Sometimes, the identification of MDAC has been critical as have the many fossil seep carbonates (Cavagna *et al.*, 1999; Campbell *et al.*, 2002).

7.3.1 Ancient deep-water fossil reefs, Algerian Sahara

Exposed fossil analogues to deep-water carbonate mounds and Norwegian deep-water coral reefs are found in the Algerian Sahara desert. After a team of German researchers visited the Ahnet Sedimentary Basin, they published a paper titled: "The World's most spectacular carbonate mounds (Middle Devonian, Algerian Sahara)" (Wendt *et al.*, 1997). From their photographs it is seen that these features are really spectacular. The haystack-shaped mounded structures are remarkably similar to some of the present, classical Norwegian deep-water coral reefs. The Ahnet reefs consist of lithified clay with fossil animals of Devonian age (380 million YBP). The animals included rugose corals, octocorals, sponges, mussels, and trilobites. They span a period of about 200,000 years. Based on the fact that there are no fossil traces of plants or algae, the structures must have been formed in the aphotic ocean zone, at an estimated water depth of about 400 m, which is a similar depth to those found living off Norway and Ireland today (Wendt *et al.*, 1997).

As they form in lines and because of the presence of framboidal pyrite, barites, and apatite in particular zones of the structures called "Neptunian dykes", the researchers concluded that they formed at the periphery of the Ahnet sedimentary

Figure 7.3. A photograph of the remarkable fossil deep-water carbonate reefs or mounds located in the Ahnet sedimentary basin, Algerian Sahara. They were, understandably, published as "The World's most spectacular carbonate mounds" (Wendt *et al.*, 1997), mainly because of their pristine detailed intactness and their undisturbed appearence. They were previously buried (and conserved) in sediments, but have been naturally exhumed over the last million or so years (photo courtesy of Bernd Kaufman).

basin over deep-seated faults where mineralised porewater flowed through to the seabed (Wendt *et al.*, 1997). Similar mounds have also been found farther to the west, in the Mader Basin, Eastern Anti-Atlas, Morocco (Kaufmann, 1998). These are the Kess Kess fossil mud mounds. Normally, fossil seep-related structures are only found underground, as buried features, seen on seismic records, and penetrated by drilling. But, fortunately, some rare, spectacular cases are exposed in arid and vegetation-free zones on the surface, as described below.

7.3.2 Kess-Kess formations, Morocco

Up to 55 m high, conical fossil carbonate formations occur in the eastern Anti-Atlas mountains, southern Morocco (Belka, 1998). These Kess-Kess carbonate mud mounds are of Early Devonian age and are possible to see only because they occur in a desert landscape, without vegetation. Their impressive characteristic features can be seen in Figure 5.1. The name Kess-Kess comes from the conical dishes used to serve the Moroccan national meal couscous. In all, there are 54 such structures within the Hamar Lagdhad area. Some have grown into groups and L-shaped structures. The number refers to centres of growth within these groupings. The Kess-Kess structures have individual diameters of up to 120 m and an average height of 20 m–45 m.

The fossils within the Kess-Kess mounds have an exceptionally high biodiversity, where tabular corals dominate "auloporids, thamnoporids, and facosites" (Brachert *et al.*, 1992). Trilobites and crinoids are found all over the conical structures. There are also solitaire, rugose corals, brachiopods, ostracods, gastropods, and pelecypods. Even though there are markings from worms and sponges, there is no sign of algae. This indicates that the structures have been growing at depths greater than 100 m (Belka, 1998). A special character of these mounds is the numerous Neptunian dikes. These are fracture systems, which have been filled with other sediments and minerals. It is suspected that these dikes represent spring conduits (seeps) during the formation of the mounds, and that the seeping fluids were of a hydrothermal nature (Belka, 1998). Another intersecting fracture pattern is also evident on these fossil carbonate mounds. The fractures were caused by tectonic forces, which were active before, during, and after their formation. In contrast to the Ahnet Basin deep-water reefs, the Kess-Kess mounds do not contain any geochemical traces of hydrothermal activity or methane-charged porewater (Belka, 1998). Thus, it is inferred that they formed as a result of springs along the fracture systems beneath the mounds.

7.4 SEAMOUNTS AND CARBONATES

Isolated mid-ocean volcanic islands like Pitcairn in the Pacific Ocean and Tristan da Cunha in the South Atlantic would be called seamounts had they been submerged beneath sea-level. Seamounts, guyots, and atolls are generally circular or conical in shape, and are perhaps (if they were visible) some of the most spectacular topographic features on the surface of our planet. They are all a result of poorly

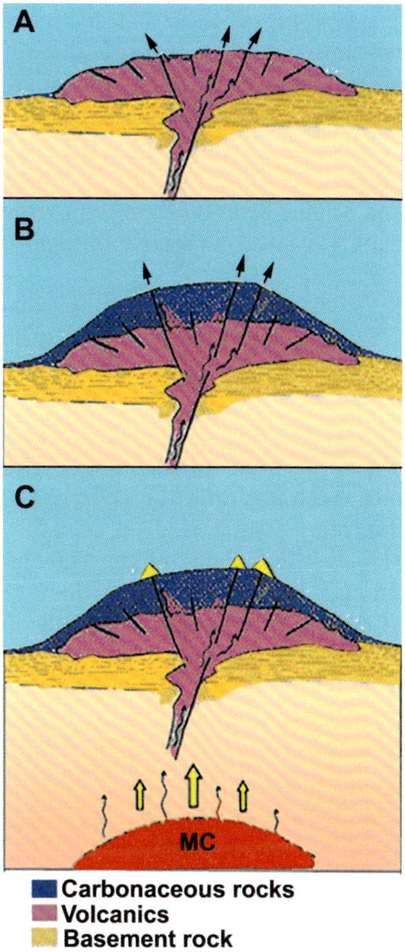

Figure 7.4. A conceptual sketch illustrating the formation of the Kess-Kess reefs, Morocco. These reefs are suspected to be fossil deepwater or cold-water coral reefs (carbonate mounds). They formed on the seabed about 380 million years ago. German researchers have documented a link with volcanic gases and mineralised porewater that seeped up through the seafloor (black arrows in A and B), above a cooling magma chamber (MC), only shown in the bottom drawing (C). The sketch shows how the reefs (yellow triangles in C) probably developed above surfacing faults and fissures. Note that carbonaceous rocks (in blue) deposited a cap over the volcanic rocks before the reefs developed (adapted from Belka, 1998).

understood "mid-plate" volcanic processes. According to recent estimates made by Mukhopadhyay *et al.* (2002) there may be as many as 2 million seamounts in the world's oceans; somehow they seem to be associated with plate flanks and also shear zones in oceanic plates. Many reports have noted the relatively low density of rock samples from seamounts. In many cases this is attributed, at least in part, to the presence of vesicles. Menard (1964) reported two examples: the first was a fragment of relatively unaltered vesicular pahoehoe lava dredged from Jasper Seamount, Baja California with a density of "only slightly more than $2\,g/cm^3$". The second was highly vesicular basalt from 3,500 m on the steep-sided Popcorn Ridge, west of Baja California. This was decrepitating when brought on deck, and "breaking with sharp snapping sounds because of the expansion of gas in the vesicles". The low density and highly porous nature of these rocks suggests that they came from magmas containing large amounts of gas (Judd and Hovland, 2007).

It seems probable that hydrothermal activity continues for some time also after volcanic activity has ceased. In a review of the biology of seamounts, Rogers (1994) noted that, because most of them are of volcanic origin "it is of no surprise that some are associated with hydrothermal venting". He identified five with hydrothermal activity: Loihi (off Hawaii), Axial Seamount, Red Volcano (one of the Larson Seamounts, 30 km east of the East Pacific Rise), Piip (Peepa) in the Bering Sea, and Marsili Seamount (in the Tyrrhenian Basin, north of Sicily in the Mediterranean Sea). However, he clearly thought that this was only the tip of the iceberg:

"The presence of hydrothermal vents on mid-oceanic seamounts and off-axis seamounts may indicate that hydrothermal emissions are more common in the deep sea than previously thought" (Rogers, 1994).

More recently, it has been observed with modern heatflow measurements that certain portions of some seamounts represent recharge areas, where seawater flows into the seabed and the underlying porous oceanic crust. This may account for the fluid plumes indicated on acoustic transects, rising into the water high above the Hancock Seamount of the Hawaiian chain (Hovland, 1988). Harris *et al.* (2002) simulated buoyancy-driven flow, then estimated that there are 15,000 seamounts of up to 2 km height, and that the mass water flux through them is 10^{14} kg per year. This is similar to estimates of the mass flux at ridge axes and through ridge flanks. It therefore seems that seamounts are important sites for different types of seabed fluid flow at different stages in their evolution: volcanic activity during, and for some time after formation, then hydrothermal activity, and finally the cycling of seawater in through the base and out into the relatively porous oceanic crust. The recent discovery of corals and other prolific sessile organisms on seamounts could be linked to active, internal fluid flow processes, as previously suggested by Rougerie and Waulthy (1988).

7.5 "LIVING FOSSIL" STRUCTURES?

7.5.1 Stromatolites and microbialites

Stromatolites and microbialites are so-called biosedimentary structures, which consist of inorganic material (sediment grains and exhaled, precipitated minerals) and bacterial cells. Tucker and Wright (1990) gave the following definition of a stromatolite: "A stromatolite is an organosedimentary structure produced by the sediment trapping, binding, and/or precipitation activity of micro-organisms, primarily the cyanobacteria." But the author of the book *Geobiology*, Kurt Konhauser (2007), says their main characteristics are local laminations:

"Distinct laminations arise when some form of recurring disruption, due to variations in sedimentation rate, salinity, seasonality, etc., causes a hiatus in the normal accumulation of biomass.

Clearly there are many more such examples of microbial niches in thermal spring settings, and hence, it should not come as a surprise that, variations in microbial communities give rise to textural variations in sinters or travertines simply because different microorganisms interact uniquely with the waters in which they bathed" (Konhauser, 2007).

These structures are important features: stromatolites are thought to have been responsible for the production of the atmosphere's free oxygen, and thus the evolution of animal life on Earth. They were widely distributed in Proterozoic seas and were still abundant in the Early Palaeozoic, but since then they have steadily declined (Playford *et al.*, 1976). Stromatolites are important biogenic carbonate structures, but are not traditionally regarded as a true fluid flow feature. However, field examples provide several lines of evidence supporting the concept that there may be a relationship, mainly with groundwater flow (Judd and Hovland, 2007).

7.5.2 Are they seep bioherms?

In 1988, we first suggested that stromatolites might have been formed as a consequence of seepage of mineralised water from the floor of lakes or the sea (Hovland and Judd, 1988; Judd and Hovland, 2007). This model is now partly supported by the microbial scientific community, although the physical and chemical ways in which it works is not yet fully understood, as mentioned by Konhauser (2007):

"Understanding modern sinter (or travertine) formation has important implications for the rock record. As the pioneering work by Walter *et al.* (1972) suggested, it may be possible to interpret some distinctive ancient stromatolite morphologies from modern thermal spring analogs, such as extant coniform stromatolites, with the Precambrian stromatolitic form *Conophyton*. Modern coniform stromatolites, dominated by *Phormidium* species, are restricted to continuously submerged, low energy geothermal pools. This environmental setting has also been proposed for the origins of *Conophyton* stromatolites, suggesting that similar cyanobacterial species may have been involved in their formation. For this to be useful, however, a general framework illustrating the association between the principal types of thermal spring biofacies and the dominant mat-forming communities still need to be developed. Unfortunately, this is where gaps in our knowledge emerge" (Konhauser, 2007).

Spring-controlled bioherms, Corinth

Along one of the major fault zones in the Gulf of Corinth, Greece, there are some strange occurrences of bioherms and reefs. Even though these occurrences have no known coral species embedded in their structure, they are beyond any doubt associated with freshwater seeps, as explained by Portman *et al.* (2005):

"The limited areal extent of the bioherms and their close association with karstified fault scarps suggest that they formed in shallow sea water where freshwater submarine springs delivered $CaCO_3$ saturated water that promoted rapid calcification of cyanobacteria. Rapid calcification and strong degassing of CO_2 from the spring water resulted in disequilibrium stable isotope compositions for the calcites."

Put in other words, these reef structures would, according to the authors, never have formed without being controlled by the seepage of CO_2-charged porewater, as also documented by Andrews et al. (2007).

Stromatolites, Lake Tanganyika

The largest of the African rift lakes and also the longest lake in the world is the 670 km long Lake Tanganyika. Its waters are rather alkaline, with a pH of between 8.6 and 9.2 (World Lake Database). Several seeps and warm submarine springs are known to occur along the shores of this lake, perhaps suggesting ongoing serpentinisation in the crust below the lake. Thus, at Cape Banza, a French team of geologists found hydrothermal water flowing through several orifices at a rate of $2 L \cdot s^{-1}$ to $3 L \cdot s^{-1}$ and a temperature of $103°C$ (TANGANYDRO Group, 1992). The temperature and abundant bubbles indicates local boiling conditions. There are multiple-orifice aragonite chimneys, up to 0.7 m high. Furthermore, groups of stromatolites grow along a cliff down to a depth of 30 m at the Luhanga hydrothermal field. Cohen et al. (1997) studied these modern stromatolites, but only in the

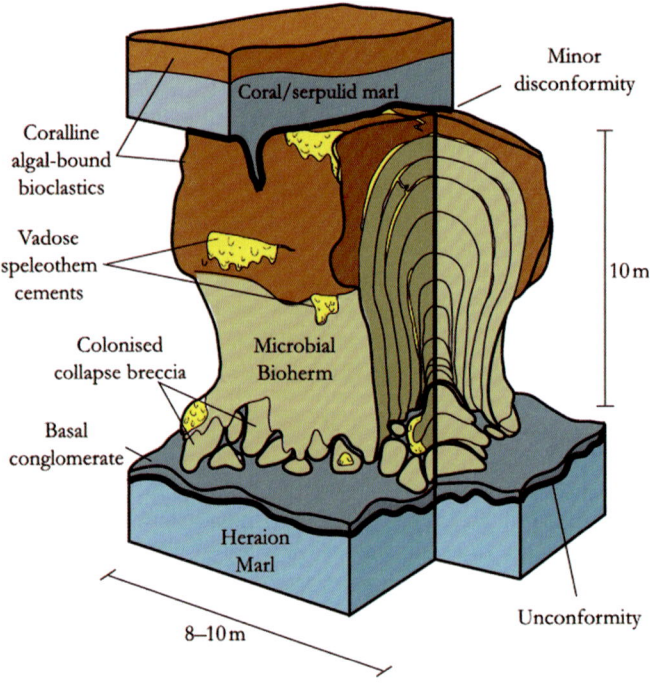

Figure 7.5. Sketch of submarine spring-controlled growth of bioherms (Portman et al., 2005), showing the principal elements described in the text. Note that the coral/serpulid marl above the minor disconformity has variable thickness and limited spatial extent, probably in part determined by the size of the bioherms and possibly by the nature of seeping fluids (adapted from Portman et al., 2005).

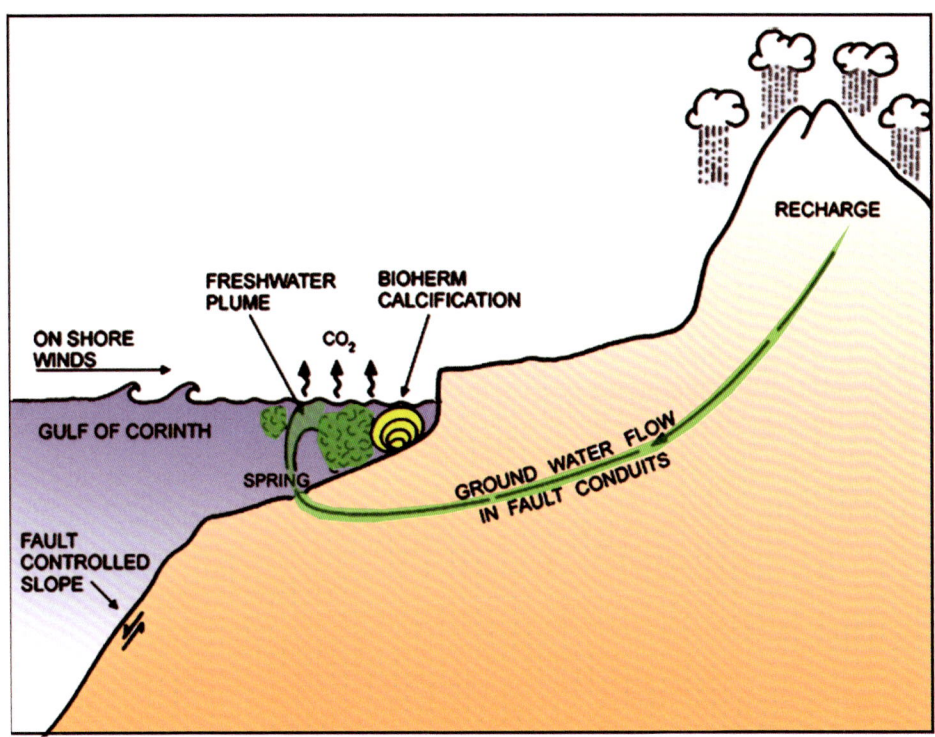

Figure 7.6. Sketch showing the suggested growth of bioherms, according to Portman *et al.* (2005). Fresh groundwater flow is focused by fault conduits in the Perachora karst aquifer emerging as shallow offshore submarine springs. Spring flow emerges as a plume in shallow shelf water that then disperses and mixes with the ambient seawater, degassing CO_2 at the surface. The shallow (fault-controlled) shelf becomes brackish, with freshwater perhaps pinned onshore by winds and waves. The brackish conditions and calcite-supersaturated water allow growth of distinctive microbial bioherms (adapted from Portman *et al.*, 2005).

context of palaeoclimate; they were not particularly concerned with their geobiology. Even so, they noticed that some of the lake water was "ground-water input from small hydrothermal springs" (Judd and Hovland, 2007).

Microbialites in Pavillion Lake, Canada

Similarly, in a very narrow (800 m) and long (5.8 km) lake in Marble Canyon, British Columbia, Canada (Pavillion Lake) there are microbialites. Because surface streams do not enter this remarkably clear lake, karst hydrology dominates. This means that whereas most other lakes rely on riverine (surface) water flow, this lake acquires water through cracks and crevices in its sides and lake bottom (i.e., groundwater flow, seepage, and springs). The microbialites were described by Laval *et al.* (2000) as being up to 3 m high, occurring along the lakesides in clusters aligned roughly perpendicular to the shoreline. They generally occur at three depths: shallow

(~10 m), intermediate (~20 m), and deep (>30 m). The shallow ones range in height from several centimetres to a few decimetres and comprise interconnected clusters of discrete round aggregates of calcite grains covered by photosynthetic microbial communities and their calcified remains. At intermediate depth, large microbialite domes (<3 m high) consist of closely spaced aggregate clusters with a preferred orientation forming vertically ribbed structural components reminiscent of cones and leaves (Laval et al., 2000). In deeper waters the structures are similar, but the individual cones and leaves are larger (20 cm–35 cm in height). Laval et al. (2000) noted that the cones "often have one or more internal conduit up to 5 mm in diameter." They concluded:

> "Based on their appearance and the presence of internal conduits, it is probable that the distribution of the intermediate to deep, cone-topped microbialites correspond to regions of groundwater seepage into the lake."

However, they also noted the presence of calcified microbial fossils within internal conduits. It seems likely that calcification is a consequence of microbial activity below surface bacterial mats (Judd and Hovland, 2007).

Microbialites, Lake Van

The largest microbialites ever found, however, occur in the alkaline Lake Van in eastern Turkey. Here, Kempe et al. (1991) described "enormous (~40 m high) tower-like microbialites". This is a remarkable lake, with a high pH (9.7–9.8) and a salinity of 21.7‰. Mantle-derived gas enters the lake together with other minerals and fluids by hydrothermal, mainly diffusive, seepage through the lake bottom. This fluid flow actually accounts for at least 0.04%–0.06 % of the total global helium flux (Kipfer et al., 1994). Because the chemistry of the lake is similar to that of the Pre-Cambrian ocean, Kempe et al. (1991) speculated that the Lake Van microbialite structures are analogues to Pre-Cambrian stromatolites. If these structures can be proved to be fluid flow-related, then this, by analogy, more or less proves that stromatolites are seepage-related (Judd and Hovland, 2007).

Taupo Volcanic Zone, New Zealand

Another indication that stromatolites may rely on fluid flow is the discovery of so-called microstromatolites in hot volcanic pools of the Taupo Volcanic Zone. In Inferno Crater and Champagne Pool, clusters of microstromatolites, up to 10 mm high, grow on small islands and on twigs (Jones et al., 1997). The spring water of the pools has abundant CO_2 bubbles and is of neutral chloride composition, but is enriched in silica, Au, Ag, As, and Sb.

These field examples of stromatolites and microbialites suggest a strong link with fluid flow. Undoubtedly, future research will tell us the true importance of fluid flow for the formation of stromatolites and microbialites, and thus also their importance for the oxygen in our atmosphere.

7.6 SUMMARY

This chapter has dealt with a host of ancient fossil analogues to the deep-water coral reefs and carbonate mounds found living today. Perhaps the best analogues to the Norwegian *Lophelia* reefs are the spectacular Devonian (380-million-year-old) Saharan fossil reefs. Although they did not have scleractinia at that time, the reef structure and evidence within the reefs suggest a seepage relationship (i.e., support for the hydraulic formation model). The best analogy for the modern giant carbonate mounds off Ireland and Mauritania are perhaps the Kess-Kess Mounds in Morocco. These have also been suggested as seep-related. So, the question arises: "Is the present the key to the past or is it the other way round?" My own conclusion is that we really do not know how and why ancient structures or modern ones formed. Thus, there is a lot of future research to be carried out on the modern occurrences. The chapter ends with a review of both modern and nascent stromatolitic, possibly seep-related structures (such as in Pavillion Lake, Canada) and also the truly seep-related microbialites, like those found in Lake Van, Turkey.

8

Competing theories

Although a decade of thorough studies in this area provided remarkable insight on mound processes, the question of their origin and stabilization over geological times remains still widely open.

Maignien *et al.* (2006)

Interestingly, and in contrast to terrestrial hydrocarbon degraders which tend to be metabolically versatile, ... their marine counterparts are mostly highly specialized obligate hydrocarbon utilizers, ...

Yakimov *et al.* (2007)

8.1 INTRODUCTION

Because the Porcupine Basin is where most detailed scientific work has been done on deep-water corals and live carbonate mounds, including drilling through a mound, this is where I will start the analysis and discussion. The intention with this chapter is to critically review some of the hypotheses proposed to explain the formation of contemporaneous deep-water coral reefs and carbonate mounds. We would all like to know why they only grow in very specific locations, how they manage to select their sites, and to survive, even through some of the most dramatic climate changes our planet has experienced (i.e., the shift from glaciation, or Arctic climate, to interglacial, or temperate climate, and back again). Perhaps there is no straightforward answer, or perhaps there is a myriad of (confusing) answers.

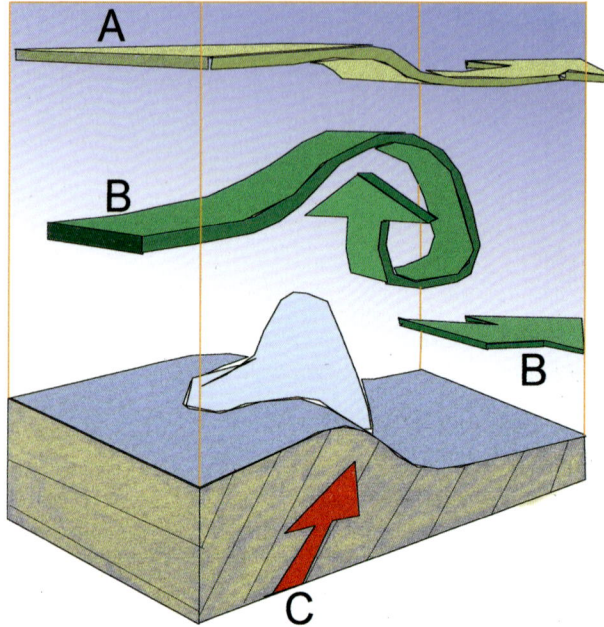

Figure 8.1. A sketch that illustrates the hydraulic theory for deep-water coral reefs and giant carbonate mounds. Arrow A: Atlantic water or any other major current that flows over the feature. Arrow B: Turbulent eddies may form around the feature as a consequence of local topography; they tend to increase the concentration of nutrients. Arrow C: Advection (migration, or seepage) of porewater and dissolved gases from the sub-surface. According to the hydraulic theory, this component is the decisive factor, as it brings a stable flux of nutrients to filter-feeders on the reef. If this component (Arrow C) is turned off, there is a high likelihood that the reef or mound will die (from Hovland and Mortensen, 1999).

8.2 REEFS THAT DIDN'T DROWN

It is only about 60 years since geologists interpreted all carbonate occurrences in the ocean and in the geological record as a clear sign of features that had formed and resided in shallow waters:

> "The 19th-century pioneers were impressed with what might best be called 'reef power', the astounding capability of reefs to grow upward and outward and to repair damage by wind and tides. The realization (after World War II) of the rapidity of the Holocene rise of sea level made reefs and platforms all the more impressive, and death by drowning became ever more difficult to imagine" (Schlager, 1981).

Since this 1981 paper, geologists have had to accept the fact that live and freshly formed deep-water carbonates occur, even in cool waters of high latitudes and at water depths up to several thousand metres, both of which are environments where calcium carbonate ($CaCO_3$) was assumed to be unstable and to dissolve. However, nature has proved to us (as it does time and again) that our conceptions and theories are wrong: the mere fact that deep-water coral reefs occur is something we have to acknowledge and positively accept as reality, and then we have to re-write our conceptual theories to find an explanation we can live with and test in future field campaigns.

8.3 A NULL HYPOTHESIS

My starting statement or *null hypothesis* in the following discussion and analyisis is: *There is a **specific reason** why a deep-water coral reef or carbonate mound forms at a certain location on the seafloor*, a statement most of us can agree is scientifically sound. Here is a review of possibilities that can form explanations and that undoubtedly will lead to new, hopefully more focused questions. Consider the analogy with a tree growing in the Sahara desert. Our null statement is similar to that above (i.e., there is a specific reason it grows there). It demonstrates how the answer can be rather complex:

— a bird happened to drop a seed there;
— the water table happened to be shallow there; or even
— some other, unknown factor caused it to grow there.

The main factors which have been used for explaining deep-water corals and carbonate mounds can be grouped into three categories:

(A) *water-related factors*, such as current velocity, water mass (type), and other physical properties (temperature, salinity, density, pH, etc.);
(B) *sub-stratum-related factors*, such as sediment/rock type, fluid flow (seepage of exotic fluids), and other physicochemical factors;
(C) *combined factors*, where one or more factors from the categories in A and B combine to make a complex set of factors.

In the following, I will review and discuss some of the published case studies, rather than performing a premature classification analysis. This is because the variability in methodologies used to study deep-water coral reefs and carbonate mounds so far are not standardised at all, and we are thus lacking a proper basis for systematic classification and sorting of factors pertinent to these structures.

8.3.1 Water-related factors

Significance of the MOW

Several researchers have focused their attention on specific water masses, their characteristic salinities, densities, and temperatures. It was early suggested that water

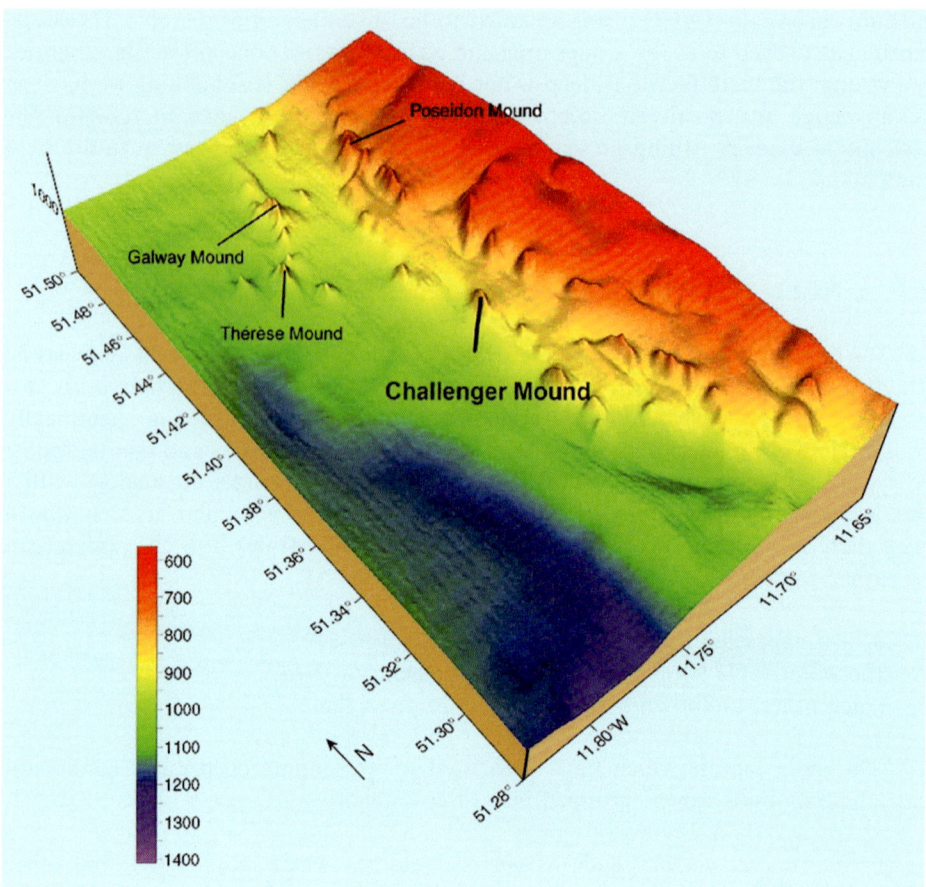

Figure 8.2. A fine perspective digital terrain model of the Belgica Mounds viewed from the southwest. Note the channel at the foot of the westernmost mounds. This image is from the Preliminary Results Report of IODP Expedition 307 (based on AWI, Alfred Wegener Institute, multibeam bathymetry, Beyer *et al.*, 2003; Ferdelman *et al.*, 2006).

streaming out from the Mediterranean may have an impact on coral reefs farther north (Freiwald, 2002). Mediterranean Outflow Water (MOW) is a dense (high-salinity) body of water (water mass) that exits the Straits of Gibraltar and dips underneath low-density Eastern North Atlantic Water (ENAW). The reason ENAW is of lower density is its lower salinity. Even though it is normally cooler than MOW, its salinity is significantly lower, and therefore it is less dense (the difference in density is actually up to $2\,\text{kg}\,\text{m}^{-3}$, according to Chen and Hsu, 2005). The high-density MOW continues as a bottom-hugging current, which follows the seafloor topography on a northerly heading, always trying to bend to the right, because of the influence of the Coriolis force (motion on a revolving globe).

Between these two vertically divided water masses there is a distinct thermocline and pycnocline (i.e., large temperature and salinity gradients), providing a major boundary in the water column, dividing the MOW (lower) and the ENAW (upper) water masses. It was Grousset *et al.* (1998) who first became aware of the potential significance of these conditions and their influence on marine sedimentary conditions. De Mol *et al.* (2002) and White (2003) looked into the dynamical significance of MOW with respect to the giant carbonate mounds, as we shall see in Section 8.5. In a later paper by another group of researchers, it was determined that:

> "The depth range of the seabed coral banks coincides with the Mediterranean Outflow Water which may control indirectly the coral distribution" (De Mol *et al.*, 2005).

Internal breaking waves

Frederiksen *et al.* (1992) found that the deep-water coral reefs off the Faroe Islands were concentrated in a particular depth interval. This instigated some research into the possibility that so-called "internal waves" were breaking against the seafloor and causing sediment resuspension and nutrient concentration within this zone. Internal wave motions are associated with a density-stratified system in which wave motion prevails on an isopycnal surface (of constant density):

> "Within the body of a stratified lake or ocean, all lines of constant density have periodic wave motion. In a continuous stratification system, these waves are called internal waves, whereas they are regarded as interfacial waves in a system with step stratification" (Chen and Hsu, 2005).

The stratification causing these waves off the Faroe Islands occur between the cold and high-density NADW (lower) and the warmer and lighter ENAW (upper) water masses. Similar breaking waves can also occur between MOW (lower) and ENAW (upper) in the Porcupine Basin.

The effect on the resuspension of sediments when internal waves break against an underwater slope can be dramatic (i.e., the turbulence and impact can cause a large resuspension of material):

> "Recent observations have also revealed that large oceanic internal waves on the continental shelf are capable of mixing sediment vertically. This has important ramifications in sediment transport, mobility of chemicals and certain biological contaminants once deposited in deeper water and adsorbed by fine sediments" (Chen and Hsu, 2005).

For explaining higher nutrient concentrations in the near-bottom water column, this is potentially a very important factor. Therefore, it merits being taken into account when dealing with deep-water coral reefs and carbonate mounds, in general.

Enhanced tidal currents

In areas of high relief, like the regions in the Porcupine Seabight, currents may be affected dramatically. When the water flow is forced round corners or is forced to run over escarpments and ridges, it has to speed up. This effect produces enhanced currents. For the Porcupine Seabight, De Mol *et al.* (2002) produced a very informative map that shows where there are enhanced tidal currents according to numerical modelling of seafloor topography and current directions. Thus, enhanced tidal currents occur at the Hovland Mounds. The map (Figure 8.3) also shows regions where internal waves may break against the slope. This actually occurs at the location of the Belgica Mounds. Such maps should be constructed for all areas where there are deep-water coral reefs, as this may explain parts of the physical environment experienced by the organisms. However, such investigations have also to be backed up with the more time-consuming, cumbersome, and costly verification of bottom current measurements.

Water density

A very intriguing and specific scientific result was found by Dullo *et al.* (2008), when they sampled and analysed seawater at numerous deep-water coral reefs and carbonate mounds. Their results showed that live corals all plotted at an S_T value of $27.5\,kg/m^3$, whereas dead coral remnants could be found at other density values. To them, this suggests hydrographic control on deep-water coral ecosystems. This water density dependency is either favouring nutrient concentration at this water density level or that it occurs on the reefs during distinct periods of the year, and supports spawning of coral larvae. This latter conclusion is supported by the larvae having a similar specific density as that measured for the water (27.5):

> "Our study shows that cold-water corals in the North Atlantic tolerate a wide range of environmental conditions. However, our data indicate that living coldwater coral reefs occur almost exclusively within the density envelope of $27.35-27.65\,kg/m^3$. This may favour nutrient concentration and possibly supports larval drift along the European continental margin" (Dullo *et al.*, 2008).

8.4 A HYDRODYNAMICAL CASE STUDY

Dorschel *et al.* (2007) studied the sediment facies distribution and current pattern (hydrodynamics) at Galway Mound, one of the Belgica Mounds, west of Ireland. Quite rightly, they state:

> "For the first time, bottom current data have been recorded at six locations over a mound thus allowing an interpretation of the local flow field to be made" (Dorschel *et al.*, 2007).

Indeed, even though many researchers have stated that currents, eddies, and/or internal breaking waves are decisive factors determining where deep-water coral reefs

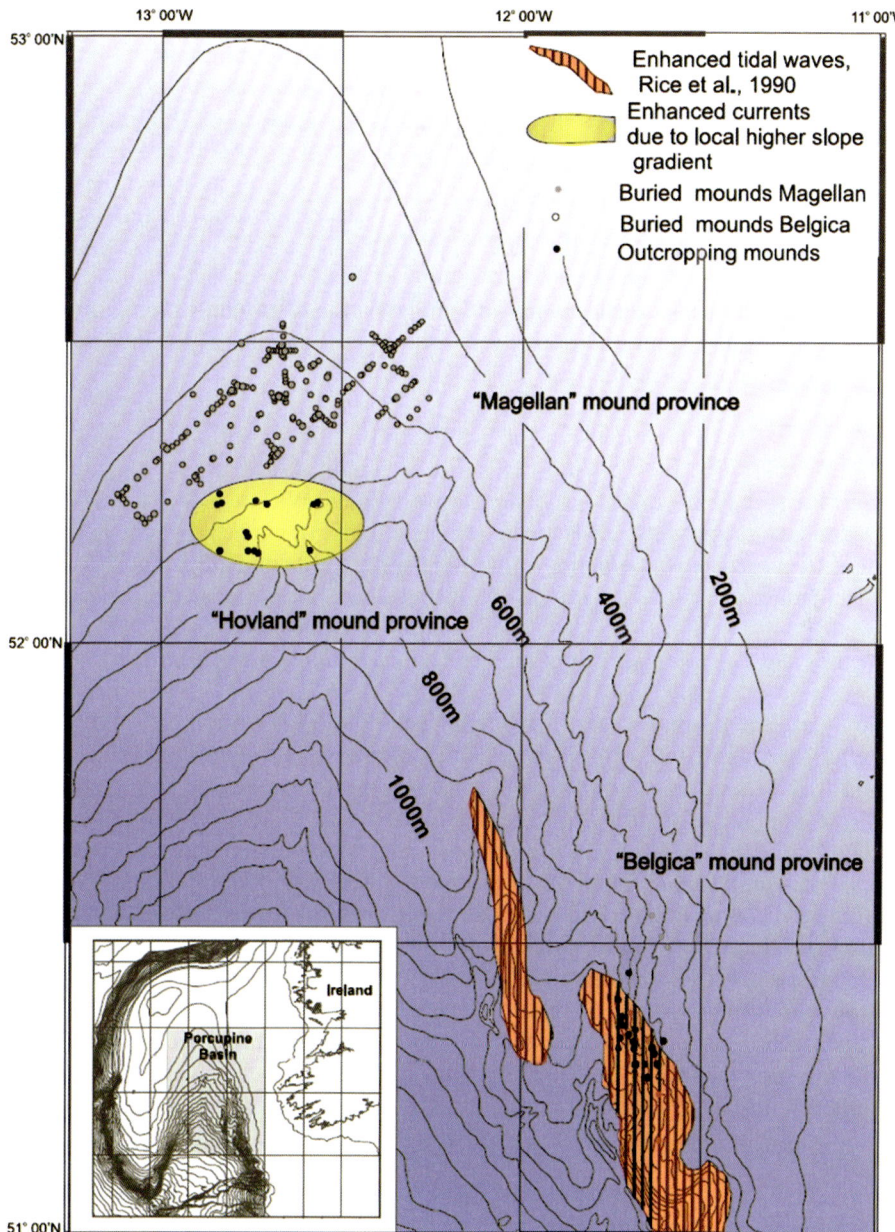

Figure 8.3. Map from the Porcupine Seabight showing the locations of the three groups of mounds: Magellan, Hovland, and Belgica. Most Magellan Mounds are buried and extinct, whereas some of the easternmost Belgica Mounds are partly buried below drift sediments; the Hovland Mounds and the westernmost Belgica Mounds are active. Note the locations of enhanced tidal waves (Rice *et al.*, 1990) and the location of enhanced currents due to the high slope gradient (courtesy De Mol *et al.*, 2002).

grow, this is actually the first time data supporting such statements have been presented. The crux of this study is based on the following rationale:

> "... this study investigates the hydrographical, biological and sedimentological factors at a key site (Galway Mound in the Belgica mound province) in the Porcupine Seabight. It focuses particularly on the local flow field over the Galway Mound and correlates it to coral facies distribution and surface sediment distributions" (Dorschel *et al.*, 2007).

After their current measurements and analyses, the researchers, not surprisingly, concluded that the strongest currents were found at the summit of the mound, where the corals were most prolific:

> "The areas with high densities of living corals coincide with areas of enhanced bottom currents. Living corals are most abundant at the mound summit where current speeds are highest and on the west flank. Within those thickets coarse sediments accumulate. Due to the sediment baffling capacity of the corals and lower energetic environments within the thickets, silts and clay deposited during slack water are protected against re-suspension" (Dorschel *et al.*, 2007).

What they did not address, however, were the two more interesting questions:

— Why are the mounds there in the first place?
— Why are there some prominent mounds in the group of mounds surrounding Galway Mound (e.g., the Challenger Mound) that are dead?

The dead mounds lack a prolific growth of coral, despite there being the same strong currents streaming over them, just as over Galway Mound. In other words, the researchers did not address the null statement, and therefore, unfortunately, their results from this research cannot be used to address these two questions.

This case clearly demonstrates the reason, in my view, why water-related (hydrodynamic) factors are of relatively little interest when it comes to answering the fundamental question of why the carbonate banks form where they do. These factors are too general for a wide area and too indiscriminate towards single (local) features. In my view, the next case brings us one step closer to asking the more interesting questions and measuring the more interesting parameters.

8.5 EXTERNAL OR INTERNAL CONTROL?

After having studied the Belgica, Magellan, and Hovland Mounds of the Porcupine Seabight using a variety of instrumentation, including high-resolution seismics and sediment coring, Ben De Mol *et al.* (2002) reviewed the possible answers to the null statement. Although they could not make any firm conclusion, they performed a full argumentation for and against four different scenarios. Their main conclusion boiled down to the urgent need for scientific drilling to resolve the issue of what triggered mound formation. The subsequent IODP Expedition 307, which drilled the dead Challenger Mound was a direct result of this work.

Some of their argumentation is provided here:

"But if the present-day distribution of living *Lophelia* and *Madrepora* coral concentrations and outcropping coral banks correlates well with areas of favourable environmental conditions (strong currents, slow sedimentation rate), it is still not understood why coral banks started to develop in such large numbers on specific sites" (De Mol *et al*., 2002).

Well, there is *no* answer in their data, as they admit. However, there are some pertinent observations: "Observations can be summarised as follows:

— Within each province mound initiation occurred rapidly and more or less simultaneously.
— There are no observations of several start-up phases within the same area.
— Mound initiation seems to be restricted to well-delineated areas—the presently defined mound provinces.
— In all three provinces, the start-up phase occurred after an erosional event, possibly during a period of non-deposition" (De Mol *et al*., 2002).

In addition, I will add to this list their observation of seafloor depressions (moats and channels), especially associated with the Hovland and Magellan Mound provinces. These observations were listed early in their article (on their p. 221) but, for some unknown reason, were deleted from the final list, above. I intend to come back to this important point.

Basically, they concluded that the event triggering the start-up phase of the giant carbonate mounds in the Porcupine Basin could actually have been a "basin-wide, environmental event", which may have been of geological (endogenic) or of oceanographic (exogenic) nature. After having reviewed the possibility of methane seeps and gas hydrates, etc. being of crucial importance as suggested earlier (Hovland *et al*., 1994, Henrict *et al*., 1998, 2001), they review some other exogenic possibilities. These other possibilities include

(1) the onset of special water masses flowing across the Porcupine Seabight; and also
(2) enhanced currents by seafloor topography and breaking internal waves.

Although these models are sound for explaining the general construction of the three groups of mounds, they hardly explain certain detailed observations. The argumentation that larvae of *Lophelia pertusa* and *Madrepora oculata* originating from the Mediterranean first settled on the Porcupine Seabight, before they travelled on further to the north, is sound and without problems. Also the reasoning that MOW, which has a higher salinity and density than the ENAW mass, may be nutrient rich and may cause enhanced current velocities and breaking internal tidal waves, are also sound factors. They can help to explain why the mounds have been thriving there for so long, even throughout the dramatic Quaternary period.

This is actually one of the most remarkable aspects of the Porcupine Seabight giant carbonate mounds: that they managed to not only sustain, but apparently thrive, through the entire Quaternary glacial/interglacial seesawing periods. During the last 600,000 years, there have been five dramatic sea-level reductions and increases, in harmony with dramatic climatic variations, whereby water either moved onto land (in the form of great ice sheets during glacials) or back into the oceans (during interglacials, as at present). The difference in sea-level during glacial (low-stand) and interglacials (high-stand) is a staggering 120 m. This is up to one-fifth of the entire water column where the Porcupine mounds are located. These changes must have significantly affected ocean current directions and speeds, as well as water temperatures, density, and salinities. As there are no reported signs that the mounds switched off during glacials, to me this fact alone speaks of their robustness. Deep-water corals must be much more robust with respect to general changes in environmental conditions than we have ever envisioned.

Although researchers know where corals originated (the Mediterranean) and where they spread to after the LGM (last glacial maximum), there are three succinct and unique facts they do not comment upon. Let us now look at them:

(A) Why are some of the great mounds still thriving today, whilst others have succumbed and become buried in layered sediments (Magellan Mound province)?
(B) Why are some mounds in the Belgica Mound province still lush and prolific (Theresa and Galway Mounds), whereas others have died (Challenger Mound)?
(C) Why do some of the mounds in the Hovland and Magellan Mound provinces have moats, whereas others do not, and why is there a distinct channel below the Belgica Mound province.

In a contest of ideas, like this, the objective is to find the model that explains the most with the least effort, or in the simplest way (i.e., by using Occam's razor: *entia non sunt multiplicanda praeter necessitatem*, "entities should not be multiplied beyond necessity"). In my view, this is where models relying on hydrodynamics alone fail: they explain some of the items, but not the other related characteristic and unique details.

In any respect, I fully agree with De Mol *et al.* (2002) in their final statement:

"Detailed dating of the base of the coral mounds is obviously required but difficult to achieve. The faunal diversity on the coral banks and the occurrence of typical Mediterranean species also illustrates the close relationship between the coral banks and the MOW (source for corals, enhanced currents) but do not provide conclusive evidence regarding the origin of the coral banks in the Porcupine Basin. Only scientific drilling will provide such conclusive evidence" (De Mol *et al.*, 2002).

Sec. 8.5] External or internal control? 189

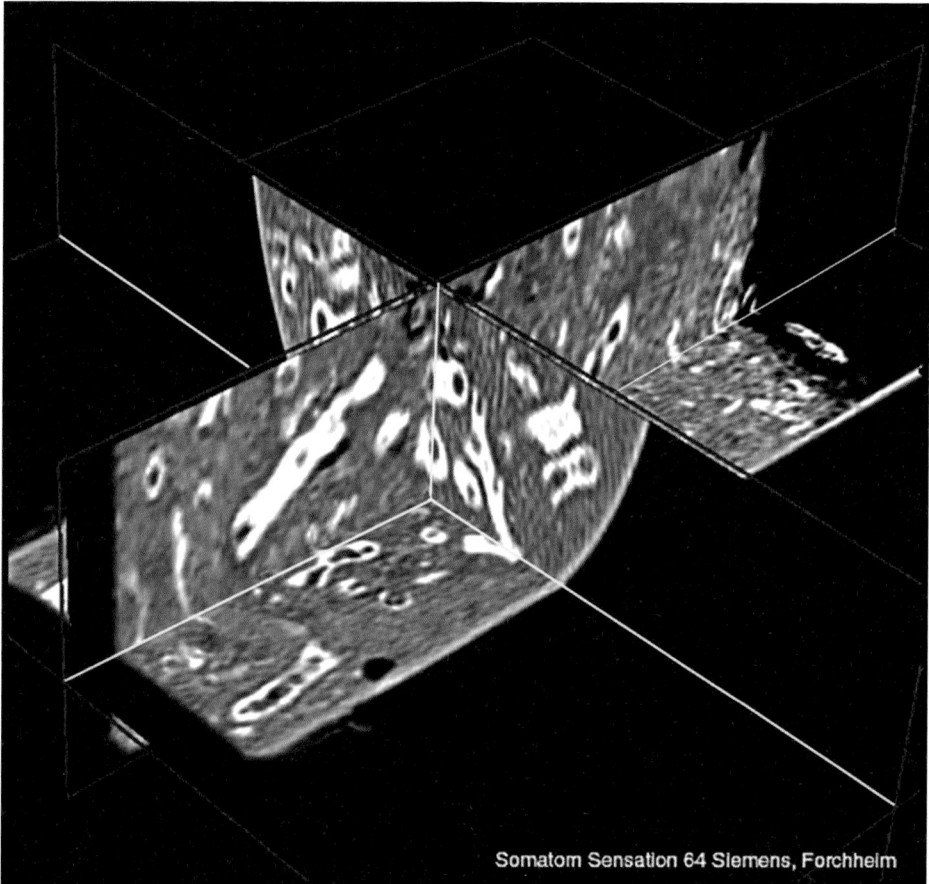

Figure 8.4. A scanned image though a soil core of the Challenger Mound (IODP Leg 307). It is an example of a CT scan from Hole U1317E (Siemens Somatom Sensation 64). This core is 10 cm across. The image clearly shows how *Lophelia* skeletal remains form a dense network, which helps to stabilize soft sediments within the giant carbonate mounds (Ferdelman *et al.*, 2006).

But, let us review the three succinct details where the hydrodynamic model fails to come up with an explanation. Perhaps the hydraulic theory (described well in Hovland and Risk, 2003) can provide some hints?

8.5.1 A case of life and death

Let us consider the first question: Why are some of the great mounds still thriving today, whilst others have succumbed and become buried in layered sediments? In the Magellan Mound Province there is a 200 m wide, living mound, Perseverance Mound, which is apparently still growing (it is about 30 m proud of the flat seafloor, Figure

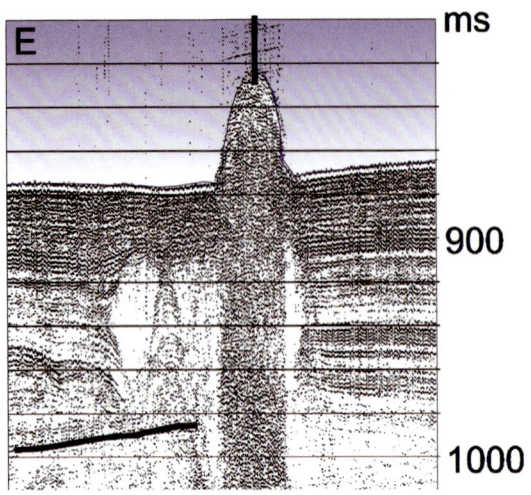

Figure 8.5. One of the Magellan Mounds: Perseverance Mound has survived to the present, in spite of all the other mounds being buried. Note that this carbonate mound is located on the shoulders of its predecessors. The slightly dipping black line at the lower left is the boundary between layered upper sediments and older strata (interpreted by and courtesy of De Mol et al., 2002).

8.5). This is despite all other surrounding Magellan Mounds being extinct as they are buried under at least 6 m of sediments (de Mol et al., 2002). From the seismic image, this mound can be seen rooted on top of buried mounds. Surely there must be a very local explanation, which is specific only to this mound and not to the others. It is futile attempting to answer this question armed only with a hydrodynamic theory. However, by applying the hydraulic theory, it becomes rather simple, as explained below.

Why are some of the Belgica Mounds prolific (e.g., Thérèse and Galway Mounds), capped with live cold-water corals, whereas others have apparently died (Challenger Mound)? Here, once again hydrodynamical models are of no avail. Current velocities become just as enhanced over dead mounds as over live ones: physical laws are indiscriminate of the biological character of the substratum. What is unique about the live ones? The answer is again local and must involve endogenic processes (i.e., seepage), as explained in the last of these cases.

Moats, channels, and subcrop topography

De Mol et al. (2002) do not attempt to address the importance of moats and channels, even though these features are pronounced and unique characteristics of both Hovland Mounds, some of the Magellan Mounds, and the Belgica Mounds (channel). Let us start with the Magellan Mounds. About 95% of these are buried below ground (i.e., they are extinct). But even so, from the seismic images, moats can clearly be seen at the base of many of these mounds. Whilst they were living, perhaps 100,000 years ago, they were located inside symmetric and asymmetric depressions, just like the current live Hovland Mounds, and on a smaller scale like the deep-water coral reefs inside pockmarks at Kristin.

Judd and Hovland (2007) have discussed the nature of some moats on the seafloor. Although most researchers take it for granted that moats occur as a

Sec. 8.5] External or internal control? 191

Figure 8.6. An artist's impression of the *Seaway Commander* performing ROV-based documentation of the HRC in 1993. An early interpretative sketch of the Haltenpipe reefs made after initial sampling and inspection in 1993 (Hovland and Thomsen, 1997). A sketch and photomontage showing the location of the Haltenpipe reefs located on Palaeocene subcropping sediments of high organic content. The nearby pockmarks bear witness of the flow of nutrient-rich fluids into the water column.

consequence of local seabed erosion due to currents passing over an obstruction, this does not necessarily have to be true. Another possibility is that they represent pockmark depressions (i.e., local erosion of the seafloor by the action of vertically migrating porewaters and gases), a purely hydraulic and endogenic cause. Not only

192 Competing theories   [Ch. 8

does this model explain the moats, it also explains where some of the reliable nutrients that feed into and keep the mounds growing and alive originate, despite dramatic climate and other environmental changes.

Finally, the question of the channel: at the base of the Belgica Mounds there is a pronounced elongate depression, a channel (De Mol *et al.*, 2002). From one of the seismic images it can be seen that the overburden thickness is much less inside the channel than anywhere else on the seafloor. Therefore, the effect of this channel is that it reduces the pressure above migrating fluids, and therefore represents "a zone of least resistance" for any fluids that are in the process of migrating upwards from depth, through the sediments, to the water column (see, e.g., Figures 8.2 and 8.3). Thus, the channel may act as a broad conduit for nutrients feeding into the water column immediately at the base of the deepest Belgica Mounds.

In a more recent analysis by De Mol *et al.* (2007), they subjected five of the deepest (950 m) Belgica Mounds to scruitiny. They used seismic profiling, sidescan sonar, swath bathymetry, and video imagery, and placed special emphasis on the lush Thérèse Mound. Their main conclusions from this study did not differ much from their previous one, which more or less confirmed the rooting of the structures:

"Seismic evidence suggests that the start-up of the coral bank development was shortly after a major erosional event of Late Pliocene–Quaternary age. The coral bank geometry has been clearly affected by the local topography of this erosional base and the prevailing current regime" (De Mol *et al.*, 2007).

The only new aspect here is that there seems to be a close link between the "topography of the erosional base" and the location of carbonate mounds. This fact strongly suggests that control is from endogenic factors. However, the most impressive result is the confirmation that these mounds have withstood dramatic climate changes and changes in all environmental parameters, such as temperature, prevailing current direction, current speed, salinity, water density, pressure, and tidal regime. The fundamental, logical interpretation is that these deep-water corals are more or less independent of climate change, and must therefore be relying on specific local conditions (i.e., they are most probably relying on a stable and consistent nutrient source, which comes from within the sedimentary basin ...).

8.6 A QUESTION OF GEOPHYSICAL INTERPRETATION, LINEAR REEFS

In the endeavour to understand how theories and models for deep-water coral reefs and giant carbonate mounds develop and are promoted, I here analyse some statements made on the basis of geophysical and geological data at some coral reefs off Norway.

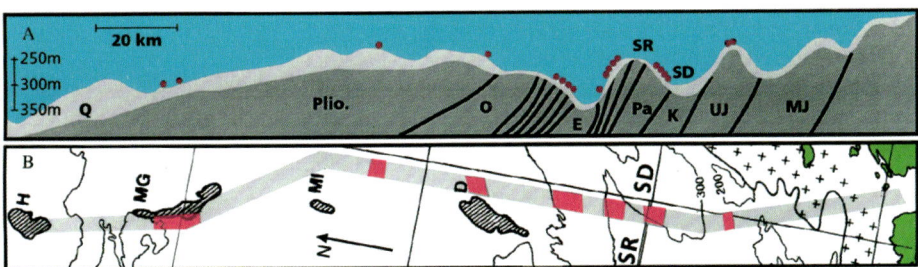

Figure 8.7. A vertical profile (upper, A) and map representation (lower, B) of the entire 200 km long Haltenpipe route, from shore (right) to the Heidrun field (left). Red indicates the locations of *Lophelia* reefs. Q = Quaternary sediments. The sedimentary rocks are, from left to right, Plio = Pliocene, O = Oligocene, E = Eocene, Pa = Palaeocene, K = Cretaceous, UJ = Upper Jurassic, MJ = Middle Jurassic. The hydrocarbon fields are H = Heidrun, MG = Midgard, MI = Mikkel, D = Draugen. SR and SD refer to the Sula Ridge and the Sula Deep, respectively (based on Hovland *et al.*, 1998).

8.6.1 Case 1: the Sula Ridge reefs

The Sula Ridge, off Mid-Norway, is a well-known flat-iron geological structure, which means that it stood up well against severe erosion by glaciation during the last glacial (Weichselian[1]) period. The other Mesozoic sedimentary rocks in the area were more severely eroded. Topographically, it is therefore the highest regional area, but even if the rocks consist of sandstones of Palaeocene age, they have not been eroded as much by ice as the Eocene rocks to the northwest and Cretaceous rocks to the southeast (Figure 8.7). The bedding planes of the Sula Ridge sandstone dip towards the west, as can be seen on seismic lines running perpendicular to the structure (also seen on Figure 8.7).

Although there is a relatively thin layer of Quaternary clays covering this part of the Norwegian Continental Shelf, there are portions, especially on the harder Palaeocene rocks, that protrude through the young sediments. This can clearly be seen in Figure 8.7, where the Paleocene bedding planes are exposed on the southeast side of the Sula Ridge, and to some degree on top of the ridge.

In their paper of 2005, Freiwald *et al.* (2005) made a very strong point of proving that coral reefs located on the summit of the Sula Ridge form along iceberg ploughmark ridges, in agreement with the bio-expectation that reefs form on hard and high ground. However, by inspecting their data (especially their fig. 5, which points at a feature representing "presumed ploughmark topography"), any geophysiscist will start to wonder if they are right. The case is that in the topographical relief of the Sula Ridge (see Figure 8.8), there are at least three sets of linear topographical elements. These are

(1) iceberg ploughmarks, which are seen as curvilinear lines in the upper left corner of Figure 8.8;

---

[1] The Weichselian was the last glacial period, lasting from 120,000 YBP until 10,000 YBP.

(2) ice-flute or current-flute marks, which are relatively broad lines running northeast–southwest; and
(3) the strike of exposed Palaeocene sandstone bedding planes, which show up as relatively thin, straight, parallel lines, also running in a northeast–southwest direction.

As with most geophysical interpretation and mapping, it takes at least two or three independent datasets to perform an unambiguous interpretation. This is because geophysical mapping relies on remote-sensing data. As shown in Figure 8.9, it is equally plausible that the coral reefs of the Sula Ridge formed their conical structures along underlying exposed Palaeocene bedding planes as along more or less fictive iceberg ploughmarks.

Geophysical interpretations can only be verified by laborious sampling and detailed analyses, and do not by themselves serve as proof. They can merely be used as indicators of factors that can be further investigated by other means. In this case, it might seem immaterial whether coral colonies construct their reefs on top of iceberg berms or on exposed bedding planes. But this is not so, because if the latter is the case they may be dependent on some nutrients coming from the beds of dipping Palaeocene rocks, in contrast to the other case where they are dependent on a hard

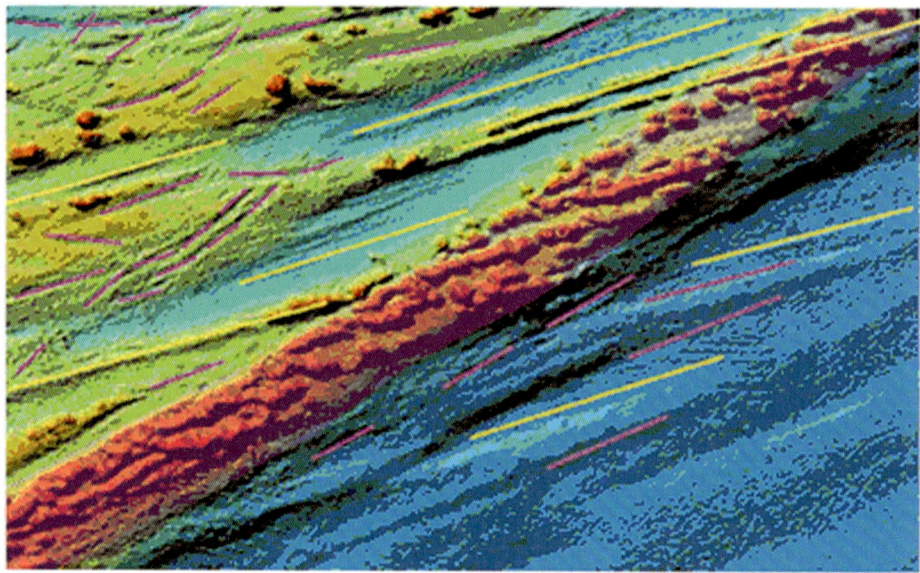

Figure 8.8. A relief map of parts of the Sula Ridge (NGU/IMR). Coral reefs show up as red haystacks, which merge into linear structures on the very summit of the ridge. The purple lines represent iceberg ploughmarks and ice-flute marks caused by previous ice contact and movement. The yellow straight lines represent exposed Palaeocene bedding planes. On the very summit of the Sula Ridge, it is impossible to discern any exposed bedding planes, flute marks, or iceberg ploughmarks because of the high density of coral reefs (map from *www.mareano.no*).

Sec. 8.6] A question of geophysical interpretation, linear reefs 195

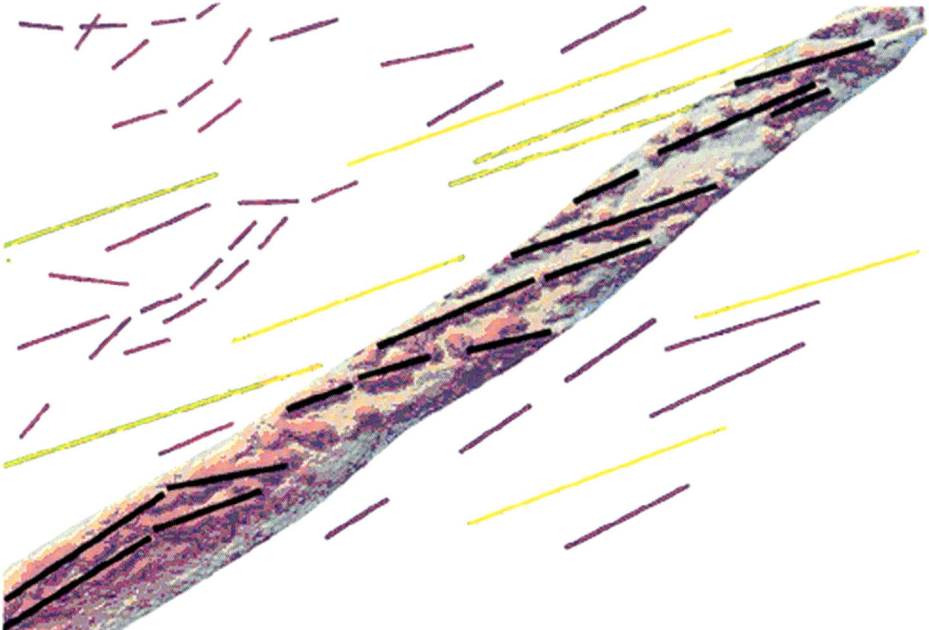

Figure 8.9. Same as previous figure, but now only showing the lineations discussed in Figure 8.8, where purple lines represent iceberg ploughmarks and flute marks, whereas yellow lines represent exposed Palaeocene bedding planes. On the summit of the ridge, neither ploughmarks, flute marks, nor exposed bedding planes can be interpreted, due to the extremely high density of coral reefs. However, by judging from the direction of the coral reef lineations (black lines), they trend more along the direction of exposed Palaeocene bedding planes (yellow lines) than along the directions of the iceberg ploughmarks (purple lines).

sub-stratum presented by boulders associated with iceberg ploughmarks. Therefore, this discussion can be put on the back burner until more evidence has been put forward (e.g., after a detailed geochemical study, where both seawater and the sub-stratum have been sampled and analysed).

Between the SR and the HR (see Figure 4.18), there is an area with numerous iceberg ploughmarks. Using Freiwald *et al.*'s logic, these should have been preferentially colonised by DWCRs. By just looking at the relief map of that area, it is easy to see that this is certainly not the case. The coral reefs of that area are not aligned along iceberg ploughmarks, as can also be seen from Figure 8.10.

8.6.2 Case 2: the Træna Deep reefs

A case in which there is no correlation between coral reef alignment and linear topographical seafloor is shown in Figure 8.10. The aligned features are ice gouge flutes (caused by grounded shelf ice, of different generations and directions of

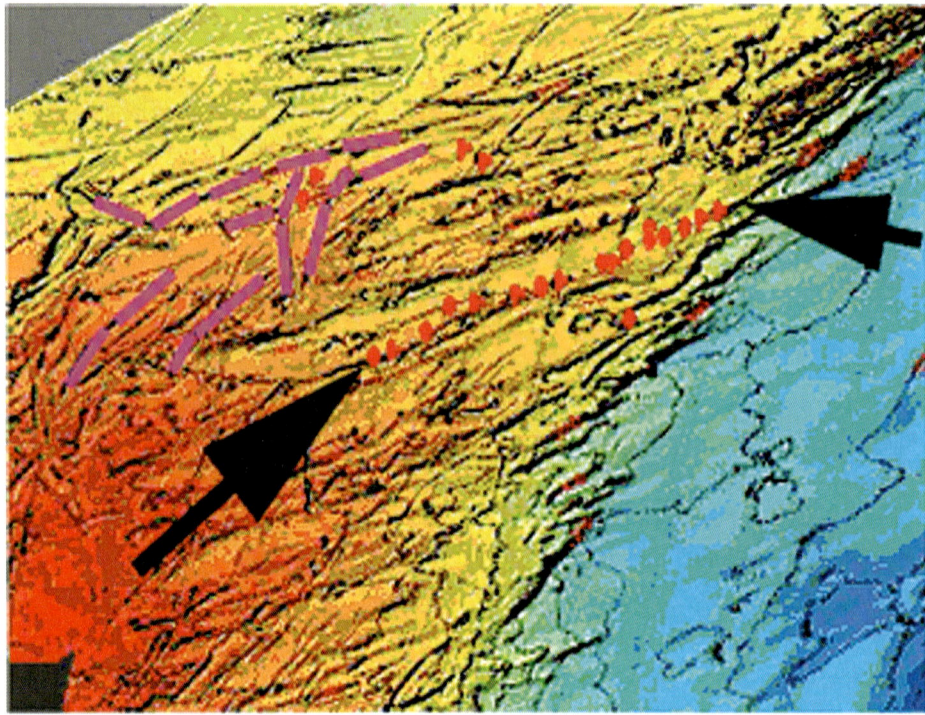

Figure 8.10. Relief map of the area between HR and SR, see previous overview map. The two black arrows are located at either end of a string of large (haystack) coral reefs. The red dots show some of the large coral reefs within the area. Close inspection shows that only a very low percentage of the reefs are associated with iceberg ploughmarks (some of which are highlighted by purple lines). This begs the question as to what causes the marked alignment (between the black arrows), if it is not following a ploughmark? We could perhaps speculate that the alignment is along a sub-surface bedding plane, which is invisible on this dataset, but which could be imaged on a shallow-seismic record across the line, perhaps combined with a geochemical survey.

movement). However, hidden immediately beneath a relatively thin layer of Quaternary deposits (mainly clay) there are sub-cropping Mesozoic rocks. The coral reefs in this area (shown by two arrows) are clearly elongated along a direction that is not due to topographical flutes. Could their aligned direction be caused by the general current direction (the coastal current) or could it be due to the alignment of sub-cropping sedimentary bedding planes? Only seismic data and current data can resolve this question. However, one aspect remains true: there is no alignment with topographical ploughmark or flutemark features (i.e., contradictory of some of the statements found in Freiwald *et al.*, 2005).

Although this example does not provide any clue as to what the reefs are aligned along (current direction or sub-surface bedding planes), it clearly demonstrates that the reefs do not align along obvious iceberg ploughmarks.

Sec. 8.6] **A question of geophysical interpretation, linear reefs** 197

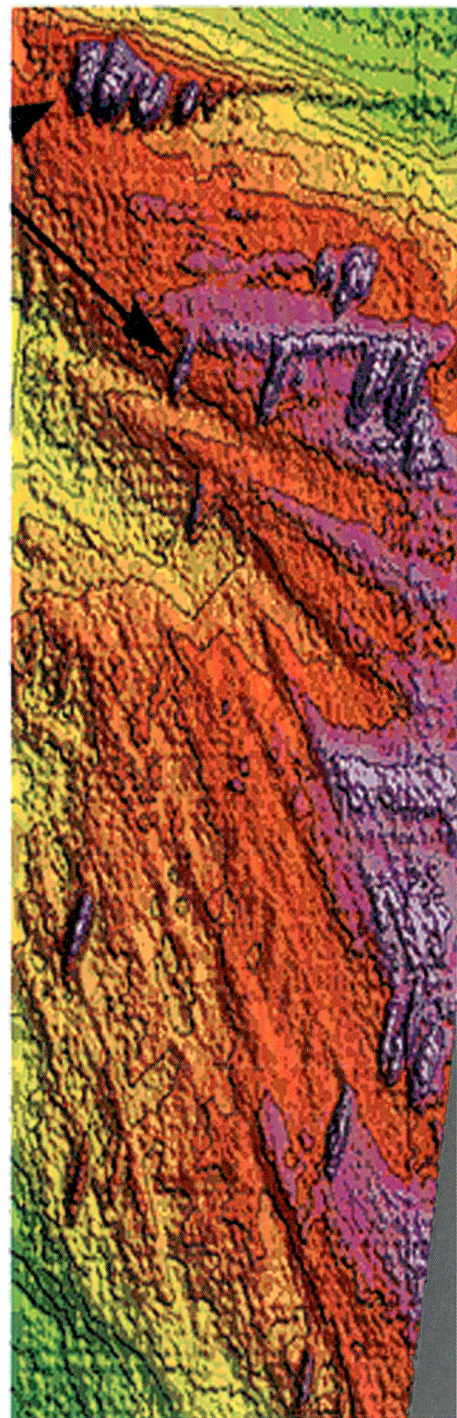

Figure 8.11. On this map over the general Træna Bank area, there are no obvious iceberg ploughmarks. Even so, there are linear coral reefs (two are arrowed), apparently aligning along the prevailing current direction of the area (see text for discussion).

8.6.3 Case 3: other linear reefs

Rezak *et al.* (1985) published a spectacular sidescan sonar image of reefs growing along a linear trend in the Gulf of Mexico (see Figure 6.3). These reefs are reported to have formed on top of a surfacing deep-seated fault, near the Flower Garden Banks. Because there is no known topographical fault scarp on the seafloor, their occurrence must be caused by seepage of fluids along the fault-line, which either produces primary (food) production or a suitable (cemented) sub-stratum for the corals.

8.7 REEFS ASSOCIATED WITH LOCAL DEPRESSIONS

8.7.1 Moats and other depressions

I have discussed the moat association with reefs in the Porcupine Basin and pointed out the similarity with pockmarks. At several other reef occurrences we have seen that deep-water coral reefs are directly associated with local, often circular, depressions in the seafloor (i.e., pockmarks, Judd and Hovland, 2007). The first such reefs to be found were in the Gulf of Mexico (Moore and Bullis, 1960). Then, we found some at the Kristin field in the Norwegian Sea (Hovland, 2005). However, there are also reefs located on the shoulders or rims of pockmarks, such as those reported by Sumida *et al.* (2004) off Brazil. Another such case, briefly mentioned in Section 4.2.2 (see Figure 4.13), is from near the HRC. In Figure 8.12, some more examples are shown from this area off mid-Norway. In addition, there is a recently reported coral occurrence associated with a depression formed around a hydrothermal vent off Japan (see Figure 8.13).

A coral reef with hydrothermal input

Hirayama *et al.* (2007) were the first to report on microbial communities associated with a shallow submarine hydrothermal system occurring within a coral reef (Figure 8.13):

> "The main hydrothermal activity occurred in a craterlike basin (depth, $\sim$23 m) on the coral reef seafloor. The vent fluid (maximum temperature, $>52°$C) contained 175 mM H_2S, and gas bubbles mainly composed of CH_4 (69%) and N_2 (29%). A liquid serial dilution cultivation technique targeting a variety of metabolism types quantified each population in the vent fluid and in a white microbial mat located near the vent. The most abundant microorganisms cultivated from both the fluid and the mat were autotrophic sulfur oxidizers, including mesophilic *Thiomicrospira* spp. and thermophilic *Sulfurivirga caldicuralii*. Methane oxidisers were the second most abundant organisms in the fluid; one novel type I methanotroph exhibited optimum growth at 37°C, and another novel type I methanotroph exhibited optimum growth at 45°C."

This case is a very good example of how the local environment is modified by input from the sub-stratum. Probably there would be a coral reef in this area, even without

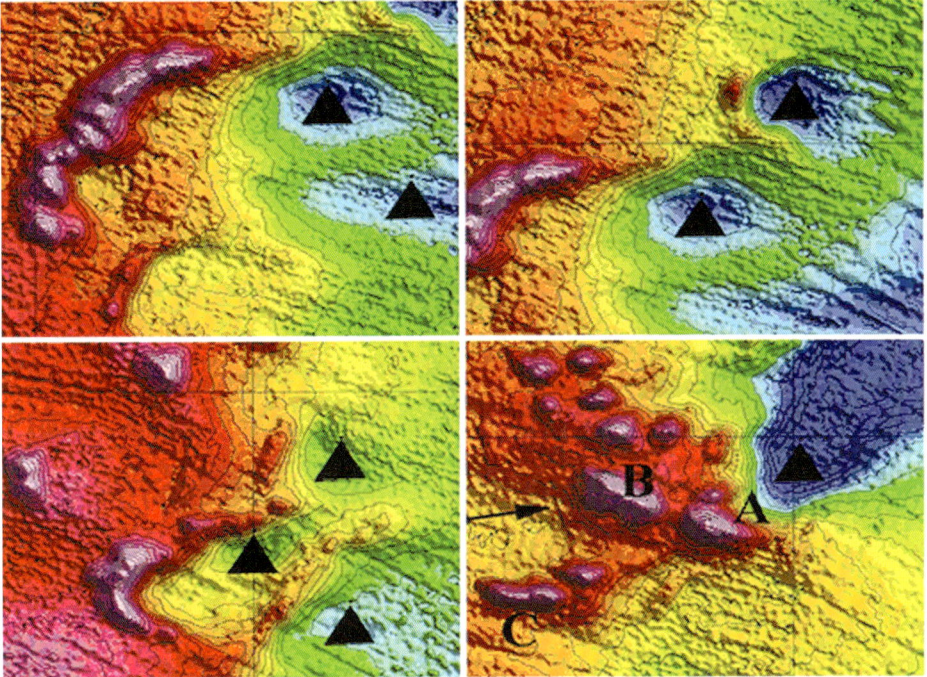

Figure 8.12. Four detailed topographical maps from the Haltenpipe Reefs, where the lower right is from the HRC (Haltenpipe Reef Cluster). The triangles denote the centres of pockmark craters formed where Cretaceous sedimentary rocks sub-crop. Note how the coral reefs seem to occupy the stable seafloor adjacent to these craters. In three of the four images, the coral reefs actually form semi-circular structures around the seepage depressions. Is this coincidental?

the hydrothermal waters seeping through, but would it be as prolific and diverse? At least there would not be any chemosynthetically associated microbes there. Figure 8.13 shows the startling vent site as imaged by the authors with shaded relief bathymetry.

8.8 NEW SUPPORT FOR THE HYDRAULIC MODEL

In 2002, Statoil supported research into the search for bacteria in Norwegian coral reefs through the VISTA project. This is a cooperation between Vitenskapsakademiet (Norwegian Academy of Science, Oslo) and Statoil. It was Sigmund Jensen, at the Institute of Biology (University of Bergen, Norway), under the leadership of Professor Nils-Kåre Birkeland, who received financial support for this work. New sediment, water, and organic samples were acquired from the reefs at the Kristin field. Both the bivalve *Acesta excavata* and sponges were part of the organic sampling. It took about 3 years, and close cooperation with other institutions in Norway, the UK,

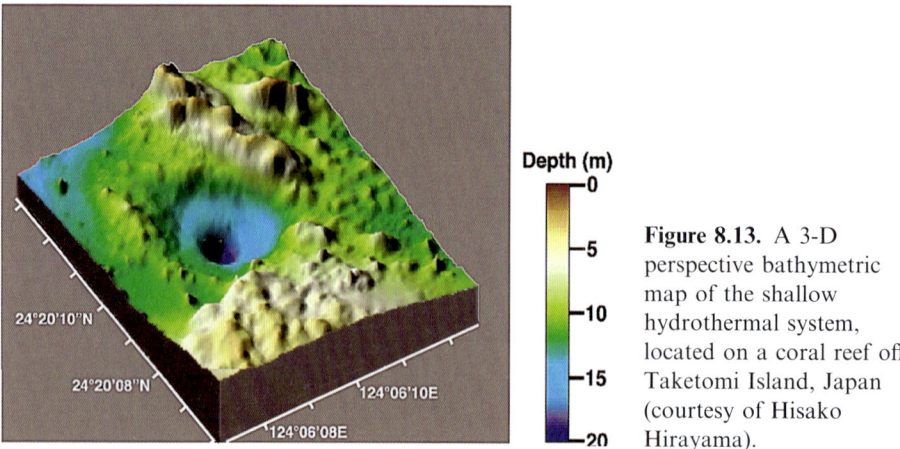

Figure 8.13. A 3-D perspective bathymetric map of the shallow hydrothermal system, located on a coral reef off Taketomi Island, Japan (courtesy of Hisako Hirayama).

and France before this work started to provide some exciting results. New samples from both the Kristin reefs and HRC were also analysed.

The breakthrough came when Sigmund found methanotrophs in sponges, sediments, and water samples. Furthermore, some strange bacteria were found in the gills of *Acesta excavata*. Although the first firm results have just been accepted for publication, the following is from the abstract of the first evidence of methane sequestration by organisms at a Norwegian coral reef:

"Recent molecular studies of the coral-associated microflora have revealed the presence of *Proteobacteria* (*Alpha, Gamma, Delta*), *Acidobacteria, Nitrospira, Firmicutes, Bacteriodetes*—and additionally, *Planctomycetes, Verrucomicrobia, Actinobacteria* and *Crenarchaea* in ambient reef-water and sediment.

In this study we investigated methanotrophic communities associated with deep-water coral reefs. Methane was explored as a carbon source fuelling bacterial growth. The bacterial commuinty of a sediment sample was analysed using stable isotope probing (SIP).

Further PCR amplification of fraction seven DNA and cloning of genes diagnostic of bacterial taxonomy (16S rRNA; 71 clones), methane uptake (*pmoA*; 42 clones) and processing (*mxaF*; 48 clones), resulted in a large portion of this fraction's 16S rRNA clones (56%) and *pmoA* clones (100%) linked with a *Methylomicrobium* sp." (Jensen *et al.*, 2008b).

In the same way as stable isotope studies and results provided proof that pockmarks were indeed formed by gas seepage (Hovland *et al.*, 1985), I think that modern microbial and stable isotope analyses can provide the necessary results that will prove a central role of fluid flow (hydraulic processes) in the foodchain of some deep-water coral reefs and carbonate mounds.

Obligate hydrocarbon-eating marine bacteria (OHCB)

According to Yakimov *et al.* (2007):

> "The introduction of oil or oil constituents into seawater leads to successive blooms of a relatively limited number of indigenous marine bacterial genera—*Alcanicorax, Marinobacter, Thallassolituus, Cycloclasticus, Oleispira* and a few others (the OHCB [Obligate hydrocarbonoclastic bacteria])—which are present at low or undetectable levels before a natural or man-made 'polluting' event. The types of OHCB that bloom depend on the latitude/temperature, salinity, redox, and other prevailing physical-chemical factors. These blooms result in the rapid degradation of many oil constituents, a process that can be accelerated further by supplementation with limiting nutrients."

However, Yakimov *et al.* (2007) also admit that the hydrocarbons in the marine environment are not only sourced from pollution:

> "Despite the ubiquity of hydrocarbons in marine systems—originating from natural seeps of oil and natural gas deposits, marine oil transport accidents and deliberate discharges, and from biomass and biological processes—true marine hydrocarbon-degrading microbes were only discovered relatively recently. Interestingly, and in contrast to terrestrial hydrocarbon degraders which tend to be metabolically versatile and utilize a large range of organic substrates, their marine counterparts are mostly highly specialized obligate hydrocarbon utilizers, the so-called marine 'obligate hydrocarbonoclastic bacteria' (OHCB)" (Yakimov *et al.*, 2007).

In essence, this revelation is remarkable and very important; it means that the ocean is pre-charged with armies of bacteria, drifting all over the world's oceans, just waiting for an opportunity to bloom. There is only one substance that triggers them, and that is the world's next most common fluid after water: namely, hydrocarbons. Thus, when a nearby mud volcano erupts on the seafloor, or a pockmark pops, or a shallow gas pocket releases a burst of hydrocarbons (methane, ethane, propane, butane, pentane, hexane, and higher hydrocarbons), or even when a hydrothermal (serpentinisation-associated) warm or hot vent erupts, the hydrocarbon plume spreads through the water column and these creatures bloom. They are obligate, which means that they are very fussy; they will only eat hydrocarbons, their sole fodder! This, then, is not a mechanism that was devised over the last 200 years (since the anthropogenic industrial revolution)—no, this must be a very ancient mechanism, nearly as old as the Blue Planet itself. In conclusion, the very existence of the OHCBs is more or less proof not only of the facts that

(1) nature takes good care of its own pollutants;
(2) natural cleaning of the oceans by OHCB has been around for millions of years;

but also that

(3) natural oil pollution is endemic to the ocean, no matter where on the planet you happen to be.

In other words, the process of hydrocarbon exudation through the seabed, streaming (pluming) through the water column, and spreading out onto the ocean surface is as ancient and as natural as the ocean itself, and remarkably echoes the conclusions, however strange this may sound, made over 250 years ago, by a bishop, Erich Pontoppidan:

> "The North Sea's fattyness is, after its saltiness, a peculiar property, ... it would be assumed that in the ocean as on land there exists, here and there, seepages of running oily liquids or streams of petroleum, naphta, sulphur, coal-oils and other bituminous liquids" Pontoppidan (1752).

8.9 GEODIVERSITY AND BIODIVERSITY, A LINK?

In their paper titled "Geosphere–biosphere coupling; cold seep related carbonate and mound formation and ecology", van Weering *et al.* (2003b) presented the following puzzling and thought-provoking comment:

> "The acceptance of a simple model of mound formation is unsatisfactory because of the observed variation in size, morphology and the prolific number of mounds with their sometimes very localised clustering."

Reading between the lines, what they probably mean is: why should such a simple process as fluid flow through the seafloor provide the basis for such diverse and complex biological structures observed around the globe (carbonate mounds and deep-water coral reefs)? Well, the answer is relatively simple: Considering how many different types of fluid flow there are, mixtures of gases and liquids combined with different flux rates, current regimes, water masses, and other physical variables, it is not at all strange that there may be innumerable different biological expressions on the seafloor. When the renowned mathematician Christopher Zeeman was interviewed by Justin Mullins (2007) on the beauty of simple proofs, he answered:

> "If a proof is very short compared to the statement, you admire its beauty, but other proofs are very long and you admire them for getting to the result after such a tortuous path."

Likewise, I believe that even though the hydraulic theory is simple in its idea, in some regions we are going to prove it only through very arduous and tortuous pathways, as the linking telltale signals may be feeble, spasmodic, and perhaps complexly interlinked with other factors.

When it comes to discussing which factors may be responsible for the occurrence of deep-water corals in the GoM, Schroeder *et al.* (2005) state:

"... coral distribution is potentially regulated by a number of biological (e.g., recruitment, food availability) and physical (e.g., current regime, temperature) processes and mechanisms. Since our understanding of deep-water coral distribution, abundance, and ecology in the GoM is minimal, it is not yet known which of these processes or mechanisms are most influential."

Therefore, the stage is open, even for the simplest of hypotheses, as long as it satisfies the majority of the pertinent observations.

John Woodside, of Vrije University, Amsterdam and others (2006) also discuss this theme:

"Differences in sites of fluid emissions are determined more by differences in their geological environment than by their stage of development. Thus, brine rich fluid seeps and brine pools are often found in association with underlying Messinian evaporites; and gas hydrates have not been found at these sites. Factors contributing to the geological control of seeps and mud volcanoes include permeability and width of escape channels, availability of fluids and fluidity of mud, temperatures and salinity of fluids, gas concentrations and fluxes to the water column, presence or not of gas hydrates and evaporitic formations, and degree of overpressure driving the emissions/eruptions" (Woodside *et al.*, 2006).

They actually mean that the geodiversity of mud volcanoes and seeps mostly depend on the general geological conditions such that each seep region could be unique because of the multitude of variables and possible combinations involved!

Mathematics of diversity

The fact that deep-water coral reefs and carbonate mounds seem to differ from area to area and from region to region, as clearly demonstrated by the examples in this book, is a very intriguing and probably important trait, which may even represent the decisive clue for why they occur where they do. Each deep-water coral reef and carbonate mound (feature) seems more or less to be unique, which means that it is very difficult to predict where to find new reefs or mounds on previously unmapped seafloor. This fact alone indicates that there may be a multitude of factors involved in determining where and how reefs and mounds may form. This suspicion is based on the mere mathematics of having more than two variables that are of importance. If there is a set of five or six important variables (perhaps ranging from type of seeping fluids, to current velocity and turbulence), then the number of different combinations to be made between these variables will increase by the multiple of the number of variables (i.e., from 120 possible permutations for 5 variables, to 720 permutations for 6 variables). The enigmatic Darwin Mounds, for example, may rely on a dense water input from the sub-surface, combined with a certain set of primary producers and currents (i.e., let us say three main determining factors). The number of permutations of how each of these three variables influence the form (or nature) of the reef is $3! = 1 \times 2 \times 3$ (i.e., 6 supposedly different reef expressions). If there was one more variable that was dominant, the number of reef types would increase by a factor of 4. This may suggest that the reefs and mounds are a product of many different variables, where one variable may dominate in one setting and another dominates in another setting, thus perhaps dramatically influencing the appearance and composition of the reef or mound.

8.10 SUMMARY AND A RE-ITERATION OF THE HYDRAULIC THEORY

This chapter contains a discussion on the various theories for why and how deep-water coral reefs and carbonate mounds build up. There are many indications pointing at a close link with the focusing and strengthening of current velocities (special eddies and internal breaking waves); that is, the hydrodynamic (external) explanation model. However, there are equally many indications pointing at a close link with sub-surface (internal) conditions (i.e., in support of the hydraulic explanation model). The chapter also contains a brief section on the very high variability in species richness, diversity, and the immense variability in physical parameters and geological settings associated with the reefs and mounds. This also indicates that the reefs and mounds are not fully dependent on either an external or an internal vector—but perhaps combinations of both.

Because more than 60% of the specific deep-water coral reefs and carbonate mounds described and discussed in this book may rely upon seeping fluids from the sub-stratum, it is pertinent to re-iterate the hydraulic theory, originally put forward as a possibility in 1990 (Hovland, 1990b), and which has been modified ever since. The basis of the hydraulic theory is the over-arching and fundamental fact that most of the seafloor is ductile and compressible, especially those areas that contain buried gas pockets and (commercial) gas reservoirs. Thus, large portions of the seafloor can be divided into either hydraulically active or hydraulically passive (inert). The former is evidenced by fluid flow features such as pockmarks and gas flares, whereas the latter is devoid of fluid flow features (Judd and Hovland, 2007). The theory is also based on another important fact, that of primary producers (bacteria and archaea) relying on chemical gradients for their subsistence. Locations with focused and even diffuse fluid flow through the seafloor contain chemical gradients and are therefore ideal habitats for primary producers. At some places, the bacteria can actually be seen as bacterial mats on the seafloor. But, most often, the chemical gradients and occurrence of primary producers can only be documented by laborious sampling and laboratory analysis (as was done on IODP Leg 307 on the Challenger Mound scientific drilling site). The seeping fluids thus fertilise the sediments and water masses by hosting not only bacterial mats, but also by introducing primary producers into the sediments, the porewater system of the upper seafloor, and eventually into the water column. Thus, the second trophic level will respond and grow on the basis of the abundance of primary producers. This theory states that a stable nutrient source may exist at or near (even below) the deep-water coral reefs and carbonate mounds, and that they can therefore survive even the harshest winter (nutrient-poor) months and even dramatic climate changes (glaciations and sea-level changes). Conversely, when and if the seepage dies out or becomes exhausted, the deep-water coral reefs and carbonate mounds may lose a very important growth vector, likely leading to its extinction.

9

An unintended extinction?

Subsurface fluid flow is a key area of earth science research, because fluids affect almost every physical, chemical, mechanical, and thermal property of upper crust.

Tryon and Brown (2001)

*So roll up and see
How they rape the universe
How they've gone from bad to worse
Who are these men of lust, greed, and glory?
Rip off the masks and let's see.
But that's not right—oh no, what's the story?
There's you and there's me
That can't be right!!*

Crime of the Century, Supertramp, 1974

Warm water streamed from every orifice and crack in the sea floor over a circular area about 100 meters in diameter. The temperature of the water was highly variable, but the maximum was about 17 degrees C. The organisms were quite selective. They choked the warmest vents. In some cases mussel reefs actually channeled the flow of water forming conduits themselves.

Edmond and Von Damm (1978)

9.1 THREATS TO DEEP-WATER CORAL REEFS

There is no doubt that deep-water coral reefs and carbonate mounds represent very valuable habitats for numerous species, besides the corals themselves. Also, for fish

Figure 9.1. The Late Devonian Kess-Kess fossil carbonate mud mounds, which resemble the currently live Porcupine Basin giant mud mounds. Is this desolation the seascape of the future? (photo courtesy of Jean-Pierre Henriet).

they represent shelter and nursing homes for juveniles, as one of the few comparative studies of on-reef vs. off-reef fish counts for deep-water coral reefs documents:

> "Fish species richness and abundance was greater on the reef than surrounding seabed. In fact, 92% of species, and 80% of individual fish were associated with the reef. The present data indicates that the reefs have a very important functional role in deepwater ecosystems as fish habitat" (Costello et al., 2005).

There seem to be two main threats to the future wellbeing of the world's deep-water coral population:

(1) Mechanical stress induced by human activities.
(2) Climate change and ocean acidification.

In addition, for each individual coral reef, there is always the possibility of dramatic environmental changes by natural causes, such as nearby underwater avalanches (burial), and in the case of the hydraulic theory being viable that the seepage or venting is naturally depleted or exhausted, and ends up getting "turned off". However, these natural causes will not be discussed further. The most pressing questions are what humans can do to ensure the future survival of the reefs and giant carbonate mounds.

9.1.1 Mechanical disturbance and burial

According to the most recent authoritative, published assessment of the situation (Roberts *et al.*, 2006), the main threats are

 (i) bottom trawling
 (ii) hydrocarbon drilling and seabed mining
 (iii) ocean acidification.

Roberts *et al.* (2006) state:

> "Compared with widespread evidence for physical damage to reef structures from bottom trawling, there is little evidence that hydrocarbon exploitation substantially threatens cold-water coral ecosystems. *L. pertusa* colonizes North Sea oil platforms and seems to have a self-seeding population, despite proximity to drilling discharge. Greatest concern is over the potential for drill cuttings to smother reef fauna, but such effects would be highly localized when compared with the extent of seabed affected by bottom trawling."

To this assessment it may be added that, whereas the fishing industry tends to operate in "blind" mode when it comes to what seafloor it operates over, the hydrocarbon industry operates with much more caution, partly preventing damage to sensitive and costly equipment and structures, and partly because of law enforcement (at least in American, UK, and Norwegian waters). Thus, no drill site is drilled without a proper pre-drilling assessment of the seafloor, whereby any physical obstructions or sensitive sessile organisms, including chemosynthetic fauna and coral reefs, are documented beforehand. The problem with bottom trawling is the insensitiveness to what is down there (i.e., indiscriminate obliteration on the seafloor). As long as such bottom trawling is legal practice, all sessile organisms in the world are threatened with extinction, at least those living at water depths shallower than 2,000 m.

When it comes to threats by human-induced silting, this can also be a real problem for deep-water coral reefs. However, judging from our own observations, especially from fields where the coral reefs thrive, despite them being located in heavily silting natural environments, such as those at Kristin and Morvin fields, offshore mid-Norway (see Chapter 3), the stony corals seem to cope well with silting events. Therefore, unless human-caused silting during construction (excavation) on the seafloor and during submarine drilling operations deposits masses of sediments over the reef, causing total or partial burial, this threat seems to be of minor concern.

The modern re-discovery of deep-water coral reefs has been very important in bringing about public awareness and also action to protect these seafloor habitats, as pointed out here:

"For several hundreds of years fishers and marine biologists have been pulling scraps of cold water coral up from the depths of the sea. Both groups had theories for how the coral was part of the ecosystems in the ocean, but what these entities even looked like in their full glory was unknown. In 1982 the Norwegian oil company Statoil, took the first ever colour stills, and black and white videos of cold water coral reefs, just off Haltenbanken. The visual image of these cold water corals were to have a profound effect upon the people who observed them, in part determining where the oil company laid gas pipes, and finally leading to protection. The combined efforts of biologists, fishers and NGOs, and the repercussion via media, on public opinion and policy-makers, played an important part in triggering and developing legislation" (Armstrong and van der Hove, 2007).

9.1.2 Climate change and acidification

The main threat, when it comes to climate change, is the consequences of the ocean taking up huge amounts of CO_2 from the atmosphere:

"Acidification is the big elephant in the room.
Reef building would grind to a halt, with grievous implications. If CO_2 emissions are not curtailed, we'll eventually see reefs dominated by sea anemones and algae" (Stone, 2007).

According to J. M. Guinotte *et al.* (2006) and Roberts *et al.* (2006), human-induced changes in seawater chemistry will alter the distribution of deep-water scleractinian corals. This is mainly because of the acidification of the seawater caused by a higher uptake of CO_2 by the oceans:

"Perhaps the most insidious threat to cold-water coral reef ecosystems is from ocean acidification. There is general consensus that atmospheric carbon dioxide levels are rising sharply, and modeled scenarios suggest that this could be the cause of the greatest increase in ocean acidification over the last 300 million years. Current research predicts that tropical coral calcification would be reduced by up to 54% if atmospheric carbon dioxide doubled. In addition to the effect acidification could have on coral calcification, modeling studies predict that the depth at which aragonite dissolves could shallow by several hundred meters, thereby raising the prospect that areas once suitable for cold-water coral growth will become inhospitable" (Roberts *et al.*, 2006).

See also Sabine *et al.* (2004) and Turley *et al.* (2006).

Carol Turley and her team (2007) looked into this and came up with the following disturbing conclusions:

"Increased greenhouse gases are causing the oceans to warm and become more acidic at unprecedented rates. The paleo record tells us that scleractinians have survived several mass extinction events, but in all cases it took several millions of years to recover. It now seems likely that perturbations in the carbon cycle, most likely resulting in ocean acidification, has played a fundamental role in all major mass extinctions of the Scleractinia. However, the extremely rapid release of anthropogenic CO_2 from fossil fuel deposits is unprecedented in geological history and risks fundamentally perturbing deep-water coral ecosystems before the scientific community has begun to map and understand them" (Turley *et al.*, 2007).

Thus, the main question, now, is very urgent: Is acidification of seawater (from a pH of 8.1 today, to a pH of 7.8 in one century) going to have any effect on the livelihood of deep-water coral reefs? The answer to this question is, like most of the processes in nature, very complex and difficult. However, because the question has been raised, and the answer is not a straightforward "no", then, we need to seriously look into it and perform some focused investigations.

We are here talking about the complex carbonate cycle in a global perspective. According to James Kasting and David Catling (2003):

"The most important part of the carbon cycle in terms of long-term climate is the inorganic carbon cycle, sometimes called the carbonate–silicate cycle. CO_2 dissolves in rainwater to form carbonic acid (H_2CO_3), which is a weak acid, but when it acts over long timescales, it is strong enough to dissolve silicate rocks. For illustrative purposes, we use the simplest silicate mineral, wollastonite ($CaSiO_3$), to represent all silicate rocks. The products of silicate weathering, including calcium (Ca^{2+}) and bicarbonate (HCO^-) ions and dissolved silica (SiO_2), are transported by streams and rivers to the ocean. There, organisms, such as foraminifera, use the products to make shells of calcium carbonate ($CaCO_3$). Limestone is the commonly preserved form of calcium carbonate. Other organisms such as diatoms and radiolarians make shells out of silica. When these organisms die, they fall into the deep ocean. Most of the shells redissolve, but a fraction of them survive and are buried in sediments on the seafloor. The combination of silicate weathering plus carbonate precipitation can be presented chemically by $CO_2 + CaSiO_3 \rightarrow CaCO_3 + SiO_2$.

If silicate weathering and carbonate precipitation were the only reactions occurring, all of Earth's CO_2 would eventually wind up in the carbonate rock reservoir and the planet would become uninhabitable. Fortunately, another part of the cycle exists. As we know from the theory of plate tectonics, the seafloor is not static. Rather, it is continuously created at the mid-ocean ridges, and it is subducted at certain plate boundaries when the denser oceanic plate dives beneath the less dense continental plate. When this happens, the overlying carbonate sediments are carried down to depths where the temperatures and pressures are much greater. Under these conditions, carbonate minerals recombine with SiO_2 (which by this time is the mineral quartz) to reform silicate minerals, releasing CO_2 in the process. This reaction is termed carbonate metamorphism. The CO_2 released from carbonate metamorphism makes its way back to the surface and re-enters the atmosphere by way of volcanism, thereby completing the carbonate–silicate cycle. This cycle replenishes all the CO_2 in the combined atmosphere–ocean system on a timescale of approximately half a million years.

As silicate weathering is the loss process for atmospheric CO_2, CO_2 concentrations should tend to fall as T_s [ocean surface temperature] rises and CO_2 should increase as T_s falls. The response time of this feedback loop is that of the carbonate–silicate cycle—hundreds of thousands to millions of years. It is thus too slow to counteract human-induced global warming, but fast enough to have a dominating effect on the billion-year timescale of planetary evolution" (Kasting and Catling, 2003).

So, again, this is a worrying situation with respect to the rising CO_2 accumulation caused by humans—or is it? Why the relatively rare mineral wollastonite was used by Kasting and Catling in this modelling is uncertain. The iron and magnesium silicates kaolinite, pyroxenes, and hornblende would be more natural to model. Perhaps there is a need to aggravate the perceived situation?

9.2 A GLIMMER OF HOPE?

9.2.1 Biological response

To some researchers, the acidification scenario discussed in Section 9.1 may well be overly pessimistic. Because we do not really have a proper understanding of the ocean's buffer dynamics when it comes to natural CO_2 sequestration and its balance with carbonate in the ocean, it has been discussed, amongst others, by Chin and Yeston (2007):

> "... prediction of atmospheric CO_2 content several centuries from now is severely hampered by the multitude of poorly understood feedback mechanisms. The most important of these is probably the interaction between atmospheric CO_2 and marine calcification. In a nutshell, the amount of CO_2 absorbed by the ocean depends on the quantity depleted through calcium carbonate incorporation into the skeletons of calcifying organisms such as foraminifera and coccolithophorids. This bioactivity is a function of the alkalinity and the pH of the ocean, which in

turn depend largely on the partial pressure of atmospheric CO_2, as well as the type of calcifying organism (*E. huxleyi* and *O. universa*). Although the carbonate chemistry of the ocean is well known, the response of different species to changes in pH and alkalinity is incompletely understood, and large differences exist between the species that have been studied. Ridgwell *et al.* (2007) have performed model calculations for a range of calcifying behaviors. They find that the strength of CO_2 calcification feedback is dominated by the assumption of which species of calcifier contributes most to carbonate production, and that ocean CO_2 sequestration could reduce the atmospheric fossil fuel CO_2 burden by 4% to 13% in the year 3000."

This means that the situation is not as dramatic as stated by others.

There is evidence that the scleractinia corals may have certain feedback mechanisms by which they can cope with acidification of the ocean by CO_2. Stolarski *et al.* (2007) recently found that a solitary azooxanthellate scleractinian coral, a *Coelosmilia* sp., had a calcite skeleton which, during evolution, turned into an aragonitic skeleton. It resembles the currently common solitary coral *Desmophyllum*, which like all other known scleractinian corals have skeletons of aragonite:

"It has been suggested that hypercalcifying organisms, including corals, are sensistive to the Mg/Ca ratio of seawater, which has changed through geologic history in response to variations in the plate tectonic cycle. According to this model, hypercalcifying aragonite-producing organisms (including corals) flourish during periods in which seawater has a Mg/Ca ratio greater than 2, whereas hypercalcifying calcite-producing organisms flourish when the Mg/Ca ratio is less than 2 (in the modern ocean Mg/Ca = 5.2). The *Coelosmilia* sp. lived in the Late Cretaceous when the inferred Mg/Ca ratio of seawater was below 2. Our findings may thus appear to support the recently proposed idea that seawater composition can even change the skeletal mineralogy of scleractinians. However, other aragonitic scleractinians lived at about the same time as the *Coelosmilia* sp. specimens studied here, and other studies have shown that the chemical and isotopic composition of scleractinian skeletons is under strong biological control. Therefore, it seems more likely that the capability of scleractinians to produce either aragonitic or calcitic skeletons is genetically determined. In any case, skeletal mineralogy can no longer be considered a conservative feature among scleractinians throghout their evolution" (Stolarski *et al.*, 2007).

This could mean that *Lophelia pertusa* and other reef-building deep-water corals may be robust enough to fight back in case ocean water slightly changes its composition, including its acidity, or perhaps its Mg/Ca ratio.

9.2.2 Climate change?

The reason for the recent climatic change, to a warmer future, is also currently debated. Perhaps even the warming of the atmosphere and ocean is not totally anthropogenically induced, as discussed by one of the world's leading oceanographers, Wallace ("Wally") Broecker and his colleague, Thomas Stocker. They have a view which is different to most researchers of the current climate change debate. Because they find similar patterns of atmospheric CO_2 rise in a previous interglacial period, they conclude that the present one is "natural":

> "It is evident that MIS 11 [MIS = Marine Isotope Stage] was a long interglacial that was not brought to an end by the precession cycle nor will the Holocene be (MIS 1). Hence, the cause for the CO_2 rise during the last 8,000 years was 'natural' and not 'anthropogenic'" (Broecker and Stocker, 2006).

However, as this stance by Broecker and Stocker is controversial, and the objective of this book is not to discuss the complexities of climate change, reference is made to recent books on this topic (i.e., Walker and King, 2007).

Another reason for being sceptical about the climate change numeric-modellers and the general circulation models used by the IPCC (Intergovernmental Panel for Climate Change) researchers, is that, whereas CO_2 is an important radiative gas, it is not the most important. Methane and water vapour are more radiative and therefore more important for the atmosphere. In this book, I have referred to methane seeps and methane concentrations in the water column and shown that very little is in fact known about the methane system of the Earth (Etiope et al., 2007). Then add to this our evolving awareness and knowledge on the interaction between ocean water and the upper mantle via the oceanic crust (i.e., the worldwide serpentinisation process) mentioned earlier. It provides vast amounts of hydrogen, carbon dioxide, and dissolved hydrocarbons that end up in the ocean and atmosphere.

These facts highlighting our lack of basic knowledge about the fluids of our planet should warn against being too assertive as to how the atmospheric climate works. But, perhaps even more paradoxical, is a similar fact: namely, that we do not even know from where the Earth's water originated, as indicated by Steven D. Jacobsen and Suzan van der Lee (2006):

> "... The mass fraction of liquid water on Earth (0.025 wt% H_2O) is still on the order of ten times less than the water content of mid-ocean ridge magmas (0.2–0.3 wt% H_2O), and about half of what model 'enriched mantle' sources contain (~0.04 wt% H_2O). Perhaps rather than asking from where did Earth's water originate, we should be asking to where has Earth's water gone?"

Thus, there is still hope that the predictions of Roberts et al. (2006), that the end is near for deep-water coral reefs, fail—and, instead, that these hidden living structures may even out-live human beings and live for another million years, and on ...

10

Conclusions

> *Don't walk behind me,*
> *I may not lead.*
> *Don't walk in front of me,*
> *I may not follow.*
> *Walk beside me,*
> *that we may be as one.*
>
> Ute, Zona (1994)

10.1 SUMMARY OF MAIN CONCLUSIONS

The true size (20 m–200 m wide, 2 m–40 m high), structure, and composition of deep-water (cold-water) coral reefs were first documented off Norway and on the Rockall Plateau, off Ireland during the period 1940–1990. The review of descriptions and occurrences of such reefs includes the entire continental European margin, from off Gibraltar to the west of Ireland, and high north to the Barents Sea, off Norway. The reefs occur along the shoulder of the continental slope, at specific locations on the continental shelves and plateaus, along portions of the Norwegian coastline, including some inlets and fjords. In the fjords (including off southwest Sweden), they tend to form on top of morainic ridges (sills). Occurrences of similar reefs off Brazil, off west and east USA, off Canada, off the Congo and in the Mediterranean Sea are also briefly reviewed and discussed. So is the occurrence of small colonies (not reefs) consisting of deep-water corals. The deepest *Lophelia* occurrence is thus found at 1,800 m near the Mid-Atlantic Ridge.

The first documentation of live, giant carbonate mounds was done in the Porcupine Basin, off Ireland. Although similar fossil structures had previously been found in the geological record, both by drilling and in exposures, their origin of formation ("driving mechanism") had never been documented or proved. The live structures reviewed here are up to 200 m high and up to a couple of kilometres across.

From scientific drilling through one of these structures, the Challenger Mound off Ireland (IODP Leg 307), it is known that they consist of unlithified, consolidated sediments that are stabilized (into a mound shape) by the dead framework of deep-water coral remnants (mainly *Lophelia pertusa* skeletons). Besides the Porcupine Basin occurrence, live, giant carbonate mounds are also described from off Namibia and off Morocco. It is suspected that similar mounds occur at many places along continental slopes around the world.

Two main theories for why deep-water coral reefs and giant carbonate mounds form at specific locations of the seafloor have been presented. These are

(1) Hydrodynamic, "external"-based theories (high water current velocities, breaking internal waves, upwelling water masses, specific water masses and currents, and ambient water density forcing)
(2) Sub-stratum, "internal"-based theories (seepage of nutrients from underground reservoirs of mineralized water, including freshwater, and gases, including hydrocarbon gases such as methane, ethane, and propane and other gases, such as carbon dioxide, hydrogen, and hydrogen sulphide).

Although there is a shortage of measurements of physical parameters such as current directions and velocities, turbulence, turbidity (including observation of breaking internal waves), nutrient amount in the near reef and mound vicinity, and long-term monitoring of parameters, the author bases his main conclusions as to why they form where they do on indirect evidence. It is concluded that most of these reefs and mounds rely on a stable (year-round) food source, which is dependent on primary producers (mainly bacteria) that live on seeping fluids originating from the sub-surface (locally or regionally). However, in some locations, such as in high current velocity environments (i.e., narrow inlets and fjord threshold settings), a reliable (year-round) nutrient provision is suspectedly made by strong, shifting tidal currents.

The future threat to deep-water corals is mainly posed by mechanical stress and destruction induced by bottom trawling, and other industry-imposed mechanical disturbance. Thus, the only way to stop total destruction of many of those reefs and mounds already mapped is by strict governmental and intergovernmental regulation of both fishing and other industrial activity (hydrocarbon and other mineral extraction). In addition, there is the threat to these structures posed by a possible lowering of the pH value of ocean water, by climate change brought on by rising atmospheric CO_2 load. However, such acidification of water at depths where most of the deep-water corals reside is regarded as being rather speculative, at least until further investigation has been done.

Appendix A

Some additional photographs and images

Figure A.1. A photograph showing the base of Reef B on the HRC. Note the absence of live stony corals. The dead *Lophelia* colonies have been re-colonised by a variety of animals, including abundant yellow sponges (*Phakellia* sp.?). Also notice the tail of the tusk (*Brosme brosme*) at left, apparently feeding undisturbed by the presence of the ROV.

Figure A.2. Some of the strange animals encountered on the Sula Ridge, during ROV surveys in 1992, were these indigo-coloured anchor animals, Echiurans (probably of the species *Bonellia viridis*). They are detritovorous, meaning that they eat detritus on the seafloor. Four individuals can be seen feeding. Upon disturbance (i.e., when the ROV comes too close), they rapidly fold back and retract into a small hole in the ground. This burrow is at the end of the long, thin anchor foot. The eastern slope up towards the Sula Ridge is rather special in that there are rocks of Palaeocene and Cretaceous age scattered around: these are sedimentary rocks of high organic content and could be the reason for the high abundance of sponges and other animals coating the rocks, as seen here.

Appendix A: Some additional photographs and images 219

Figure A.3. A large group of Norway redfish (*Sebastes viviparus*) over a block of largely dead *Lophelia*. This video-grabbed image was acquired in May 1991, and could be showing fish congregating for spawning (at least one of the fish in the centre has a wide stomach). This species gives birth to live sprats (the eggs are hatched inside their bodies).

Figure A.4. Large blocks of dead *Lophelia* colonies are favoured locations for many types of sponges, as seen here. Note also numerous shrimp (their eyes glow red) and rice coral (*Primnoa resedaeformis*). This photograph is from HRC.

Appendix A: Some additional photographs and images 221

Figure A.5. A very rare image of a dark-violet, unknown coral species. The video footage is from the northward-facing side of Reef A of the HRC (Hovland and Mortensen, 1999). The poor resolution of this "vintage" colour recording prevents proper species identification. In addition, numerous organisms are seen, ranging from sponges to feather stars and shrimps.

Figure A.6. A video-grabbed image from one of the pockmark reefs at Kristin, taken in April 2007. This is "heaven" for marine biologists: name the fish and the other macro-fauna!

Appendix A: Some additional photographs and images 223

Figure A.7. One of the early photographs taken on the HRC, this is from the summit of Reef A in 1993.

224 Appendix A: Some additional photographs and images

Figure A.8. Here is some detail from the HRC in 1997. Two anemones (*Bolocera thuediae*?) and a crab, *Lithodes maja*, between them. They are living on dead *Lophelia* (species identification by Sofie Vandendriessche, 2002).

Figure A.9. A beautiful *Paragorgia arborea* living in 200 m water depth inside a gulley between skerries (small islands) about 15 km west of Brønnøysund (town in mid-Norway). The photo was taken in April 2007.

226 Appendix A: Some additional photographs and images

Figure A.10. At the base of *Lophelia* reefs, large sea urchins, such as this 10 cm wide specimen of the species *Cidaris cidaris* can be found. This is from the HRC in 1993 (species identification by Sofie Vandendriessche, 2002).

Appendix A: Some additional photographs and images 227

Figure A.11. A fine photo of the common saithe (*Pollachius* sp.), about 0.8 m in length, as it searches for shrimp or other food, above a lush colony of *Lophelia pertusa*. In the background lies a Norwegian redfish (*Sebastes viviparus*), tactically associated with the red rice coral (*Primnoa resedaeformis*).

Figure A.12. On these reefs, which are up to 24 m high, coral colonies grow up and outwards and often form overhangs. When the mechanical strain is too large, they topple over and fall down the slope. This close-up is from the upper portion of the toppled block in Figures 4.11 and 4.12. Note the blue clayey sediments still intact amongst the dead *Lophelia pertusa* network. A squat lobster, probably of the species *Munida sarsi*, is seen defending its small sedimentary territory.

Appendix A: Some additional photographs and images 229

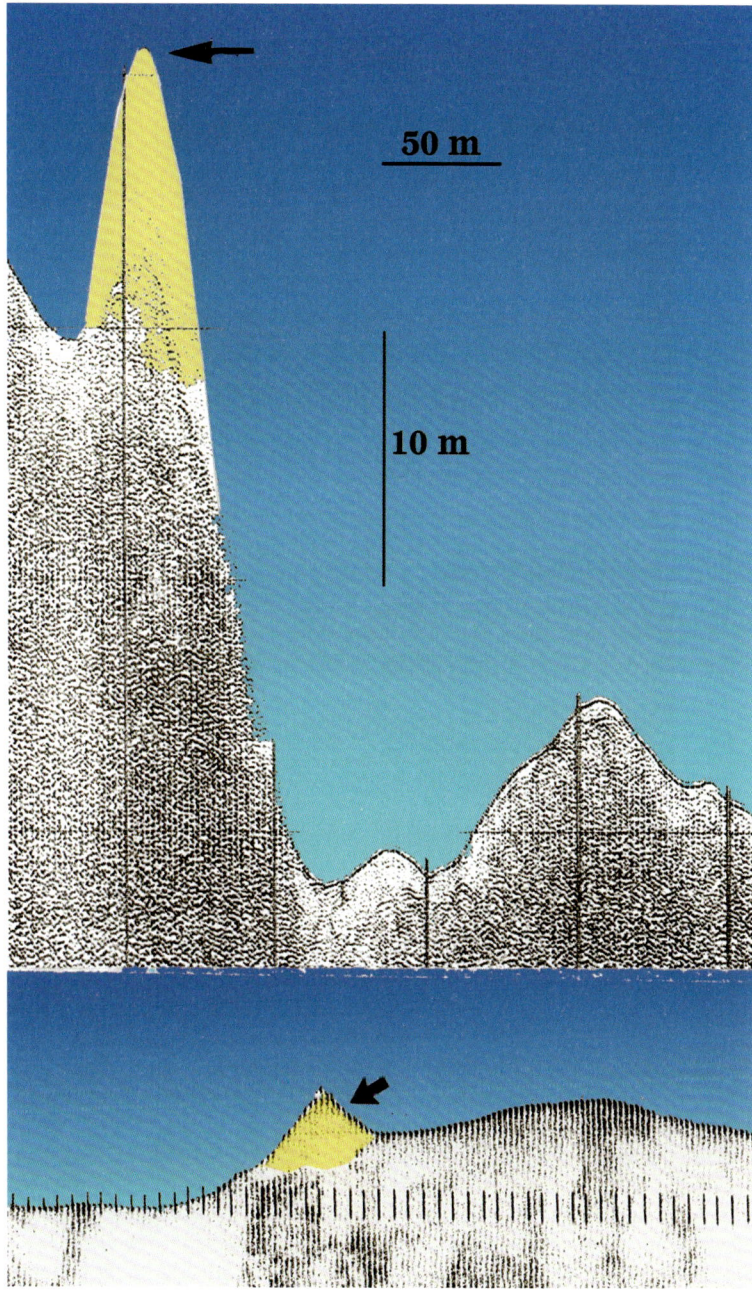

Figure A.13. A coloured version of the original monochrome sub-bottom profiler (top) and sidescan sonar (bottom) recordings of the Fugløy reef of 1982 (Hovland and Mortensen, 1999). Black arrows point at the reef summit (upper) and the characteristic cone shape of this reef (lower).

Figure A.14. Two video-grabbed images from the centre of complex pockmark G11, at Nyegga (see pp. 24–31), at a water depth of 750 m and ambient water temperature of −0.7°C (minus point-seven degrees C!). They show the occurrence of rare and beautiful sea lilies (stalked crinoids). Stalked crinoids mostly inhabit deep water and are therefore difficult for ordinary underwater enthusiasts to observe. These animals are neither abundant nor familiar today. However, they dominate the Palaeozoic fossil record of echinoderms and shallow marine habitats until the the Permo-Triassic extinction, when they suffered a near complete extinction. Stalked crinoids attach to the sea bottom by stalks, which lead up to what is known as the calyx, the base of the pentameral system of feeding arms (the "feathers"). These feathers have a dual purpose: they are used to filter food (via ciliated grooves, ambulacral canals) and to "fly" with. The animals are extremely elegant when they decide to fly off into the water column (perhaps searching for better feeding locations, if and when seeps turn on and off). The lower image shows a detail of the upper one (the yellow central rectangle in the upper image). Note how the crinoid's stalks are tickling (feeling) the back of a sleeping eelpout fish (which is probably just conserving energy in the extremely cold Norwegian Sea water). This scene is one of utter serenity and peace from the deep, dark, largely unknown Nordic ocean.

Appendix A: Some additional photographs and images 231

Figure A.15. These two high-resolution underwater photographs were taken at the Kristin A2 reef in 2004. The lower one is a detail of the upper overview photo (yellow lines define the field of view). Note the abundance of articulating ("breathing") *Acesta excavata* bivalves living on and within the mature 2 m wide *Lophelia* colony. Note also the abundance of shrimp (seen with shining eyes in the lower image).

Appendix B

Epilogue

The more we learn about the ocean, the more intriguing the story gets: it's like exploring a hidden valley on a newly discovered planet. For the last 18 years, I have had the privilege of being a member of the Environmental Protection and Safety Panel of the IODP (Integrated Ocean Drilling Program), where we discuss "big" science (geology, geophysics, biology, climatology, chemistry, and physics) with some of the world's best scientists. As the workings of our planet are slowly revealed by mapping, sampling, measuring, and drilling, the clearer it becomes that the main answers to the hidden secrets (origin of life, etc.) lie in the interaction and exchange of material between the mantle and the deep ocean. The heat systems at the plate boundaries, where water circulates into the porous oceanic crust and interacts with the warm underlying mantle substance (peridotite), are as yet barely understood. Even though these heat systems (hot vents) are buried below kilometres of sediments, they still work, by circulating fluids and refining minerals. Some of the recent discoveries are like scenes from science fiction or fairytales (e.g., the "Lost City" discovery of 2000).

The ocean contains "everything", in much the same way as described by the little boy Azaro in Ben Okri's wonderful book *Songs of Enchantment* from Africa:

"I looked, and saw them again. I saw them in the revelations of moonlight. I saw their hidden and glorious radiance. I stared in trembling wonder at the mighty procession of wise spirits from all the ages, from eras past and eras to come. I watched the glorious stream of hierophants and invisible masters with their caravans of eternal delights, their floating pyramids of wisdom, their palaces of joy, their windows of infinity, their mirrors of lovely visions, their dragons of justice, their lions of the divine, the unicorns of mystery, their crowns of love-won illumination, their diamond sceptres and golden staffs, their hieratic standards and their shining thyrsi of magic ecstasy. I gazed at the royal and serene spirits from higher realms that restore balances. They were continuing their majestic procession to the great meeting-place in the mind and dreams of the world. They were moving temporarily from their adventures of infinity to our earthly realm which for centuries has cried out for more vision, more transformation, and the birth of a new cycle of world justice" (Okri, 1993).

To end this book, what is more appropriate than to re-iterate what I think is the greatest discovery in our time: the first sitings of the hot vents on the EPR (East Pacific Rise) in 1977:

"*Alvin* sinks passively in the ocean's water column at a rate of 30 to 35 meters per minute. Thus it settled to a depth of 2,500 meters in one and a half hours. At that point, 100 meters from the bottom, it jettisoned a set of weights and gained neutral trim. We moved downward until we hovered above the sea floor, which proved to be a gentle slope. For about half an hour we cruised about, trying to find the target. Each of us had a Plexiglas porthole to look through. Then we stopped to collect some rocks. As our pilot employed the submarine's mechanical arm to wrestle with a basalt pillow (lava) a couple of large purple sea anemones engaged our attention. Only when our gaze finally wandered away from them did we realize that the water within the range of our lights was shimmering. The hastily measured temperature of the water was five degrees above the ambient water temperature (2.05 degrees C). With all thoughts of rocks forgotten, we captured a sample of the water and then continued on our course upslope. Here we came on a fabulous scene.

The typical basaltic terrain at the ridge axis is bleak indeed. Monotonous fields of brown pillows are cut by faults and fissures. One must examine several square meters to find a single organism. Yet here was an oasis. Reefs of mussels and fields of giant clams were bathed in the shimmering water, along with crabs, anemones and large pink fish. The remaining five hours of 'bottom time' passed in something close to frenzy" (Edmond and Von Damm, 1978).

This sole discovery is parallel to the moon landing, and the first human step on another heavenly body, even though it has never been touted as such. Since 1977, up to one hundred different hot vents have been studied. Not only have new animal species been found, but a whole new group of animals, the chemosynthetic and those that are independent of sunlight and a changing climate; these newly discovered communities will survive even the harshest of climate changes, human-made or natural.

Martin Hovland
Stavanger, February 2008

References

Aharon, P.; Socki, R. A.; and Chan, L. (1987). Dolomitization of atolls by sea water convection flow: Test of a hypothesis at Niue, South Pacific. *Journal of Geology*, **95**, 187–203.

Akhmetzanov, A. M.; van Weering, Tj.; Kenyon, N. H.; and Ivanov, M. (1998). Carbonate mounds and reefs at the Rockall Trough and Porcupine margins. *IOC (Intergovernmental Oceanographic Commission, UNESCO) Workshop Report No. 143*, pp. 45–46.

Al-Aasm, I. S.; and Vernon, M. (2007). Waulsortian-like mounds of the Mississippian Pekisko Formation, northwestern Alberta: Petrographic and chemical attributes. *Marine and Petroleum Geology*, **24**, 616–631.

Andrews, J. A.; Portman, C.; Rowe, P.; Leeder, M. R.; and Kramers, J. D. (2007). Sub-orbital sea-level change in early MIS 5e: New evidence from the Gulf of Corinth, Greece. *Earth Planet. Sci. Lett.*, **259**, 457–468.

Armstrong, C. W.; and van der Hove, S. (2007). The formation of policy for protection of cold-water coral off the coast of Norway. *Internal Report*. University of Tromsø.

Anderson, D. L. (2007). Looking glass worlds. *Elements*, June, 160.

Bacon, C.; and Lanphere, M. A. (1990). The geologic setting of Crater Lake, Oregon. In: Drake, E. T.; Larson, G. L.; Dymond, J.; and Collier, R. (eds), *Crater Lake: An Ecosystem Study*. American Association for the Advancement of Science, San Francisco, pp. 19–27.

Ballard, R.D.; and Grassle, J.F. (1979). Return to oases of the deep. *National Geographic Magazine*, **156**, 680–705.

Baraza, J.; Ercilla, G.; and Nelson, C. H. (1999). Potential geologic hazards on the eastern Gulf of Cadiz slope (SW Spain). *Marine Geology*, **155**, 191–215.

Barry, J.M.; and Kochevar, R.E. (1998). A tale of two clams: Differing chemosynthetic life styles among vesicomyids in Monterey Bay cold seeps. *Cahiers de Biologie Marine*, **39**, 329–331.

Belenkaia, I. (2000). Gas-derived carbonates: Reviews in morphology, mineralogy, chemistry and isotopes (data collected during the TTR programme cruises during 1995–1999). *Sixth International Conference on Gas in Marine Sediments, September 5–9*, Abstracts Book, VNIIOkeangeologia, St. Petersburg, Russia, pp. 9–10 (abstract).

Belka, Z. (1998). Early Devonian Kess-Kess carbonate mud mounds of the Eastern Anti-Atlas (Morocco), and their relation to submarine hydrothermal venting. *J. Sed. Res.*, **68**, 368–377.

Berryhill Jr, H. L. (1987). Late Quaternary facies and structure, northern Gulf of Mexico. *Am. Assoc. Petrol. Geologists Studies in Geology*, **23**.

Bett, B.J. (2001). UK Atlantic Margin Environmental Survey: Introduction and overview of bathyal benthic ecology. *Continental Shelf Research*, **21**, 917–956.

Bett, B.J.; and Cruise Participants (1999). *RRS Charles Darwin Cruise 112C, 19 May–24 June 1998*. Atlantic Margin Environmental Survey: Seabed survey of deep-water areas (17th round tranches) to the north and west of Scotland. Southampton Oceanography Centre Cruise Report No. 25, 171 pp.

Bett, B. J.; Billett, D. S. M.; Masson, D. G.; and Tyler, P. A. (2001). *RRS Discovery Cruise 244, 7 July–10 August 2000*. A multidisciplinary study of the environment and ecology of deep-water coral ecosystems and associated seabed facies and features (The Darwin Mounds, Porcupine Bank and Porcupine Seabight). Southampton Oceanography Centre, Cruise Report No. 36, 108 pp.

Beyer, A.; Schenke, H. W.; Klenke, M.; and Niederjasper, F. (2003). High resolution bathymetry of the eastern slope of the Porcupine Seabight. *Marine Geology*, **198**, 27–54.

Bennett, I. (1971). *The Great Barrier Reef*. Landsdowne Press, p. 183.

Blinova, V.; and Stadnitskaia, A. (2001). Composition and origin of the hydrocarbon gases from the Gulf of Cadiz mud volcano area. In: Akhamanov, G.; and Suzyumov, A. (eds), *Geological Processes on Deep-water European Margins: International Oceanographic Commission Workshop Report No. 175 on the International Conference and Ninth Postcruise Meeting of the Training Through Research Programme, Moscow-Mozhenka, Russia, 28 January–2 February 2001*. UNESCO, Paris, pp. 45–46.

Blondel, Ph.; and Murton, B.J. (1997). *Handbook of Seafloor Sonar Imagery*. Wiley/Praxis, Chichester, UK, 314 pp.

Bøe, R. (2007). *Gasslekkasjer fra havbunnen utenfor Vesterålen* (Gas leakage from the seafloor off Vesterålen). Available at www.mareano.no, November, 2007 [in Norwegian].

Bouriak, S.; Vanneste, M.; and Soutkine, A. (2000). Inferred gas hydrates and clay diapers near the Storegga Slide on the southern edge of the Vøring Plateau, offshore Norway. *Marine Geology*, **163**, 125–148.

Braechert, T. C.; Buggisch, W.; Flügel, E.; Hüssner, H. M.; Joachimski, M. M.; Tourneur, F.; and Waliser, O. H., 1992. Controls of mud mound formation: The Early Devonian Kess-Kess carbonates of the Hamar Laghdad, Antiatlas, Morocco. *Geologische Rundschau*, **81**, 15–44.

Breeze, H.; Davis, D. S.; Butler, M.; and Kostylev, V. (1997). Distribution and status of deep-sea corals off Nova Scotia. *Marine Issue Comm.*, Spec. Publ. 1, Ecology Action Centre.

Broecker, W. S.; and Stocker, T. F. (2006). The Holocene CO_2 rise: Anthropogenic or natural? *EOS Trans. Am. Geophys. Union*, **87**(3), 27.

Brøgger, W. C. (1901). Om de senglaciale og postglaciale nivåforandringer I Kristianiafeltet (Molluskfaunan). *Norges Geologiske Undersøgelse*, No. 31, 183–185 [in Norwegian].

Brooks, J. M.; Kennicutt II, M. C.; Fay, R. R.; McDonald, T. J.; and Sassen, R. (1984). Thermogenic gas hydrates in the Gulf of Mexico. *Science*, **223**, 696–698.

Brooks, J. M.; Cox, H. B.; Bryant, W. R.; Kennicutt, M. C.; Mann, R. G.; and MacDonald, I. R. (1986). Association of gas hydrates and oil seepage in the Gulf of Mexico. *Organic Geochemistry*, **10**, 221–234.

Brown, M. A.; and Dodd, J. R. (1990). Carbonate mud bodies in Middle Mississippian strata of southern Indiana and northern Kentucky: End members of a Middle Mississippian mud mound spectrum? *Palaios*, **5**, 236–243.

Bryn, P.; Solheim, A.; Berg, K.; Lien, R.; Forsberg, C.F.; Haflidason, H.; Ottesen, D.; and Rise, L. (2003). The Storegga Slide Complex: Repeated large scale sliding in response to climatic cyclicity. In: J. Locat and J. Mienert (eds), *Submarine Mass Movements and Their Consequences*. Kluwer Academic, Dordrecht, the Netherlands, pp. 215–222.

Bugge, T. (1983). *Submarine Slides on the Norwegian Continental Margin, with Special Emphasis on the Storegga Area*. Continental Shelf Institute (IKU), Trondheim, Norway, Publication 110, 152 pp.

Bugge, T.; Knarud, R.; and Mørk, A. (1984). Bedrock geology on the mid-Norwegian continental shelf. In: A.M. Spencer (ed.), *Petroleum Geology of the North European Margin*. Norwegian Petroleum Society/Graham & Trotman, London, pp. 271–283.

Buhl-Mortensen, L. (2007). Skal bli best på bunnen (Shall be best on the bottom). *VG Helg* (Norwegian Newspaper supplement), 6 April 2007, pp. 22–25.

Buhl-Mortensen, L.; and Mortensen, P. B. (2005). Distribution and diversity of species associated with deep-sea gorgonian corals off Atlantic Canada. In: Freiwald, A.; Roberts, J. M. (eds). *Cold-water Corals and Ecosystems*. Springer-Verlag, Berlin, pp. 849–878.

Bünz, S.; Mienert, J.; and Berndt, C. (2003). Geological controls on the Storegga gas-hydrate system of the mid-Norwegian continental margin. *Earth Planetary Sci. Lett.*, **209**(3–4), 291–307.

Cairns, S. D.; and Chapman, R. E. (2001). Biogeographic affinities of the North Atlantic deepwater scleractinia. In: Willison, J. H. M.; Hall, J.; Gass, S. E.; Kenchington, E. L. R.; Butler, M.; and Doherty, P. (eds), *Proceedings of the First International Symp. on Deep-Sea Corals*, Ecology Action Centre and Nova Scotia Museum, Halifax, Canada, pp. 30–57.

Campbell, K. A.; Farmer, J. D.; and Des Marais, D. (2002). Ancient hydrocarbon seeps from the Mesozoic convergent margin of California: Carbonate geochemistry, fluids and palaeoenvironments. *Geofluids*, **2**, 63–94.

Cann, J.; and Morgan, J. (2002). Secrets of the Lost City. *Geoscientist*, **12**(11), 4–7.

Cavagna, S.; Clari, P.; and Martire, L. (1999). The role of bacteria in the formation of cold seep carbonates: Geological evidence from Monferrato (Tertiary, NW Italy). *Sedimentary Geology*, **126**, 253–270.

Chanton, J. P.; Martens, C. S.; and Kelley, C. A. (1989). Gas transport from methane-saturated, tidal freshwater and wetland sediments. *Limnology and Oceanography*, **34**, 807–819.

Chanton, J. P.; Martens, C. S.; and Paull, C. K. (1991). Control of pore-water chemistry at the base of the Florida escarpment by processes within the platform. *Nature*, **349**, 229–231.

Charlou, J. L.; and Donval, J.-P. (1993). Hydrothermal methane venting between 12°N and 26°N along the Mid-Atlantic Ridge. *J. Geophys. Res.*, **98**, 9625–9642.

Charlou, J. L.; Philippe, J.-B.; Donval, J. P.; Knoery, J.; Foucher, J.-P.; Pellé, H.; and the MEDINAUT Scientific Party (2000). High methane concentrations in plumes and brines associated with mud volcanoes of the Eastern Mediterranean Sea (MEDINAUT Diving Cruise, November/December 1998). *VIth International Conference on Gas in Marine Sediments, St Petersburg, September 2000*, p. 15 (abstract).

Chen, C. Y.; and Hsu, J. R. C. (2005). Interaction between internal waves and a permeable seabed. *Ocean Engineering*, **32**, 587–621.

Childress, J. J.; Fisher, C. R.; Brooks, J. M.; Kennicutt II, M. C.; Bidigare, R.; and Anderson, A. E. (1986). A methanotrophic marine molluscan (bivalvia, Mytilidae) symbiosis: Mussels fuelled by gas. *Science*, **233**, 1306–1308.

Chin, G.; and Yeston, J. (2007). Choose calcifiers with care. *Science*, **317**, 572.

Clari, P. A.; Gagliardi, C.; Governa, M. E.; Ricci, B.; and Zuppi, G. M. (1988). I calcari di Marmorito: Una testimonianza di processi diagenetici in presenza di metano. *Bollettino del Museo Regionale di Scienze Naturali Torino*, **5**, 197–216 [in Italian].

Cohen, A. S.; Talbot, M. R.; Awramik, S. M.; Dettman, D. L.; and Abell, P. (1997). Lake level and paleoenvironmental history of Lake Tanganyika, Africa, as inferred from late Holocene and modern stromatolites. *Geological Society of America (Bulletin)*, **109**, 444–460.

Colman, J. G.; Gordon, D. M.; Lane, A. P.; Forde, M. J.; and Fitzpatrick, J. J. (2005). Carbonate mounds off Mauritania, Northwest Africa: Status of deep-water corals and implications for management of fishing and oil exploration activities. In: Freiwald, A.; and Roberts, M. (eds), *Cold-water Corals and Ecosystems*. Springer-Verlag, Heidelberg, pp. 417–441.

Costello, M. J.; McRea, M.; Freiwald, A.; Lundälv, T.; Jonsson, L.; Bett, B. J.; van Weering, T. C. E.; de Haas, H.; Roberts, M. J.; and Allen, D., 2005. Role of cold-water coral *Lophelia pertusa* coral reefs as fish habitat in the North East Atlantic. In: Freiwald, A.; and Roberts, M. (eds), *Cold-water corals and Ecosystems*. Springer-Verlag, Heidelberg, pp. 771–805.

Croker, P. F. (1992). Seabed mounds (carbonate knolls) in the Porcupine Basin, W. Ireland. *Annual Irish Geological Research Meeting* (unpublished abstract).

Croker, P.; and O'Loughlin, O. (1998). A catalogue of Irish offshore carbonate mud mounds. *Conference on Carbonate Mud Mounds and Cold Water Reefs, University of Gent, Belgium, February 7–11, 1998.*

Croker, P. F.; Shannon, P. M. (1987). The evolution and hydrocarbon prospectivity of the Porcupine Basin, offshore Ireland. In: Brooks, J.; and Glennie, K. W. (eds), *Petroleum Geology of NW Europe*. Graham & Trotman, London, pp. 633–642.

Cunha, M. R.; Hilário, A. M.; Teixeira, I. G.; and all the Shipboard Scientific Party aboard the TTR-10 Cruise (2001). The faunal community associated to mud volcanoes in the Gulf of Cadiz. In: Akhamanov, G.; and Suzyumov, A. (eds), *Geological Processes on Deep-water European Margins: International Oceanographic Commission Workshop Report No. 175 on the International Conference and Ninth Post-cruise Meeting of the Training Through Research Programme, Moscow-Mozhenka, Russia, 28 January–2 February 2001.* UNESCO, Paris, pp. 61–62 (abstract).

Dando, P. R. (2001). *A Review of Pockmarks in the UK Part of the North Sea, with Particular Reference to Their Biology*. Strategic Environmental Assessment SEA2, Technical Report TR_001, Department of Trade and Industry, London, 21 pp.

Dando, P. R.; and Southward, A. J. (1986). Chemoautotrophy in bivalve molluscs of the genus *Thyasira*. *Journal of the Marine Biological Association of the United Kingdom*, **66**, 915–929.

Davis, A. S.; Clague, D. A.; Bohrson, W. A.; Dalrymple, G. B.; and Greene, H. G. (2002). Seamounts at the continental margin of California: A different kind of oceanic intraplate volcanism. *Geol. Soc. Amer. Bull.*, **114**, 316–333.

De Cock, H. (2005). 3D-seismische studie van koralbaanken aan de oostelijke rand van het Porcupine bekken. MSc. thesis, Ghent University, Belgium [in Dutch].

De Haas, H.; and Shipboard Scientific Party (2006). Seismic and sedimentological investigations of the carbonate mounds and mud volcanoes at the Pen Duick Escarpment and SE Gulf of Cadiz. *Cruise Report R.V. Pelagia Cruise M2006*.

De Mol, B.; Van Rensbergen, P.; Pillen, S.; Van Herreweghe, K.; Van Rooij, D.; McDonnell, A.; Ivanov, M.; Swennen, R.; and Henriet, J. P. (2002). Large deep-water coral banks in the Porcupine Basin, southwest of Ireland. *Marine Geology*, **188**, 193–231.

De Mol, B.; Henriet, J.-P.; and Canals, M. (2005). Development of coral banks in Porcupine Seabight: Do they have Mediterranean ancestors? In: Freiwald, A.; and Roberts, M. (eds), *Cold-water Corals and Ecosystems*. Springer-Verlag, Heidelberg, pp. 515–533.

De Vogelaere, A.; Burton, E. J.; Trejo, T.; King, C. E.; Clague, D. A. (2005). Deep-sea corals and resource protection at the Davison Seamount, California, U.S.A. In: Freiwald, A.; and Roberts, M. (eds), *Cold-water Corals and Ecosystems*. Springer-Verlag, Heidelberg, pp. 1189–1198.

Dons, C. (1927). *Sjøen* (The Sea). J. W. Cappelens Forlag, Oslo, Norway, 108 pp. [in Norwegian].

Dons, C. (1944). Norges korallrev. *Norsk Vidensk Selsk Trondheim Forh.*, **16A**, 37–82 [in Norwegian].

Dorschel, B.; Hebbeln, D.; Rüggeberg, A.; and Dullo, C. (2007). Carbonate budget of a cold-water coral carbonate mound: Propeller Mound, Porcupine Seabight. *Int. J. Earth Sciences*, **96**, 73–83.

Dullo, W.-Chr.; Flögel, S.; and Rüggeberg, A. (2008). Cold-water coral growth in relation to the hydrography of Celtic and Nordic European Continental Margin. Accepted for publication in *Marine Ecological Progress Series*.

Dymond, J.; Collier, R. W.; and Watwood, M. E. (1989). Bacterial mats from Crater Lake, Oregon and their relationship to possible deep-lake hydrothermal venting. *Nature*, **342**, 673–675.

Eberli, G.; Swart, P.; and Malone, M. (1996). *Ocean Drilling Program Leg 166, Preliminary Report: The Bahamas Transect*. ODP, Texas A&M University, College Station, TX.

Edmond, J. M.; and Von Damm, K. (1978). Hot springs on the ocean floor. *Scientific American*, June, 70–85.

Egorov, A. V.; Crane, K.; Rozhkov, A. N.; and Vogt, P. R. (1999). Gas hydrates that outcrop on the sea floor: Stability models. *Geo-Marine Letters*, **19**, 68–75.

Etiope, G.; Martinelli, G.; Caracausi, A.; and Italiano, F. (2007). Methane seeps and mud volcanoes in Italy: Gas origin, fractionation and emission to the atmosphere. *Geophys. Res. Lett.*, **34**, L14303, doi:10.1029/2007GL030341.

Feary, D. A.; Hine, A. C.; James, N. P.; and Malone, M. J. (2004). Leg 182 synthesis: Exposed secrets of the Great Australian Bight. In: Hine, A. C.; Feary, D. A.; and Malone, M. J. (eds), *Proceedings of the Ocean Drilling Program: Scientific Results Vol. 182*. ODP, Texas A&M University, College Station, TX.

Ferdelman, T. G.; Kano, A.; Williams, T.; Henriet, J.-P.; and the Expedition 307 Scientists (2006). *Expedition 307 Summary: Proceedings of the Integrated Ocean Drilling Program, Vol. 307*, doi:10.2204/iodp.proc.307.101.2006.

Fisher, C. R.; MacDonald, I. R.; Macko, S.; Nelson, K.; Sassen, R.; Joy, M.; Carney, R.; McMullin, E. (1998). Colonization of gas hydrates by metazoans. *Conference Proceedings. Gas hydrates: Resources? Hazards? Origins? Geoscience '98, Keele University, UK, April 1998* (abstract).

Flemings, P. B.; Liu, X.; and Winters, W. J. (2003). Critical pressure and multiphase flow in Blake Ridge gas hydrates. *Geology*, **31**, 1057–1060.

Flügel, H. J.; and Langhof, I. (1983). A new hermaphroditic pogonophore from the Skagerrak. *Sarsia*, **68**, 131–138.

Forsberg, C. F.; Planke, S.; Tjelta, T. I.; Svanø, G.; Strout, J. M.; and Svensen, H. (2007). Formation of pockmarks in the Norwegian Channel. *Proceedings of the Sixth Int. Offshore Site Investigation and Geotechnics Conf., 11–13 September, London*, pp. 221–230.

240 References

Fosså, J. H.; and Alvsvåg, J. (2003). Kartlegging og overvåking av korallrev (Mapping and monitoring coral reefs). In: Sjøtun, K.; and Dahl, E. (eds), *Havets Miljø*, **2**, 62–67 [in Norwegian].

Fosså, J. H.; and Mortensen, P. B. (1998). Artsmangfoldet på Lophelia-korallrev og metoder for kartlegging og overvåkning (Biodiversity on Lophelia reefs and methods for mapping and monitoring). *Fisken og Havet*, **17**, 1–95 [in Norwegian].

Fosså, J. H.; Mortensen, P. B.; and Furevik, D. M. (2000). Lophelia-korallrev langs norskekysten: Forekomst og tilstand (Lophelia coral reefs along the Norwegian coast: Occurrence and conditions). *Fisken og Havet*, **2**, 94 [in Norwegian].

Foubert, A.; Maignien, L.; Beck, T.; Depreiter, D.; Blamart, D.; and Henriet, J.-P. (2005). Pen Duick Escarpment on the Moroccan Margin: A new mound lab? *Geosphere–Biosphere Coupling Processes: The TTR Interdisciplinary Approach towards Studies of the European and North African Margins*. IOC (Intergovernmental Oceanographic Commission, UNESCO) Workshop Report No. 197, p. 17.

Foubert, A.; Depreiter, D.; Beck, T.; Maignien, L.; Pannemans, B.; Frank, N.; Blamart, D.; and Henriet, J.-P. (2008). Carbonate mounds in a mud volcano province off north-west Morocco: Key to processes and controls. *Marine Geology*, **248**(1–2), 74–96.

Frederiksen, R.; Jensen, A.; and Westerberg, H. (1992). The distribution of the scleractinian coral *Lophelia pertusa* around the Faroe Islands and the relation to internal tidal mixing. *Sarsia*, **77**, 157–171.

Freiwald, A. (1995). *Deep-water Coral Reef Mounds on the Sula Ridge, mid-Norway Shelf*. Universität Bremen, Field Report, Cruise 24/95, R/V Victor Hensen, 13 pp.

Freiwald, A. (2002). Reef-forming cold-water corals. In: Wefer, G.; Billett, D.; Hebbeln, D.; Jørgensen, B. B.; Schlüter, M.; and van Weering, T. (eds), *Ocean Margin Systems*. Springer-Verlag, Berlin, pp. 365–385.

Freiwald, A.; and Roberts, J. M. (2005). *Cold-water Corals and Ecosystems*. Springer-Verlag, Berlin, 1200 pp.

Freiwald, A.; Henrich, R.; and Pätzold, J. (1997). Anatomy of a deep-water coral reef mound from Stjernsund, west Finnmark, northern Norway. In: James, N. P.; and Clarke, J. A. D. (eds), *Cool-Water Carbonates*. Society of Sedimentary Geologists (SEPM), Special Publication No. 56, pp. 140–161.

Freiwald, A.; Wilson, J. B.; and Henrich, R. (1999). Grounding Pleistocene icebergs shape Recent deep-water coral reefs. *Sedimentary Geology*, **125**, 1–8.

Freiwald, A. (2002). Reef-forming cold-water corals. In: G. Wefer, D. Billett, D. Hebbeln, B.B. Jørgensen, M., Schlüter, T. Van Weering (eds), *Ocean Margin Systems*. Springer-Verlag, Berlin, pp. 365–385.

Freiwald, A.; Hühnerbach, V.; Lindberg, B.; Wilson, J. B.; and Campbell, J. (2002). The Sula Reef Complex, Norwegian Shelf. *Facies*, **47**, 179–200.

Fryer, P.; Saboda, K. L.; Johnson, L. E.; Mackay, M. E.; Moore, G. F.; and Stoffers, P. (1990). Conical Seamount: SeaMARC II, Alvin submersible, and seismic-reflection studies. *Proceedings of the Ocean Drilling Programme: Initial Report No. 125*, pp. 69–80.

Früh-Green, G. L.; Connolly, A. D.; Plas, A.; Kelley, D. S.; and Grobéty, B. (2004). Serpentinization of oceanic peridotites: Implications for geochemical cycles and biological activity. *The Subseafloor Biosphere at Mid-ocean Ridges*. Geophysical Monograph Series No. 144, American Geophysical Union, Washington, DC, pp. 119–136.

Gallagher, J. W.; Hovland, M.; Amaliksen, K. G.; Braaten, A. M.; Jacobsen, T.; and Granli, J. R. (1991). *Shallow Gas on Haltenbanken*. Statoil Open Report No. 91023027.

Games, K. P. (2001). Evidence of shallow gas above the Connemara oil accumulation, Block 26/28, Porcupine Basin. In: P. M. Shannon, P. D. W. Haughton, and D. V. Corcoran (eds), *The Petroleum Exploration of Ireland's Offshore Basins*. Geological Society, London, Special Publication No. 188, pp. 361–373.

Gardner, J. M. (2001). Mud volcanoes revealed and sampled on the Western Moroccan continental margin. *Geophysical Research Letters*, **28**, 339–342.

Gass, S. E.; and Roberts, J. M. (2006). The occurrence of the cold-water *Lophelia pertusa* (scleractinian) on oil and gas platforms in the North Sea: Colony growth, recruitment and environmental controls on distribution. *Mar. Pollution Bull.*, **52**, 549–559.

Gay, A.; Lopez, M.; Berndt, C.; and Séranne, M. (2007). Geological control on focused fluid flow associated with seafloor seeps in the Lower Congo Basin. *Marine Geology*, **244**, 68–92.

Gleick, E. (1997). Life in the abyss: The new age of discovery. *Time*, Special Issue, 35–37.

Glenn, K. (2002). Coral reefs and hydrcarbon seeps. *Ausgeonews*, **68**, 4–7.

Grant, A. C.; and Jackson, A. E. (1994). *Seabed Mounds on Orphan Knoll: Vestiges of a Hydrocarbon Basin off Eastern Canada?* Geological Survey of Canada, Contrib. 22694.

Godø, K. (2007). Snublet over gass-indikasjoner (Stumbled across gas indications). *Harstad Tidene* (a local newspaper), October 15th, p. 1 [in Norwegian].

Goreau, N. I.; and Hayes, R. L. (1977). Nucleation catalysis in coral skeletogenesis. *Proceedings of the Third International Coral Reef Symposium, Miami*, p. 4.

Gravdal, A.; Haflidason, H.; and Evans, D. (2003). Seabed and subsurface features on the southern Vøring Plateau and northern Storegga slide escarpment. In: Mienert, J.; and Weaver, P. (eds), *European Margin Sediment Dynamics*. Springer-Verlag, Berlin, pp. 111–117.

Grousset, F. E.; Joron, J. L.; Latouche, C.; Treuil, M.; Maillet, N.; Faurgères, J. C.; Gonthier, E. (1998). Mediterranean outflow through the Strait of Gibraltar since 18,000 years B.P.: Mineralogical and geochemical arguments. *Geo-Marine Letters*, **8**, 25–34.

Guinotte, J. M.; Orr, J.; Cairns, S.; Freiwald, A.; Morgan, L.; and George, R. (2006). Will human induced changes in seawater chemistry alter the distribution of deep-sea scleractinian corals? *Front Ecol. Environ.*, **4**, 141–146.

Gunnerus, J. E. (1768). Om nogle Norske coraller. *Kongelige Norske Videnskabers Selskabs Skrifter*, **4**, 38–73.

Han, X.; Suess, E.; Huang, Y.; Wu, N.; Bohrmann, G.; Su, X.; Eisenhauer, A.; Rehder, G.; and Fang, Y. (2008). Jiulong methane reef: Microbial mediation of seep carbonates in the South China Sea. *Marine Geology*, doi:10.1016/j.margeo.2007.11.012.

Harris, R. N.; Fisher, A. T.; and Chapman, D. S. (2002). Fluid flow through seamounts: Patterns of flow and implications for global fluid flux and heat loss. *EOS Trans. Am. Geophys. Union*, Fall Meeting Supplement, Abstract T52F-05.

Haymon, R. M.; Fornari, D. J.; Von Damm, K. L.; Lilley, M. D.; Perfit, M. R.; Edmon, J. M.; Shanks III, W. C.; Lutz, R. A.; Grebmeier, J. M.; Carbotte, S. *et al.* (1993). Volcanic eruption of the mid-ocean ridge along the East Pacific Rise crest at 9°45–52′N: Direct submersible observations of seafloor phenomena associated with an eruption event in April, 1991. *Earth Planetary Science Letters*, **119**, 85–101.

Helland-Hansen, B. (1916). Havforskning (Ocean Research). In: Swanstrøm, L. (ed,), *Illustrert verdensgeografi* (Illustrated world geography), Vol. 2. Alb. Cammermeyers Forlag, Kristiania, pp. 789–820.

Henriet, J.-P. (1999). *Proposal: Porcupine- Belgica Carbonate Mud Mounds and Cold Water Reefs, Deep Biosphere–Geosphere Coupling*. Renard Centre of Marine Geology (RCMG), Universiteit Ghent, Open Report BOF99/GOA/009, 80 pp.

Henriet, J.-P.; and Dullo, C. W. (2005). *Atlantic Mound Drilling 2: Morocco Margin*. IODP 673 Proposal.

Henriet, J.-P.; De Mol, B.; Pillen, S.; Vanneste, M.; Van Rooij, D.; Versteeg, W.; Croker, P. F.; Shannon, P. M.; Unninthan, V.; Bouriak, S.; Chachkine, P.; and Porcupine–Belgica Shipboard Party (1998). Gas hydrate crystals may help build reefs. *Nature*, **391**, 648–649.

Henriet, J.-P.; De Mol, B.; Vanneste, M.; Huvenne, V.; Van Rooij, D.; and Porcupine–Belgica Shipboard Parties, 1997–1999 (2001). Carbonate mounds and slope failures in the Porcupine Basin: A development model involving past fluid venting. In: Shannon, P. M.; Haughton, P.; and Corconan, D. (eds), *The Petroleum Exploration of Ireland's Offshore Basins*. Geological Society, London, pp. 375–383.

Henriet, J.-P.; Foubert, A.; Maignien, L.; and the Proposals 573-Light, 673-Pre2 Teams, Expedition 307 Shipboard Party (2006). *Atlantic Carbonate Mounds: Challenges for Ocean Drilling, from Porcupine to the Moroccan Margin*. IOC (Intergovernmental Oceanographic Commission, UNESCO) Workshop Report No. 201, pp. 24–25.

Hirayama, H.; Sunamura, M.; Takai, K.; Nunoura, T.; Noguchi, T.; Oida, H.; Furushima, Y.; Yamamoto, H.; Oomori, T.; Horikoshi, K. (2007). Culture-dependent and independent characterization of microbial communities associated with a shallow submarine hydrothermal system occurring within a coral reef off Taketomi Island, Japan. *Applied and Environmental Microbiology*, **73**(23), 7642–7656.

Holm, N. G.; and Charlou, J. L. (2001). Initial indications of abiotic formation of hydrocarbons in the Rainbow ultramafic hydrothermal system, Mid-Atlantic Ridge. *Earth Planet. Sci. Lett.*, **191**, 1–8.

Hovland, M. (1981). Characteristics of pockmarks in the Norwegian Trench. *Marine Geology*, **39**, 103–117.

Hovland, M. (1988). Organisms: The only cause of scattering layers? *EOS Trans. Am. Geophys. Union*, **69**(31), 760.

Hovland, M. (1990a). Suspected gas-associated clay diapirism on the seabed off Mid Norway. *Mar. and Petrol. Geol.*, **7**, 267–276.

Hovland, M. (1990b). Do carbonate reefs form due to fluid seepage? *Terra Nova*, **2**, 8–18.

Hovland, M. (1991). Large pockmarks, gas-charged sediments and possible clay diapirs in the Skagerrak. *Marine and Petroleum Geology*, **8**, 311–316.

Hovland, M. (1992). Hydrocarbon seeps in northern marine waters: Their occurrence and effects. *Palaios*, **7**, 376–382.

Hovland, M. (2002). On the self-sealing nature of marine seeps. *Continental Shelf Res.*, **22**, 2287–2394.

Hovland, M. (2005). Pockmark-associated coral reefs at the Kristin field off Mid-Norway. In: Freiwald, A.; and Roberts, J. M. (eds), *Cold-water Corals and Ecosystems*. Springer-Verlag, Berlin, pp. 623–632.

Hovland, M. (2007). Discovery of prolific natural methane seeps at Gullfaks, northern North Sea. *Geo-Marine Letters*, doi:10.1007/s00367-007-0070-6.

Hovland, M.; and Croker, P. F. (1993). Fault-associated seabed mounds in the Porcupine Basin, offshore Ireland. *Proceedings of the 55th EAEG Annual Meeting, Stavanger, Norway, June 1993* (expanded abstract).

Hovland, M.; and Curzi, P. (1989). Gas seepage and assumed mud diapirism in the Italian Central Adriatic Sea, *Marine and Petroleum Geology*, **6**, 161–169.

Hovland, M.; and Indreeide, A. (1980). Mapping the seafloor offshore west coast Norway. *Hydrographic Review*, **100**, 154–162.

Hovland, M.; and Judd, A. G. (1988). *Seabed Pockmarks and Seepages: Impact on Geology, Biology and the Marine Environment*. Graham & Trotman, London, 293 pp.

Hovland, M.; and Mortensen, P. B. (1999). *Norske korallrev og prosesser i havbunnen* (Norwegian coral reefs and processes in the seafloor). John Grieg Forlag, Bergen, Norway, 160 pp. [in Norwegian with English summary].

Hovland, M.; and Risk, M. (2003). Do Norwegian deep-water coral reefs rely on seeping fluids? *Marine Geology*, **198**, 83–96.

Hovland, M.; and Svensen, H. (2006). Submarine pingoes: Indicators of shallow gas hydrates in a pockmark at Nyegga, Norwegian Sea. *Marine Geology*, **228**, 15–23.

Hovland, M.; and Thomsen, E. (1997). Cold-water corals: Are they hydrocarbon seep related? *Marine Geology*, **137**, 159–164.

Hovland, M.; Talbot, M.; Qvale, H.; Olaussen, S.; and Aasberg, L. (1987). Methane-related carbonate cements in pockmarks of the North Sea. *J. Sed. Petrol.*, **57**, 881–892.

Hovland, M.; Croker, P.; and Martin, M. (1994a). Fault-associated seabed mounds (carbonate knolls?) off western Ireland and north-west Australia. *Mar. and Petrol. Geol.*, **11**, 232–246.

Hovland, M.; Farestveit, R.; and Mortensen, P. B. (1994b). Large cold-water coral reefs off mid-Norway: A problem for pipelaying? *Proc. Oceanology Int., Brighton*, **3**, 35–40.

Hovland, M.; Mortensen, P. B.; Brattegard, T.; Strass, P.; and Rokoengen, K. (1998). Ahermatypic coral banks off mid-Norway: Evidence for a link with seepage of light hydrocarbons. *Palaios*, **13**, 189–200.

Hovland, M.; Vasshus, S.; Indreeide, A.; Austdal, L.; and Nilsen, Ä. (2002). Mapping and imaging deep-sea coral reefs off Norway, 1982–2000. *Hydrobiologia*, **471**, 13–17.

Hovland, M.; Svensen, H.; Forsberg, C. F.; Johansen, H.; Fichler, C.; Fosså, J. H.; Jonsson, R.; and Rueslåtten, H. (2005). Complex pockmarks with carbonate ridges off mid-Norway: Products of sediment degassing. *Marine Geology*, **218**, 191–206.

Hovland, M.; Fichler, C.; Rueslåtten, H.; and Johnsen, H. K. (2006). Deep-rooted piercement structures in deep sedimentary basins: Manifestations of supercritical water generation at depth? *Journal of Geochemical Exploration*, **89**, 157–160.

Hurtley, S.; and Szuromi, P. (2007). Rethinking coral composition. *Science*, **318**, 13.

Huvenne, V. A. I.; Blondel, Ph.; and Henriet, J.-P. (2002). Textural analyses of sidescan sonar imagery from two mound provinces in the Porcupine Seabight. *Marine Geology*, **189**, 323–341.

Huvenne, V. A. I.; De Mol, B.; and Henriet, J.-P. (2003). A 3D seismic study of the morphology and spatial distribution of buried coral banks in the Porcupine Basin, SW of Ireland. *Marine Geology*, **198**, 5–25.

Huvenne, V. A. I.; Beyer, A.; De Haas, H.; Dekindt, K.; Henriet, J.-P.; Kozachenko, M.; Olu-Le Roy, K.; Wheeler, A. J.; and TOBI/Pelagia 197 and CARACOLE participants (2005). The seabed appearance of different coral bank provinces in the Porcupine Seabight, NE Atlantic: Results from sidescan sonar and ROV seabed mapping. In: Freiwald, A.; and Roberts, M. (eds), *Cold-water Corals and Ecosystems*. Springer-Verlag, Heidelberg, pp. 535–569.

Ivanov, M.; Westbrook, G. K.; Blinova, V.; Kozlova, E.; Mazzini, A.; Nouzé, H.; and Minshull, T. A. (2007). First sampling of gas hydrates from the Vøring Plateau. *EOS Trans. Am. Geophys. Union*, **88**(19), 209–212.

Jacobsen, S. D.; and van der Lee, S. (2006). *Earth's Deep Water Cycle*. American Geophysical Union, Washington, DC, Geophys. Monograph No. 168, pp. vii–viii.

James, N. P. *et al.* (2000). Quaternary bryozoan reef mounds in cool-water, upper slope environments: Great Australian Bight. *Geology*, **28**, 647–650.

James, N. P. *et al.* (2004). Origin of late Pleistocene bryozoan reef-mounds, Great Australian Bight. *J. Sed. Research*, **74**, 20–48.

Jensen, P.; Aagaard, I.; Burke Jr, R. A.; Dando, P. R.; Jørgensen, N. O.; Kuijpers, A.; Laier, T.; O'Hara, S. C. M.; and Schmaljohann, R. (1992). "Bubbling reefs" in the Kattegat: Submarine landscapes of carbonate-cemented rocks support a diverse ecosystem at methane seeps. *Marine Ecology Progress Series*, **83**, 103–112.

Jensen, S.; Neufeld, J. D.; Birkeland, N.-K.; Hovland, M.; and Murrell, J. C. (submitted 2008a). Diversity of bacteria associated with the sponge *Desmacidon* sp. and the ambient environment in a deepwater coral reef at the Kristin field off the coast of Norway. *Deep-Sea Research I*.

Jensen, S.; Neufeld, J. D.; Birkeland, N.-K.; Hovland, M.; Murrell, J. C. (submitted 2008b). Methane assimilation by *Methylomicrobium* in deepwater coral reefs off Norway. *FEMS, Microbiology Ecology*.

Jerosch, K.; Schlüter, M.; Foucher, J.-P.; Allais, A.-G.; Klages, M.; and Edy, C. (2007). Spatial distribution of mud flows, chemoautotrophic communities, and biogeochemical habitats at Håkon Mosby Mud Volcano. *Marine Geology*, **243**, 1–17.

Jones, B.; Renaut, R. W.; and Rosen, M. R. (1997). Vertical zonation of biota in microstromatolites associated with hot springs, North Island, New Zealand. *Palaios*, **12**, 220–236.

Jonsson, L.; and Lundälv, T. (2001). Detailed mapping of a shallow occurrence of *Lophelia pertusa* (L.) in the Kosterfjord, Sweden. In: Willison, J. H. M.; Hall, J.; Gass, S. E.; Kenchington, E. L. R.;Butler, M.; and Doherty, P. *Proceedings of the First International Symp. Deep-Sea Corals*. Ecology Action Centre, Halifax, Nova Scotia, Canada, 210 pp.

Jørgensen, B. B. (2007). The deep subseafloor biosphere: Discovering the largest living community on Earth. *DRILLS: Integrated Ocean Drilling Program Brochure*. IODP, Washington, DC, 2 pp.

Jørgensen, N. O. (1992). Methane-derived carbonate cementation of marine sediments from the Kattegat, Denmark: Geochemical and geological evidence. *Marine Geology*, **103**, 1–13.

Judd, A. G.; and Hovland, M. (1992). The evidence of shallow gas in marine sediments. *Continental Shelf Res.*, **12**(10), 1081–1096.

Judd, A. G.; and Hovland, M. (2007). *Submarine Fluid Flow, the Impact on Geology, Biology, and the Marine Environment*. Cambridge University Press, Cambridge, UK, 475 pp.

Jung, W.-Y.; Vogt, P.; Haflidason, H.; Parsons, B. (2001). US Navy submarine NR-1 dives in the upper Storegga slide area, Norwegian margin. *European Union of Geologists 11th Conf. Proceedings* (Session RCM4) (abstract).

Karl, D. M. (1995). Ecology of free-living, hydrothermal vent microbial communities. In: D.M. Karl (ed.), *The Microbiology of Deep-Sea Hydrothermal Vents*. CRC Press, New York, pp. 35–124.

Kamenev, G. M.; Nadtochy, V. A.; and Kuznetsov, A. P. (2001). *Conchele bisecta* (Conrad, 1849) (Bivalvia: Thyasiridae) from cold-water methane-rich areas of the Sea of Okhotsk. *The Veliger*, **44**, 84–94.

Kasting, J. F.; and Catling, D. (2003). Evolution of a habitable planet. *Annu. Rev. Astron. Astrophys.*, **41**, 429–463.

Kaufmann, B. (1998). Diagenesis of middle Devonian carbonate mounds Mader basin (Eastern Anti-Atlas, Morocco). *Journal of Sedimentary Research*, **67**, 945–956.

Kelley, D. S.; Karson, J. A.; Blackman, D. K.; Früh-Green, G. L.; Butterfield, D. A.; Lilley, M. D.; Olson, E. J.; Schrenk, M. O.; Roe, K. K.; Lebon, Q. T. *et al.* (2001). An off-axis hydrothermal vent field near the Mid-Atlantic Ridge at 30°N. *Nature*, **412**, 145–149.

Kempe, S.; Kazmierczak, J.; Landmann, G.; Konuk, T.; Reimer A.; and Lipp, A. (1991). Largest known microbialites discovered in Lake Van, Turkey. *Nature*, **349**, 605–608.

Kennicut, M. C.; Brooks, J. M.; Bidigare, R. R.; Fay, R.; Wade, T. L.; and McDonald, T. J. (1985). Vent type taxa in a hydrocarbon seep region on the Louisiana slope. *Nature*, **317**, 351–353.

Kenyon, N. H.; Akhmetzhanov, A. M.; Wheeler, A. J.; van Weering, T. C. E.; de Haas, H.; and Ivanov, M. K. (2003). Giant carbonate mud mounds in the southern Rockall Trough. *Marine Geology*, **195**, 5–30.

King Jr, D. T. (1986). Waulsortian-type buildups and resedimented (carbonate-turbidite) facies, early Mississippian Burlington shelf, central Missouri. *J. Sediment. Petrol.*, **56**, 471–479.

King, L. H.; and MacLean, B. (1970) Pockmarks on the Scotian Shelf. *Geol. Soc. Am. Bull.*, **81**, 3142–3148.

Kipfer, R.; Aeschebach-Hertig, W.; Baur, H.; Hofer, M.; Imboden, D. M.; and Signer, P. (1994). Injection of mantle type helium into Lake Van (Turkey): The clue for quantifying deep water renewal. *Earth & Planetary Science Letters*, **125**, 357–370.

Kohout, F. A. (1967). Ground-water flow and the geothermal regime of the Florida Plateau. *Transactions of the Gulf Coast Geological Society*, **17**, 339–354.

Konhauser, K. (2007). *Introduction to Geomicrobiology*. Blackwell Publishing, Oxford, UK, 425 pp.

Kuhn, T. S. (1970). *The Structure of Scientific Revolutions*. University of Chicago Press, Chicago, 210 pp.

Laberg, J. S.; and Vorren, T .O. (1993). A Late Pleistocene submarine slide on the Bear Island Trough Mouth Fan. *Geo-Marine Letters*, **13**, 227–234.

Laberg, J. S.; Berndt, C.; and Bünz, S. (2000). *Marine Geological and Geophysical Cruise in the Norwegian–Greenland Sea*. Cruise Report, *R.V. Jan Mayen*, 3–18 July. Department of Geology, University of Tromsø, Norway. 24 pp.

Laval, B.; Cady, S. L.; Pollack, J. C.; McKay, C. P.; Bird, J. S.; Grotzinger, J. P.; Ford, D. C.; and Bohm, H. R. (2000). Modern freshwater microbialite analogues for ancient dendritic reef structures. *Nature*, **407**, 626–629.

Leake, J. (1997). Oil team finds giant British coral reef. *Sunday Times*, 22nd June, p. 16.

LeDanois, E. (1948). *Les profondeurs de la mer*. Payot, Paris, 303 pp.

Le Page, J. M. (1998). Capital humain, Chômage et productivité. *Le capital Humain (OCDE), dimensions économiques et managériales*. Presses universitaires d'Angers.

Lien, R. (1983). Iceberg scouring on the Norwegian continental shelf. *Proc. Offshore Technol. Conf. (OTC), Houston, Texas*, **15**, 41–45.

Lilley, M. D.; Butterfield, D. A.; Olson, E. J.; Lupton, J. E.; Macko, S. A.; and McDuff, R. E., 1993. Anomalous CH_4 and NH_4^+ concentrations at an unsedimented mid-ocean-ridge hydrothermal system. *Nature*, **364**, 45–47.

Lindberg, B. (2004). Cold-water coral reefs on the Norwegian shelf: Acoustic signature, geological, geomorphological and environmental setting. PhD thesis, Department of Geology, University of Tromsø, Norway.

Lindberg, B.; and Mienert, J. (2005). Sedimentological and geochemical environment of the Fugløy Reef off Norway. In: Freiwald, A.; and Roberts, J. M. (eds), *Cold-water Corals and Ecosystems*. Springer-Verlag, Berlin, pp. 633–650.

Lindberg, B.; Christensen, O.; and Fosså, J. H. (2004). The geologic and morphologic setting of the Træna reef area based on high resolution acoustic data. In: Lindberg, B. (ed.), *Cold-water Coral Reefs of the Norwegian Shelf: Acoustic Signature, Geological, Geomorphological and Environmental Setting*. Department of Geology, University of Tromsø, Norway, Article V, pp. 1–18.

Lindberg, B.; Berndt, C.; and Mienert, J. (2007). The Fugløy Reef at 70°N: Acoustic signature, geologic, geomorphologic and oceanographic setting. *Int. J. Earth Sci. (Geol. Rundsch.)*, **96**, 201–213.

Linné, C. (1758). *Systemae naturae per Regna tria naturae, secundum classes, ordines, genera, species*, Tomus I. Regnum animale, Tenth Edition, Stockholm.

Lonsdale, P. (1977). Clustering of suspension-feeding macrobenthos near abyssal hydrothermal vents at oceanic spreading centres. *Deep-Sea Research*, **24**, 857–863.

Lundälv, T.; and Jonsson, L. (2000). *Inventering av Koster-Väderöområder med ROV-teknik*. Naturvårdsverket, Rapport 5079, Stockholm, Sweden.

Lundälv, T.; and Jonsson, L. (2003). Mapping of deep-water corals and fishery impacts in the north-east Skagerrak, using acoustical and ROV survey techniques. *Sixth Underwater Science Symposium*. Society for Underwater Technology, London (abstract).

MacDonald, I. (1990). Spatial ecology of hydrocarbon seeps in the northern Gulf of Mexico. Thesis. Texas A&M University, College Station, TX, pp. 135.

MacDonald, I.; and Leifer, I. (2002). Constraining rates of carbon flux from natural seeps on the northern Gulf of Mexico slope. *VIIth International Conference of Gas in Marine Sediments, Baku, Azerbaijan, October*, p. 119 (abstract).

MacDonald, I. R.; Boland, G. S.; Baker, J. S.; Brooks, J. M.; Kennicutt II, M. C.; and Bididgare, R. R. (1989). Gulf of Mexico hydrocarbon seep communities. II. Spatial distribution of seep organisms and hydrocarbons at Bush Hill. *Marine Biology*, **101**, 235–247.

MacDonald, I.; Butman, D. B.; Sager, W. W.; Peccini, M. B.; and Guinasso, N. L. (2000). Pulsed flow from a Gulf of Mexico mud volcano. *Geology*, **28**, 907–910.

MacDonald, I. R.; Leifer, I.; Sassen, R.; Stine, P.; Mitchell, R.; and Guinasso Jr, N. R. (2002). Transfer of hydrocarbons from natural seeps to the water column and atmosphere. *Geofluids*, **2**, 95–107.

Maignien, L.; Wouters, N.; and Henriet, J.-P. (2006). *Biogeochemistry of Carbonate Mounds from the Pen Duick Escarpment, Gulf of Cadiz*. IOC (Intergovernmental Oceanographic Commission, UNESCO) Workshop Report No. 201, pp. 31–32.

Maldonado, A.; Somoza, L.; and Pallarés, L. (1999). The betic orogen and the Iberian–African boundary in the Gulf of Cadiz: Geological evolution. *Marine Geology*, **155**, 9–43.

Manheim, F. T. (1967). Evidence for submarine discharge of water on the Atlantic continental slope of the southern United States, and suggestions for further research. *Transactions of the New York Academy of Sciences (Series II)*, **29**, 839–853.

Masson, D. G.; Bett, B. J.; Billett, D. S. M.; Jacobs, C. L.; Wheeler, A. J.; and Wynn, R. B. (2003). The origin of deep-water, coral-topped mounds in the northern Rockall Trough, Northeast Atlantic. *Marine Geology*, **194**, 150–180.

Mazurenko, L.; Soloviev, V.; Ivanov, M.; Pinheiro, L.; and Gardner, J. (2001). Geochemical features of gas hydrate-forming fluids of the Gulf of Cadiz. In: Akhamanov, G.; and Suzyumov, A. (eds), *Geological Processes on Deep-water European Margins: International Oceanographic Commission Workshop Report No 175 on the International Conference and Ninth Post-cruise Meeting of the Training Through Research Programme, Moscow-Mozhenka, Russia, 28 January–2 February 2001*. UNESCO, Paris, p. 50.

Mazzini, A.; Svensen, H.; Hovland, M.; and Planke, S. (2006). Comparison of strikingly different authigenic carbonates in a Nyegga complex pockmark, Norwegian Sea. *Marine Geology*, **231**, 89–102.

Medialdea, T.; Vegas, R.; Somoza, L.; Vázquez, J. T.; Maldonado, A.; Díaz-del-Río, V.; Maestro, A.; Córdoba, D.; and Fernández-Puga, M.C. (2004). Structure and evolution of the "Olistostrome" complex of the Gibraltar Arc in the Gulf of Cádiz (eastern Central Atlantic): Evidence from two long seismic cross-sections. *Marine Geology*, **209**, 173–198.

Menard, H. W. (1964). *Marine Geology of the Pacific*. McGraw-Hill, New York.

Michaelis, W. (2000). *Boreale Schwämme als marine Naturstoffquelle (BOSMAN)*. Technical Cruise Report Pos 254, University of Hamburg, 52 pp.

Mienert, J.; Posewang, J.; and Baumann, M. (1998). Gas hydrates along the north-eastern Atlantic Margin: Possible hydrate bound margin instabilities and possible release of methane. In: Henriet, J.-P.; and Mienert, J. (eds), *Gas Hydrates: Relevance to World Margin Stability and Climate Change*. Geological Society, London, Special Publication No. 137, pp. 275–291.

Mikkelsen, N.; Erlenkauser, H.; Killingley, J. S.; and Berger, W. H. (1982). Norwegian corals: Radiocarbon and stable isotopes in *Lophelia pertusa*. *Boreas*, **5**, 163–171.

Milkov, A. V. (2000). Worldwide distribution of submarine mud volcanoes and associated gas hydrates. *Marine Geology*, **167**, 29–42.

Milkov, A. V.; Vogt, P. R.; Crane, K.; Lein, A. Y.; Sassen, R.; and Cherkashev, G. A. (2004). Geological, geochemical, and microbial processes at the hydrate-bearing Håkon Mosby mud volcano: A review. *Chemical Geology*, **205**, 347–366.

Miller, J. (1986). Facies relationships and diagenesis in Waulsortian mudmounds from the lower Carboniferous of Ireland and N. England. In: Schroeder, J. H.; and Purser, B. H. (eds), *Reef Diagenesis*. Springer-Verlag, Berlin, pp. 311–335.

Minshull, T. A.; Singh, S. C.; and Westbrook, G. K. (1994). Seismic velocity structure at a gas hydrate reflector, offshore western Colombia, from full waveform inversion. *Journal of Geophysical Research*, **99**, 4715–4734.

Moore, D. R.; and Bullis, H. R. (1960). A deep-water coral reef in the Gulf of Mexico. *Bulletin of Marine Sciences*, **10**, 125–128.

Mortensen, P. B.; and Fosså, J. H. (2001). Coral reefs and other bottom habitats on the Tautra Ridge in Trondheimsfjorden. *Fisken og Havet*, **7**, 43.

Mortensen, P. B.; and Fosså, J. H. (2006). Species diversity and spatial distribution of invertebrates on *Lophelia* reefs in Norway. *Proceedings of the Tenth International Coral Reef Symp., Okinawa, Japan*, pp. 1849–1868.

Mortensen, P. B.; and Rapp, H. T. (1998). Oxygen- and carbon isotope ratios related to growth line patterns in skeletons of *Lophelia pertusa* (L) (Anthozoa: Scleractinia): Implications for determination of linear extention rates. *Sarsia*, **83**, 433–446.

Mortensen, P. B.; Hovland, M.; Brattegard, T.; and Farestveit, R. (1995). Deep water bioherms of the scleractinian coral *Lophelia pertusa* (L.) at 64°N on the Norwegian shelf: Structure and associated megafauna. *Sarsia*, **80**, 145–158.

Mortensen, P. B.; Hovland, M. T.; Fosså, J. H.; and Furevik, D. M. (2001). Distribution, abundance and size of *Lophelia pertusa* coral reefs in mid-Norway in relation to seabed characteristics. *J. Mar. Biol. Ass. UK*, **81**, 581–597.

Mortensen, P. B.; Buhl-Mortensen, L.; and Gordon Jr, D. C. (2006). Distribution of deep-water corals in Atlantic Canada. *Proceedings of Tenth Int. Coral Reef Symp., Okinawa, Japan*, pp. 1849–1868.

Mortensen, P. B.; Buhl-Mortensen, L.; Gebruk, A. V.; and Krylova, E. M. (2008). Occurrence of deep-water corals on the Mid-Atlantic Ridge based on MAR-ECO data. *Deep-Sea Res. II*, **55**, 142–152.

Morton, B.; and Morton, J. (1983). *The Sea Shore Ecology of Hong Kong*. Hong Kong University Press, 350 pp.

Mukhopadhyay, R.; Iyer, S. D.; and Ghosh, A. K. (2002). The Indian Ocean Nodule Field: Petrotectonic evolution and ferromanganese deposits. *Earth-Science Reviews*, **60**, 67–130.

References

Mullins, J. (2007). Interview: Why mathematics is beautiful. *New Scientist*, **2613**, 21 July 2007, 34.

Mullins, H. T.; Newton, C. R.; Heath, K.; and Vanburen, H. M. (1981). Moderen deep-water coral mounds north of Little Bahama bank: Criteria for recognition of deep-water coral bioherms in the rock record. *J. Sed. Petrol.*, **51**, 999–1013.

Møller, M. M.; Nielsen, L. P.; and Jørgensen, B. B. (1985). Oxygen responses and mat formation by *Beggiatoa* spp. *Applied Environmental Microbiology*, **50**, 373–382.

Naeth, J.; di Primio, R.; Horsfield, B.; Schaefer, R. G.; Shannon, P. M.; Baileey, W. R.; and Henriet, J.-P. (2005). Hydrocarbon seepage and carbonate mound formation: A basin modelling study from the Porcupine Basin (offshore Ireland). *Journal of Petroleum Geology*, **28**, 147–166.

Naganuma, T.; Otsuki, A.; and Seki, H. (1989). Abundance and growth rate of bacterioplankton community in hydrothermal vent plumes of the North Fiji Basin. *Deep-Sea Research*, **36**, 1379–1390.

Nelson, D. C.; and Fisher, C. R. (1995). Chemoautotrophic and methanotrophic endosymbiotic bacteria at deep-sea vents and seeps. In: Karl, D. M. (ed.), *The Microbiology of Deep-Sea Hydrothermal Vents*. CRC Press, Boca Raton, FL, pp. 125–167.

Neumann, A. C.; Koefoed, J. W.; and Keller, G. H. (1977). Lithoherms in the Straits of Florida. Geology, 5, 4-10.

Newman, K. R.; Cormier, M.-H.; Weissel, J. K.; Driscoll, N. W.; Kastner, M.; Solomon, E. A.; Robinson, G.; Hill, J. C.; Singh, H.; Camilli, R.; and Eustice, R. (2008). Active methane venting observed at giant seafloor pockmarks along the U.S. mid-Atlantic shelf break. *Earth and Planetary Science Letters*, **267**(1/2), 341–352.

Newton, C.R., Mullins, H.T., Gardulski, A.F., 1987. Coral mounds on the west Florida slope: Unanswered questions regarding the development of deep-water banks. *Palaios*, **2**, 59–67.

Niemann, H.; Elvert, M.; Hovland, M.; Orcutt, B.; Judd, A. G.; Suck, I.; Gutt, J.; Joye, S.; Damm, E.; Finster, K.; and Boetius, A. (2005). Methane emission and consumption at a North Sea gas seep (Tommeliten area). *Biogeosciences*, **2**, 335–351.

Nilsson, P. (1997). *Biological Values of the Kosterfjord Area: A Compilation and Analysis of Present Knowledge*. Naturvårdsverket, Rapport 4749, Stockholm, Sweden.

NOAA (2002). The state of coral reef ecosystems of the United States and Pacific Freely Associated States. Available at *http://www.nccos.noaa.gov/doucuments/status_coralreef.pdf*, and also see *http://www.gulfbase.org/reef/view.php?rid=efgb*

Nordgaard, O. (1912). Et gammelt *Lophohelia*-rev I Trondheimsfjorden. *Det Kgl. Norske Videnskabers Selskabs Skrifter*, **3**, 1–8 (In Norwegian).

Nunn, J. A.; and Harris, N. B. (2007). Subsurface seepage of seawater across a barrier: A source of water and salt to peripheral salt basins. *Geol. Soc. Am. Bull.*, **119**(9), 1201–1217.

O'Brien, G. W.; Glenn, K.; Lawrence, G.; Williams, A.; Webster, M.; Burns, S.; and Cowley, R. (2002). Influence of hydrocarbon migration and seepage on benthic communities in the Timor Sea, Australia. *APPEA Journal*, pp. 1–14.

Okri, B. (1993). *Songs of Enchantment*. Vintage, 297 pp.

Orange, D. L.; Greene, H. G.; Reed, D.; Martin, J. B.; McHugh, C. M.; Ryan, W. B. F.; Maher, N.; Stakes, D.; and Barry, J. (1999). Widespread fluid expulsion on a translational continental margin: Mud volcanoes, fault zones, headless canyons, and organic-rich substrate in Monterey Bay, California. *Geological Society of America Bulletin*, **111**, 992–1009.

O'Reilly, B. M.; Readman, P. W.; Shannon, P. M.; and Jacob, A. W. B. (2003). A model for the development of a carbonate mound population in the Rockall Trough based on deep-towed sidescan sonar data. *Marine Geology*, **198**(1–2), 55–66.

Parkes, R. J.; Webster, G.; Cragg, B. A.; Weightman, A. J.; Newberry, C. J.; Ferdelman, T. G.; Kallmeyer, J.; Jørgensen, B. B.; Aiello, I. W.; Fry, J. C. (2005). Deeep-sub-seafloor prokaryotes stmulated at interfaces over geological time. *Nature*, **436**, 390–394.

Paull, C. K.; and Dillon, W. P. (1980). Structure, stratigraphy, and geologic history of Florida–Hatteras Shelf and inner Blake Plateau. *Am. Assoc. Petrol. Geol. Bull.*, **64**, 339–358.

Paull, C. K.; Hecker, B.; Commeau, R.; Freeman-Lynde, R. P.; Neumann, A. C.; Corso, W. P.; Golubic, S.; Hook, J. E.; Sike, E.; and Curray, J. (1984). Biological communities at the Florida escarpment resemble hydrothermal vent taxa. *Science*, **226**, 965–967.

Paull, C. K.; Neumann, A. C.; am Ende, B. A.; Ussler III, W.; and Rodriguez, N. M. (2000). Lithoherms on the Florida–Hatteras slope. *Marine Geology*, **166**, 83–101.

Pfannkuche, O.; and the shipboard participants (2006). *Maria S. Merian Cruise 1 Leg 3.* Report 10.05–18.05. Alfred Wegener Institute, Bremen.

Pimenov, N.; Savvichev, A.; Lein, A.; and Ivanov, M. (2000). *Microbiology of North Atlantic cold seeps*. VIth International Conference on Gas in Marine Sediments, September 5–9. VNIIOkeangeologia, St. Petersburg, Russia, p. 111 (abstract).

Pinheiro, L.; Ivanov, M.; Sautkin, A.; Akhmanov, G.; Magelhães, V. H.; Volkonskaya, A.; Monteiro, J. H.; Somoza, L.; Gardner, J.; Hamouni, N.; and Cunha, M. R. (2003). Mud volcanism in the Gulf of Cadiz: Results from the TTR-10 cruise. *Marine Geology*, **195**, 131–151.

Pires, D. D. O. (2007). The zooxanthellate coral fauna of Brazil. In: R.Y. George and D. Cairns (eds), *Conservation and Adaptive Management of Seamounts and Deep-sea Coral Ecosystems*. Rosenstiel School of Marine and Atmospheric Science, University of Miami, pp. 265–272.

Playford, P. E.; Cockbain, A. E.; and Low, O. H. (1976). *Geology of the Perth Basin, Western Australia*. Geological Survey of Western Australia, Bulletin No. 124.

Pontoppidan, E. (1752). *Det første forsøg paa Norges Naturlige historie* (The first attempt on Norway's Natural History). A. Linde, London.

Portman, C.; Andrews, J. E.; Rowe, P. J.; Leeder, M. R.; and Hoogewerff, J. (2005). Submarine-spring controlled calcification and growth of large *Rivularia* bioherms, Late Pleistocene (MIS 5e), Gulf of Corinth, Greece. *Sedimentology*, **52**, 441–465.

Pratt, B. R. (1982). Stromatolitic framework of carbonate mud-mounds. *J. Sediment. Petrol.*, **52**, 1203–1228.

Pray, L. C. (1965). Limestone clastic dikes in Mississippian bioherms, New Mexico. *Abstr. Geol. Soc. Am.*, Special Paper 82, 154–155.

Reed, J. K. (1980). Distribution and structure of deep-water *Oculina varicosa* coral reefs off central eastern Florida. *Bull. Mar. Sci.*, **30**(3), 667–677.

Reed, J. K. (1992). Submersible studies of deep-water *Oculina* and *Lophelia* coral banks off southeastern U.S.A. *Diving for Science*, 143–151.

Reitner, J.; Peckmann, J.; and Arp, G. (2001). Polygenetic types of seepage-related carbonate deposits and mounds. *European Union of Geologists 11th Conference Proceedings (Session RCM6)* (abstract).

Reston, T. J.; Parnell, J.; Stubenrauch, A.; Walker, I.; and Perez-Gussinye, M. (2001). Detachment faulting, mantle serpentinization, and serpentinite-mud volcanism beneath the Porcupine Basin, southwest of Ireland. *Geology*, **29**, 587–590.

Rezak, R.; Bright, T. J.; and McGail, D. W. (1985). *Reefs and Banks of the Northern Gulf of Mexico: Their Geological, Biological, and Physical Dynamics*. John Wiley & Sons, New York, 259 pp.

Rice, A. L.; Thurston, M. J.; and New, A. L. (1990). Dense aggregations of a hexactinellid sponge, *Pheronema carpenteri*, in the Porcupine Seabight (northeast Atlantic Ocean), and possible causes. *Prog. Oceanogr.*, **24**, 179–196.

Ridgwell, A.; Hargreaves, J. C.; Edwards, N. R.; Annan, J. D.; Lenton, T. M.; Marsh, R.; Yool, A.; and Watson, A. (2007). Marine geochemical data assimilation in an efficient Earth System Model of global biogeochemical cycling. *Biogeosciences*, **4**(1), 87–104.

Roberts, H. H.; and Carney, R. S. (1997). Evidence of episodic fluid, gas, and sediment venting on the northern Gulf of Mexico continental slope. *Economic Geology*, **92**, 863–879.

Roberts, J. M.; Long, D.; Wilson, J. B.; Mortensen, P. B.; and Gage, J. D. (2003). The cold-water coral *Lophelia pertusa* (Scleractinia) and enigmatic seabed mounds along the north-east Atlantic margin: Are they related? *Marine Pollution Bulletin*, **46**, 7–20.

Roberts, J. M.; Wheeler, A. J.; and Freiwald, A. (2006). Reefs of the deep: The biology and geology of cold-water coral ecosystems. *Science*, **312**, 543–547.

Roberts, M. J.; and Gass, S. (2005). Looking for *Lophelia. Planet Earth*, Spring 2005, pp. 26–27. Available at *www.nerc.ac.uk*

Rogers, A. D. (1994). The biology of seamounts. *Advances in Marine Biology*, **30**, 305–350.

Rokoengen, K.; and Østmo, S. R. (1985). *Shallow Geology off Fedje, Western Norway*. IKU Report No. 85.140, 23 pp.

Rokoengen, K.; Rise, L.; Bryn, P.; Frengstad, B.; Gustavsen, B.; Nygaard, E.; and Sættem, J. (1995). Upper Cenozoic stratigraphy on the Mid-Norwegian continental shelf. *Norsk Geologisk Tidsskrift*, **75**, 88–104.

Rollet, N.; Logan, G. A.; Ryan, G.; Judd, A. G.; Totterdell, J. M.; Glenn, K.; Jones, A. T.; Kroh, F.; Struckmeyer, H. I. M.; Kennard, J. M.; and Earl, K. L. (2007). Shallow gas and fluid migration in the northern Arafura Sea (Offshore Northern Australia). *Marine and Petroleum Geology*, doi:10.1016/j.marpetgeo.2007.07.010.

Roth, A. A. (1995). Fossil reefs and time. *Origins*, **22**(2), 86–104.

Rougerie, F.; and Waulthy, B. (1988). The endo-upwelling concept: A new paradigm for solving an old paradox. *Proceedings of the Sixth International Coral Reef Symposium, Australia*, Vol. 3, pp. 1–6.

Russell, M. J.; and Hall, A. J. (1997). The emergence of life from iron monosulfide bubbles at a submarine hydrothermal redox and pH front. *J. Geol. Soc. London*, **154**, 377–402.

Russell, M. J.; Hall, A. J.; Boyce, A. J.; and Fallick, A. E. (2005). On hydrothermal convection systems and the emergence of life. *Economic Geology*, **100**, 419–438.

Sabine, C. L.; Feely, R. A.; Gruber, N.; Key, R. M.; Lee, K.; Bullister, J. L.; Wanninkhof, R.; Wong, C. S.; Wallace, D. W. R.; Tilbrook, B.; Millero, F. J.; Peng, T. H.; Kozyr, A.; Ono, T.; and Rios, A. F. (2004). The oceanic sink for anthropogenic CO_2. *Science*, **305**, 367–371.

Salas, C.; and Woodside, J. (2002). *Lucinoma kazani* n. sp. (Mollusca: Bivalvia): Evidence of a living benthic community associated with a cold seep in the Eastern Mediterranean Sea. *Deep-Sea Research I*, **49**, 991–1005.

Sars, M. (1865). *Om de I Norge forekommende fossile dyrelevninger fra Quartærperioden*. Universitetets program for første halvaar 1864, Kristiania, 134 pp. [in Norwegian].

Sassen, R.; Cole, G. A.; Drozd, R.; and Roberts, H. H. (1994). Oligocene to Holocene hydrocarbon migration and salt-dome carbonates, northern Gulf of Mexico. *Marine and Petroleum Geology*, **11**, 55–65.

Schäfer, P.; Ritzrau, W.; Schlüter, M.; and Thiede, J. (2001). *The Northern North Atlantic: A Changing Environment*. Springer-Verlag, Berlin, 500 pp.

Schlager, W. (1981). The paradox of drowned reefs and carbonate platforms. *Geological Soc. Am. Bull., Part 1*, **92**, 197–211.

Schmaljohann, R.; and Flügel, H. J. (1987). Methane-oxidizing bacteria in Pogonophora. *Sarsia*, **72**, 91–98.

Schmaljohann, R.; Faber, E.; Whiticar, M. J.; and Dando, P. R. (1990). Co-existence of methane- and sulphur-based endosymbioses between bacteria and invertebrates at a site in the Skagerrak. *Marine Ecology Progress Series*, **61**, 119–124.

Schroeder, W. W.; Brooke, S. D.; Olson, J. B.; Phaneuf, B.; McDonough III, J. J.; and Etnoyer, P. (2005). Occurrence of deep-water *Lophelia pertusa* and *Madrepora oculata* in the Gulf of Mexico. In: Freiwald, A.; and Roberts, M. (eds), *Cold-water Corals and Ecosystems*. Springer-Verlag, Heidelberg, pp. 297–307.

Shinn, E. A. (1993). *Geology and Human Activity in the Florida Keys: Energy and Marine Geology*. US Geological Survey, Reston, VA/US Department of the Interior, October.

Shipboard Scientific Party (2000). Leg 182 Summary. In: Feary, D. A.; Hine, A. C.; Malone, M. J. *et al.* (eds), *Proceedings of the Ocean Drilling Program, Initial Report No. 182*. ODP, Texas A&M University, College Station, TX, pp. 1–58.

Sibuet, M.; and Olu, K. (1998). Biogeography, biodiversity and fluid dependence of deep-sea cold-seep communities at active and passive margins. *Deep-Sea Research II*, **45**, 517–567.

Skjærseth, J. B.; Tangen, K.; Swanson, P.; Christiansen, A. C.; Moe, A.; and Lunde, L. (2004). *Limits to Corporate Social Responsibility: A Comparative Study of Four Major Oil Companies*. Fridtjof Nansen Institute, Lysaker, Norway, Report No. 7.

Soley, J. (1910). The oilfields of the Gulf of Mexico. *Scientific American*, Supplement No. 1788, April 9th.

Somoza, L. (2001). Hydrocarbon seeps, gas hydrate and carbonate chimneys in the Gulf of Cadiz: An example of interaction between tectonic and oceanographic controlling factors. *Natural Hydrocarbon Seeps, Global Tectonics and Greenhouse Gas Emissions: European Science Foundation Exploratory Workshop, August 27 and 28*, p. 18 (abstract).

Somoza, L.; Díaz-del-Río, V.; León, R.; Ivanov, M.; Fernández-Puga, M. C.; Gardner, J. M.; Hernández-Molina, F. J.; Pinheiro, L. M.; Rodero, J.; Lobato, A.; Maestro, A., Vázquez, J. T.; Medialdea, T.; and Fernández-Salas, L. M. (2003). Seabed morphology and hydrocarbon seepage in the Gulf of Cadiz mud volcano area: Acoustic imagery, multi-beam and ultra-high resolution seismic data. *Marine Geology*, **195**, 152–176.

Southward, A. J.; Southward, E. C.; Dando, P. R.; Rau, G. H.; Felbeck, H.; and Flilgel, H. (1981). Bacterial symbionts and low $^{13}C/^{12}C$ ratios in tissues of Pogonophora indicate unusual nutrition and metabolism. *Nature*, **293**, 616–620.

Stetson, T. R.; Squires, D. F.; and Pratt, R. M. (1962). Coral banks occurring in deep water on the Blake Plateau. *American Museum Novatites*, **2114**, 1–39.

Stolarski, J.; Meibom, A.; Przeniosło, S.; and Mazur, M. (2007). A Cretaceous scleractinian coral with a calcitic skeleton. *Science*, **318**, 92–94.

Stone, R. (2007). A world without corals? *Science*, **316**, 678–681.

Strømgren, T. (1971). Vertical and horizontal distribution of *Lophelia pertusa* (Linné) in Trondheimsfjorden on the west coast of Norway. *Kongelig Norske Vidensk. Selsk. Skrifter*, **6**, 1–19.

Sumida, P. Y. G.; Yoshinaga, M. Y.; Madureira, L. A. S.-P.; and Hovland, M. (2004). Seabed pockmarks associated with deepwater corals off SE Brazilian continental slope, Santos Basin. *Marine Geology*, **207**, 159–167.

Squires, D. F. (1964). Fossil coral thickets in Wairarapa, New Zealand. *J. Paleont.*, **38**, 905–915.

TANGANYDRO Group (1992). Sublacustrine hydrothermal seeps in northern lake Tanganyika, East African Rift: 1991 Tanganydro expedition. *BCREDP, Elf Aquitaine Production*, **16**, 55–81.

Taviani, M.; Freiwald, A.; and Zibrowius, H. (2005). Deep coral growth in the Mediterranean Sea: An overview. In: Freiwald, A.; and Roberts, M. (eds), *Cold-water corals and Ecosystems*. Springer-Verlag, Heidelberg, pp. 137–156.

Teichert, C. (1958). Cold- and deep-water coral banks. *Am. Assoc. of Petroleum Geologists Bulletin*, **42**, 1064–1082.

Thiel, V.; Peckmann, J.; Richnow, H.-H.; Luth, U.; Reitner, J.; and Michaelis, W. (2001). Molecular signals for anaerobic methane oxidation in Black Sea seep carbonates and a microbial mat. *Marine Chemistry*, **73**, 97–112.

Thomson, C. W. (1873). *The Depths of the Sea*. MacMillan, London, 527 pp.

Thrasher, J.; Fleet, A. J.; Hay, S.; Hovland, M.; and Düppenbecker, S. (1996). Understanding geology as the key to using seepage in exploration: The spectrum of seepage styles. In: Schumacher, D.; and Abrams, M. A. (eds), *Hydrocarbon Migration and Its Near-surface Expression*. AAPG Mem., Vol. 66, pp. 223–241.

Tjelta, T. I.; Svanø, G.; Strout, J. M.; Forsberg, C. F.; Johansen, H.; and Planke, S. (2007). Shallow gas and its multiple impact on a North Sea production platform. *Proceedings of the Sixth Int. Offshore Site Investigation and Geotechnics Conf., 11-13 September, London*, pp. 205–220.

Tréhu, K. A.; Ruppel, C.; Holland, M.; Dickens, G. R.; Torres, M. E.; Collett, T. S.; Goldberg, D.; Riedel, M.; and Schultheiss, P. (2006). Gas hydrates in marine sediments: Lesson from scientific ocean drilling. *Oceanography*, **19**, 124–143.

Tryon, M. D.; and Brown, K. M. (2001). Complex flow patterns through Hydrate Ridge and their impact on seep biota. *Geophys. Res. Lett.*, **28**(14), 2863–2867.

Tucholke, B. E.; Sibuet, J.-C.; Klaus, A.; and the Shipboard Party (2004). *Proceedings ODP: Initial Report, Leg 210*. College Station, TX, Ocean Drilling Program.

Tucker, M. E.; and Wright, V. P. (1990). *Carbonate Sedimentology*. Blackwell Scientific, Oxford, UK, 482 pp.

Turley, C.; Blackford, J.; Widdicombe, S.; Lowe, D.; Nightingale, P. D.; Rees, A. P. (2006). Reviewing the impact of increased atmospheric CO_2 on oceanic pH and the marine ecosystem. In: Schnellnhuber, H. J.; Cramer, W.; Nakicenovic, N.; Wigley, T.; and Yohe, G. (eds), *Avoiding Dangerous Climate Change*. Cambridge University Press, Cambridge, pp. 65–70.

Turley, C. M.; Roberts, J. M.; and Guinotte, J. M. (2007). Corals in deep-water: Will the unseen hand of ocean acidification destroy cold-water ecosystems? *Coral Reefs*, **26**(3), 445–448.

Vacelet, J.; Fiala-Medioni, A.; Fisher, C. R.; and Boury-Esnault, N. (1996). Symbiosis between methane-oxidizing bacteria and a deep-sea carnivorous cladohizid sponge. *Mar. Ecol. Prog. Ser.*, **145**, 77–85.

Van Dover, C. L. (2000). *The Ecology of Deep-sea Hydrothermal Vents*. Princeton University Press, Princeton, NJ, 424 pp.

Van Dover, C. L.; Aharon, P.; Bernhard, J. M.; Caylor, E.; Doerries, M.; Flickinger, W.; Gilhooly, W.; Goffredi, S. K.; Knock, K. E.; Macko, S. A. *et al.* (2003). Blake Ridge methane seeps: Chaacterization of a soft-sediment, chemosynthetically based ecosystem. *Deep-Sea Research*, **50**, 281–300.

Vandendriessche, S. (2002). The epifauna associated with cold-water coral ecosystems. M.Sc. thesis, Renard Centre of Marine Geology, Belgium.

Van Rensbergen, P.; Depreiter, D.; Pannemans, B.; Moerkerke, G.; Van Rooij, D.; Marsset, B.; Akhmanov, G.; Blinova, V.; Ivanov, M.; Rachidi, M.; Magalhaes, V.; Pinheiro, L.; Cunha, M.; and Henriet, J.-P. (2005). The El Arraiche mud volcano field at the Moroccan Atlantic slope, Gulf of Cadiz. *Marine Geology*, **219**, 1–17.

Van Rensbergen, P.; Depreiter, D.; Pannemans, B.; Moerkerke, G.; Van Rooij, D.; Marsset, B.; Akhmanov, G.; Blinova, V.; Ivanov, M.; Rachidi, M.; Magalhaes, V.; Pinheiro, L.; and Henriet, J.-P. (2007). The El Arraiche mud volcano field at the Moroccan Atlantic slope, Gulf of Cadiz. *Marine Geology*, in press.

Van Rooij, D.; Kozachenko, M.; Blamart, D.; Lekens, W.; Wheeler, A.; and Henriet, J.-P. (2000). The acoustic and sedimentological face of the sediments surrounding the Belgica Mounds Province. *EOS Trans. Am. Geophys. Union*, **81**(41), F638.

Van Rooij, D.; De Mol, B.; Huvenne, V.; Ivanov, M. K.; and Henriet, J.-P. (2003). Seismic evidence of current-controlled sedimentation in the Belgica mound province, upper Porcupine slope, southwest of Ireland. *Marine Geology*, **195**, 31–53.

Van Rooij, D.; Depreiter, D.; Bouimetarhan, I.; De Boever, E.; De Rycker, K.; Foubert, A.; Huvenne, V.; Réveillaud, J.; Staelens, P.; Vercruysse, J.; Versteeg, W.; and Henriet, J.-P. (2005). First sighting of active fluid venting in the Gulf of Cadiz. *EOS*, **86**(49), 509–511.

van Weering, T. C. E.; de Haas, H.; Lykke-Andersen, H.; and de Stigter, H. (2001). Giant carbonate mounds in the Rockall Trough, NE Atlantic Ocean. *European Union of Geologists 11th Conference Proceedings* (Session RCM6) (abstract).

Van Weering, Tj. C. E.; de Haas, H.; de Stiger, H. C.; Lykke-Andersen, H.; and Kouvaev, I. (2003a). Structure and development of giant carbonate mounds at the SW and SE Rockall Trough margins. *Marine Geology*, **198**, 67–81.

Van Weering, Tj. C. E.; Dullo, C.; and Henriet, J.-P. (2003b). An introduction to geosphere–biosphere coupling: Cold seep related carbonate and mound formation and ecology. *Marine Geology*, **198**, 1–3.

Viana, A. R.; Faugères, J. C.; Kowsmann, R. O.; Lima, J. A. M.; Caddah, L. F. G.; and Rizzo, J. G. (1998). Hydrology, morphology and sedimentology of the Campos continental margin, offshore Brazil. *Sedimentary Geology*, **115**, 133–157.

Vogt, P. R.; Crane, K.; Pfirman, S.; Sundvor, E.; Cherkis, N.; Flemming, H.; Nishimura, C.; and Shor, A. (1991). SeaMarc II sidescan sonar imagery and swath bathymetry in the Nordic basin. *EOS Transactions*, **72**, 486.

Völker, H.; Schweisfurth, R.; and Hirsch, P. (1977). Morphology and ultrastructure of *Crenothrix polyspora* Cohn. *Journal of Bacteriology*, **131**, 306–313.

Walker, G.; and King, D. (2007). *The Hot Topic*. Bloomsbury/Harcourt.

Walter, M. R.; Bauld, J.; and Brock, T. D. (1972). Siliceous algal and bacterial stromatolites in hot spring and geyser effluents of Yellowstone National Park. *Science*, **178**, 402–405.

Weaver, P. P. E.; Billett, D. S. M.; Boetius, A.; Danovaro, R.; Freiwald, A.; and Sibuet, M. (2004). Hotspot ecosystems research on Europe's deep-ocean margins. *Oceanography I*, **17**(4), 132–143.

Wegener, G.; Shovitri, M.; Knittel, K.; Niemann, H.; Hovland, M.; and Boetius, A. (2008). Biogeochemical processes and microbial diversity of the Gullfaks and Tommeliten methane seeps (northern North Sea). *Biogeosciences*, **5**, accepted for publication.

Wendt, J.; Belka, Z.; Kaufmann, B.; Kostrewa, R.; and Hayer, J. (1997). The World's most spectacular carbonate mounds (Middle Devonian, Algerian Sahara). *J. Sedimentary Research*, **67**(3), 424–436.

Westbrook, G. K.; Carson, B.; Musgrave, R. J.; Ashi, J.; Baranov, B.; Brown, K. M.; Camerlenghi, A.; Caulet, J.-P.; Chamov, N.; Clennell, M. B. *et al.* (1994). *Proceedings ODP: Initial Reports, Part 1, Cascadia Margin*, Vol. 146, 611 pp.

Wheeler, A. J.; Kozachenko, M.; Beyer, A.; Foubert, A.; Huvenne, V. A. I.; Klages, M.; Masson, D. G.; Olu-Le Roy, K.; and Thiede, J. (2005) Sedimentary processes and carbonate mound morphology in the Belgica mouond province, Porcupine Seabight,

NE Atlantic. In: Freiwald, A.; and Roberts, M. (eds), *Cold-water Corals and Ecosystems*. Springer-Verlag, Heidelberg, pp. 517–603.

Wheeler, A. J.; Beyer, A.; Freiwald, A.; de Haas, H.; Huvenne, V. A. I.; Kozachenko, M.; Olu-Le Roy, K.; and Opderbecke, J. (2007). Morphology and environment of cold-water coral carbonate mounds on the NW European margin. *Int. J. Earth Sciences*, **96**, 37–56.

White, M. (2003). Comparison of near seabed currents at two locations in the Porcupine Sea Bight: Implications for benthic fauna. *J. Mar. Biol. Ass. UK*, **83**, 683–686.

Williams, T.; Kano, A.; Ferdelman, T.; Henriet, J.-P.; Abe, K.; Andres, M. S.; Bjerager, M.; Browning, E. L.; Cragg, B. A.; DeMol, B. *et al.* (2006). Cold-water coral mounds revealed. *EOS Trans. Am. Geophys. Union*, **87**(47), 525–526.

Wilson, J. B.; and Freiwald, A. (1998). How icebergs shape deep-water coral reefs. *Conference Proceedings: Carbonate Mud Mounds and Cold Water Reefs, Gent, Belgium, February 1998*.

Wilson, J. B. (1979a). The distribution of the coral *Lophelia pertusa* (L.) [*L. prolifera* (Pallas)] in the north-east Atlantic. *Journal of the Marine Biological Association of the UK*, **59**, 149–164.

Wilson, J. B. (1979b). "Patch" development of the deep-water coral *Lophelia pertusa* (L.) on Rockall Bank. *J. Mar. Biol. Ass. UK*, **59**, 165–177.

Winn, C. D.; Karl, D. M.; and Massoth, G. J. (1986). Microorganisms in deep-sea hydrothermal plumes. *Nature*, **320**, 744–746.

Woodside, J.; Lykousis, V.; Foucher, J.-P.; Dupre, S.; Alexandri, S.; de Lange, G.; Mascle, J.; Zitter, T.; scientists of the ANAXIMANDER, MEDIFLUX, and MEDINAUT projects (2006). *Seafloor Fluid Emissions in the Eastern Mediterrranean: Variety as a Function of Environment*. IOC (Intergovernmental Oceanographic Commission, UNESCO) Workshop Report No. 201, p. 65.

Yakimov, M. M.; Timmis, K. N.; Golyshin, P. N. (2007). Obligate oil-degrading marine bacteria. *Current Opinion in Biotechnology*, **18**, 257–266.

Yusifov, M.; and Rabinowitz, P. D. (2004). Classification of mud volcanoes in the South Caspian Basin, offshore Azerbaijan. *Marine and Petroleum Geol.*, **21**, 965–975.

Zibrowius, H.; and Gili, J.-M. (1990). Deep-water scleractiniana (Cnidaria: Anthozoa) from Namibia, South Africa, and Walvis Ridge, Southeastern Atlantic. *Scientia Marina*, **54**, 19–46.

Zona, G. A. (1994). *The Soul Would Have No Rainbow If the Eyes Had No Tears*. Simon & Schuster, New York, 128 pp.

2-D hydrocarbon basin modelling, 108
2-D seismic line, 68
^{13}C-depleted carbonate, 134
δ^{13}C values, 26
62nd parallel, 40

Abandoned, 95
Absorbed, 212
Academic research, 25
Accretionary wedge, 131
Accumulation of sponges, 81
Acergy Petrel, xiii, 15
Acergy Viking, xiii
ACES, 106
Acesta, 95
   bivalves, 96
   excavata, 15
Acid, 221
Acidobacteria, 200
Acoustic
   blanking, 97
   characteristics, 49
   evidence, 65, 80
   flare, 67, 80
   plumes, 127
   reflection, 44
   shadow, 46, 49
Acoustically high reflective, 45
Active migration, 61

Adhesive substrate, 10
Adsorbed, 57
Advection, 61, 180
Africa, 129, 137
   rift lakes, 174
Agdenes, 40
Aggregates, 75
Ahermatypic, 146
   coral reefs, 78
Ahnet
   reefs, 169
   Sedimentary Basin, 169
Alberta, 40, 163, 167
Alcanicorax, 201
Alcalcaline, 174
Ålesund, 16
Aleutian
   Islands, 158
   Trench, 30
Algae, 7, 10
Algeria, 163
   Sahara, 163
Alice in Wonderland, 9
Alkalinity, 119
Alkor, 95
Along-slope gravitational sliding, 130
Alpha, 200
Aluminium cylinder, 19
Alvin, 141, 151

Ambergris
  Caye, 148
  Creek, 148
Amber head, 148
Ambient seawater, 27
American
  geologists, 144
  government, 144
  law, 145
AMS, 158
Amsterdam, 203
Anaerobic
  conditions, 43
  sediments, 43
Analogues, 163
Anaximander (mud volcanic) mountains, 54
Anchor
  animals, 218
  chains, 70
Ancient, 163
Anemones, 224
Animals, 15
Animal community, 26
Anomalies, 127
Anoxic fluids, 82
Anoxic
  methane oxidation, 33
  oxidation of methane, 33
Antarctic
  continental shelf, 55
  Intermediate Water, 142, 144
Annual temperature change, 64
AOM, 33
Aquarium, 9
Aragonite, 29, 129
Arborea, 9
Arafura Sea, 160
Armies of bacteria, 201
Armoured, 165
Archaea, 37, 75
  symbionts, 79
Architects, 7
Articulating, 96
Artificial image, 41
Artist impression, 42
  photomontage, 59
Ashmore Reef, 160
Ascending gas bubbles, 87

Asgard field, 25
Askeladden field, 14
Associated fauna, 130
Atalante mud volcano, 76
Atlantic
  Coral Ecosystem Study, 106
  Margin, 100
  Ocean, 1
Atmospheric
  CO_2 content analysis, 73
Anti-Atlas mountains, 170
Aragonite, 213
  -producing organisms, 213
  skeletons, 213
Australia, 107
Authors, 174
Autumn cruise, 86
Autotrophic, 198
Auto-precipitation, 152
Australian carbonate knolls, 160
Aquifers, 150, 175
Axial Seamount, 172
Azooxanthellate scleractinian corals, 213
Azores archipelago, 155

Bacteria, 37
  and archaea, 79
  cells, 118, 172
  growth, 85
  mats, 26
  symbionts, 79
Bacteriodetes, 200
Baffling, 164
Banda gas field, 137
Barites, 169
Baseline geochemical conditions, 56
Basin modelling, 108
Bathymetric map, 25
Beautiful photo, 90
Baja California, 171
Barbados, 75, 76
  accretionary prism, 75
Barents Sea, 16, 76
Barrier Reef, 106
Basalt pillow, 232
Basement rock, 91
Basin-wide, 187
Bathymodiolus heckerae, 52
Beaches, 148

Bear Island, 31
  Fan Slide, 31
Bearing capacity, 137
Bedrock, 44
Beggiatoa, 32
  coverage, 33
  filaments, 35
Beitstadfjord, 15
Belgica, 107, 139, 165
  Mounds, 167, 184
Belgium, 120
Belize, 147
Benthic
  fauna, 122
  foraminifera, 122
Bergen, 9
Bering Sea, 172
Beryl oil field, 94
Bicarbonate, 10, 211
Bill Bailey Bank, 100
Biodiversity, 41, 85, 160
BIOFAR programme, 100
Bioherm, 7, 152
Bioactivity, 212
Biodensity, 41
Bio-expectations, 69, 93
Biofacies, 173
Biogenic carbonate, 173
Biogeographic, 157
Biological
  contaminants, 183
  response, 212
  structure, 7
Biology, 172
Biomarkers, 168
Biomass production, 36
Biosedimentary, 172
Bishop, 15
Bituminous liquids, 202
Bivalve, 15
  association, 48
Blanking, 97
Blobs, 92
Block 6407/4, 82
Block diagram, 97
Black corals, 158
Black smoker vents, 129
Black smokers, 128

Black/White video
Blake Ridge, 52, 152
Blind, 209
Blind canyon, 151
Blocky rubble, 123
Blowout, 95
Blue Planet, 201
Bolocera thuediae, 224
Bonellia viridis, 218
Bonjardim mud volcano, 130, 132
Book, 132
Borgenfjord, 13
Bottom
  -simulating reflector, 26
  trawling, 209
Boulders, 4, 195
Brachiopods, 122
Branching skeletons, 115
Brazil, 141
Breisunddjupet, 5
  reefs, 5
Brent Spar buoy, 93
Brine seeps, 148
British Colombia, 175
British Isles, 105
Brønnøysund, 225
Brosme brosme, 4
Brucite, 129
BSR, 26
Bubble seeps, 148
Bubbling
  reefs, 98
  reef carbonates, 99
Bubira
  fields, 142
  reefs, 142
Buffering capacity, 165
Buried
  salt, 145
  mounds, 107
Burrowing
  clams, 51
  organisms, 209
Butane, 167, 201
Butterfly-shaped structures, 114
Bush Hill, 145
Builders, 7
Bright spots, 44

258 Index

C_1, 26
C^{14} and U/Th dating, 158
C_2, 56
C_3, 56
C_4, 56
C_5, 26, 56
Ca, 129
Ca^{2+}, 211
$CaCO_3$, 10, 145, 211
Calcification, 11, 174, 210
Calcified remains, 176
Calcite, 152
  skeleton, 9, 213
Calcium, 9
  carbonate, 181
  -poor, 129
Calification feedback, 213
California, 151
Calyptogena, 51
Camera
  pod, 39, 83
  recording, 83
Campos Basin, 142
  reefs, 142
Canada, 13, 166
Candidate, 133
Canyon, 151, 153
Cap rocks, 145
Capacity, 137
Cape Banza, 174
Cape Fear, 152
Captain Soley, 149
Carbon
  dating, 46
  dioxide, 44
  isotope values, 168
Carbonaceous rock, 171
Carbonates, 26
  cemented sandstone, 99
  compensation depth, 129
  knolls, 107
  rock, 26
  shoals, 103
  –silicate cycle, 211
Carbonic acid, 211
Carboniferous, 164
Carcasses, 51
Cariophyllia sp., 145

Carlos Ribeiro mud volcano, 132
Carpets of deep-water corals, 130
Cascadia Accretionary Prism, 117
CASQ core, 133
Catchphrase, 165
Causative formation mechanism, 78
Cavity filling, 168
Cay-Caulker island, 148
CCD, 129
Cell counts
Cemented sandstone, 99
Cenozoic, 119
  bryozoan mounds, 118
Centre for Geomicrobiology, 199
Century, 207
CH_4, 56, 110
  fluxes, 33
Challenger Mound, 114
Champagne Pool, 176
Channels, 192
Charlie Gibbs Fracture Zone, 156
Chart, 95
Chemical
  components, 86
  reactions, 128
  species, 59
Chemolithoautotrophic, 36, 50
Chemosynthesis, 75
  oasis, 31
  tubeworms, 31
  based fauna, 209
Chief scientist, 89
Chimney, 129, 174
Chinguetti, 137
Cidaris cidaris, 226
Circular structures
Cladorhiza, 75
Clay
  layer, 14
  -dominated till, 4
Cliff walls, 154
Climate change, 153
Climatology, 231
Clones, 200
Clusters, 94
  of nematocysts, 9
Cnidaria, 7
CO, 110

CO_2, 12, 110
  accumulation, 212
  bubbles, 176
  burden
Coal oils, 202
Coarse-grained sediments, 3
Coastline, 142
Cobbles, 4
Coccolithophorids, 212
Cod, 6
Coelenterata, 7
Coelosmilia, 213
Coherent acoustic reflections, 49
Cold seep, 30
  communities, 168
Cold War, 16
Cold-water carbonate mounds, 115
College Station, 145
Colombia, 175
Colony, 89
Colour
  film, 17
  image, 134
Combinations, 204
Combined factors, 181
COMET, 97
Common aspects, 145
Comparative analyses, 73
Complex pockmarks, 26, 93
  at Troll, 93
  G11, 26
Composition, 204
Compressible, 204
Concentration, 153
Conceptual
  diagram, 59
  model, 59
Concession Block
  184, 145
  6407/4, 79
Conclusions, 215
Concrete platforms, 94
Condensates, 70
Conduits, 192, 207
Cones, 16, 176
Cone-shaped features, 176
Congo, 156, 215
Congregating, 219

Conical
  deishes, 170
  methane-derived carbonate rocks, 93
Connemara, 107
Conophyton stromatolites, 175
Conservation, 106
Construction engineer, 7
Contaminants, 183
Continental
  crust, 128
  shelf, 18, 101
Continuous methane fluxes, 97
Controls on methane fluxes
Cool water, 7
Copenhagen, 46
Coral, 8, 210
  calcification, 210
  clump, 18
  colonies, 3
  cup, 8
  debris, 3, 81
  oasis, 39, 89
  park, 89
  reefs, 7, 87
  taxa, 160
  trees, 154
Corallite polyp, 8, 11
Corinth, 173
Coriolis force, 151
Corollary, 119
Corrosion, 27
  pits, 28
Corrosive bottom layer, 113
Cosmopolitan, 95
Couscous, 170
Co-workers, 121
Crab, 11
Cracks, 87, 126
Crater Lake, 35
Craters, 95
  -like basin, 198
Crenarchaea, 200
Cretaceous
  period, 10, 11, 46
  sedimentary rocks, 189
  subcrop basin, 56
Crevice, 175
Crime, 207
Crinoids, 89, 170, 230

Crown, 19
Cruises, 86, 95
  #259, 95
Crustacea, 11
Crux, 186
Crystalline
  basement, 108
  bedrock, 44
Crystals, 152
CT scan, 189
Cuba, 150
Cubic square cm, 118
Current
  component, 87
  direction, 82
  flute marks, 194
Curvilinear troughs, 74
Cyanobacteria, 172, 174
Cycloclasticus, 201

Damage, 73
Damon Mound salt diapir, 145
Danish, 98
  "bubbling reefs", 98
  waters, 98
Dark-violet unknown coral, 221
Darwin Mounds, 203
Database, 76
Davison Seamount, 159
Dead coral assemblages, 130, 132
Dead zone, 117
Deep nutrient-rich water, 81
Deeply sourced fluids, 127
Deep-ocean
  blooms, 36
  hydrothermal vents, 49
Deep Rover, 35
Deep-towed boomer, 47
Deep-water
  coral reefs, 4
  coral skeletons
  corals, 7
  ecosystems, 208
  fossil reefs, 169
  mound community, 115
Degassing, 133
De-glaciation, 55
Delta, 41, 146
Demersal trawling activity, 138

Demopongiae, 75
Dense aggregates, 79
Density, 181
  boundary, 120
Designers, 7
Desmacidon, 75
Desmophyllum
  cristagalli, 3, 8
  sp., 114, 158, 213
Destabilize fine-grained sediments, 78
Devonian fossil reefs, 163
Dewatering, 20
Diagenetic, 11, 168
Diagnostic, 168
Diameter, 72
Diapiric mud, 44
Diatoms, 211
Digital terrain model, 42, 60
Dipping
  Palaeocene rocks, 193
  reflections, 57
Discharge, 129
Disconformity, 174
Discontinuities
Discontinuous acoustic reflection, 50
Discovery, 141
Disequilibrium, 174
Disrupted slide material, 90
Dissociation, 27, 64
Dissolution, 128
Dissolved, 86
  chemical components, 86
  gases, 21
  silica, 211
Distribution, 67
Diversity, 203, 204
DNA analyses, 73
Dolomite, 167
Dome-like structure, 156
Doming of the seafloor, 45
Don Quixote mud volcano, 134
Dons, Carl 15
Downslope, 137
Draugen field, 25, 62
Drift sediments, 115
Drilling, 22, 95
  campaign, 153
  site survey, 22
Driving mechanism, 215

Drøbak, 14, 15
DTM, 60
Ductile, 204
DWCRs, 195

E. huxley, 213
Early Devonian age, 170
Early Pleistocene, 158
Earth, 7, 173
East Flower Garden Bank, 145
East Pacific Rise, 36, 172
Eastern Anti-Atlas, 170
Eastern Atlantic, 156
Eastern Mediterranean Sea, 54
Ebullition, 79
Echinoderms, 15
Echinoids, 122
Echiurans, 218
Echosounder, 45
  data, 87, 143
Ecological, 157
ECOMOUND, 106
Econiches, 130
Edda Fonn, xiii
Edda Freya, xiii
Eddies, 180
Eelpout, 32
Eggagrunnen, 87
Eggs, 219
El Arriche mud volcano, 133
Elephant, 210
Elongated
  depressions, 192
  ridges, 91
Emanating free gas
Embedded, 96
Emitting bubbles, 95
ENAW, 183
Endemic, 158, 160
Endodermis, 8
Endogenic, 187
Endosymbionts, 10, 31
Endosymbiotic bacteria, 30, 79
Energetic environments, 165
Energy-poor environments, 39
Enhanced
  acoustic layers, 44
  ethane/methane ratios, 117

reflections, 65
tidal currents, 184
Enigmatic
  Darwin Mounds, 203
  giant carbonate mounds, 103
Enriched
  habitat, 99
  mantle, 214
Enrichment, 93
Environment, 1
Environmental
  changes, 192
  controls on mound formation along the
    European margin, 106
  controls, 106, 110
  damage, 106
  factors, 67
  Protection and Safety Panel, 202
Epifauna, 73
  organism, 73
Episodic removal of fine-grained sediments,
    79
Epithelia, 12
Equipment, 17
Erosion of fines, 138
Ethane, 56, 167
Eunice norvegica, 11
European
  Commission, 106
  continenatal margin, 184
  group, 114
  margins, 215
  Union Fifth Framework research project,
    110
Evaporites, 203,
Evolutionary
  connection, 11
  patterns, 157
Excavation, 210
Exogenic possibilities, 187
Exotic fluids, 36
Expansion of gas, 171
Expedition 307, 115
Exploration drilling, 95
Exposed
  by erosion, 99
  carbonate rock, 147
  Palaeocene bedding, 194
Expulsion of free gas, 28

Extended tentacles, 9
Extinction, 205
Extraction, 129

Fabulous scene, 232
Facosites, 170
Fan-like, 20
Faroe
  Bank Channel, 124
  Bank, 100
  Isles, 124
  –Shetland Channel, 124
Fathoms, 154
Faults, 87
  conduits, 175
  -controlled, 175
  line, 198
Fauna, 26
  diversity, 188
Favourable topography, 67
Fe^{2+}, 36
Feather stars, 221
Features, 28
  database, 76
Fedje
  Island, 91
  reefs, 69, 91
Feedback mechanisms, 212, 213
Fertilization, 59
FeS, 128
FFI, 86
Fields of
  floating oil, 149
  giant clams, 232
Filaments 164
Filter-feeding organisms, 93
Fine-grained
  detritus, 69
  grained sand, 78
  grained sediments, 78
Fines, 138
Finland, 13
Finnmark, 16
Firm sub-stratum, 78
Firmicutes, 200
First-hand observation, 39
Fish, 7
  habitat, 208
  shoals, 127
  species richness, 208
  species, 208
Fishermen, 13, 92
Fishing lines, 82
Fissures, 126, 232
Fiuza MV fixation, 129
Fjord
  glacier, 41
  threshold, 216
Flares, 127
Flat-iron, 193
Flemish Cap, 155
Flexible pipes, 70
Floating University, 106
Floc (flocculent) material, 36
Floholman reefs, 81, 88
Florida, 150
  Escarpment, 33
  Keys, 150
  peninsula, 150
  platform, 153
  Straits, 142, 150
Flow
  field, 184
  line, 73
  of water, 207
Fluids, 11, 27
  emissions, 203
  flow related, 176
  migration pattern, 134
  sapping, 90
  seep site, 156
  seepage, 23
Fluidization, 27
Flute marks, 193
Flux, 31, 139
  measurements, 153
  of gas, 139
  rates, 53
Focused investigations, 211
Foinaven, 100
Food
  availability, 147
  chain, 37
  supply, 10
Foraminifera, 122, 211
Forest of risers, 70
Formation, 215
  mechanism, 78

Forsvarets Forskningsinstitutt, 86
Fossil
　coral reef, 19
　fuel, 213
　mound, 164
　seep, 200
　seep carbonates, 168
Foundation, 9
Fracture zones, 156
Framboidal pyrite, 169
Frame-building organism, 164
Framework-building coral species, 123
France, 29
Fredrikshavn, 99
Free hydrocarbons, 56
Free-swimming, 10
Freezing, 64
French, 16, 174
　biologists, 16, 174
　team, 174
Fresh groundwater flow, 175
Freshwater, 216
　seeps, 152
Frozen gas hydrates, 62
Fugløy, 17
　banken, 88
　reef, 17

G11, 26
Galapagos Rift, 49
Galway Mound, 110
Gas, 26
　bubbles, 21, 41, 87
　charged, 48, 65
　escape, 86
　extraction industry, 145
　hydrate pingoes, 26
　hydrates, 26, 62
　migration, 97
　plumes, 67
　seepage, 86, 127
　-bearing fluids, 127
　-charged sediments, 149
　hydrate pingoes, 28
Gastropods, 122
Gemini mud volcano, 134, 135
Genera, 158
Genetically determined, 213
Geo-biological
　process, 128
　work, 64
Geochemical Exploration and Research
　　Group, 145
Geochemistry, 26
Geodia sp, 72
Geodiversity, 203
Geological
　controls on mound formation, 106
　horizon, 114
　model, 145
Geologists, 89
Geomicrobial, 115
GEOMOUND, 106
Geophysical
　and geological data, 114
　interpretation, 192
　survey, 60
Geosphere–biosphere coupling, 202
Geosund, xiii
GERG, 145
German researchers, 171
Ghent University, 120
Giant carbonate mounds, 135
Giant concrete platforms, 94
Giant limid bivalves, 79
Giant mussel, 50
Giant protozoa, 124
Giant structures, 110
Gibraltar, 215
Ginsburg mud volcano, 132
Glacial, 130, 188
　period, 99
Glaciation, 188, 205
Glass coral, 8
Glauconitic and silty sandstone, 114
Global prokaryote profile, 118
Globe, 128
Glory, 207
Golfstrømmen, 126
GoM, 144
Gorge, 153
G.O. Sars, 82
Grand Isle, 145
Gravel, 4, 20
Gravitational sliding, 130
Gravity, 156
　corer, 16

Great Australian Bight, 118
  bryozoan mounds, 118, 119
Great Bahama Bank, 150
Great fork-beard, 101
Greatest discovery, 232
Greece, 173
Greed, 207
Green Canyon, 145
Greenland–Iceland–Norwegian Seas, 3
Grid size, 74
Grounded shelf ice, 55
Grounding icebergs, 55
Groundwater, 20, 41
  seepage, 176
Guaymas Basin, 35
Gulf of Cadiz, 129
Gulf of Corinth, 173
Gulf of Mexico, 23, 79, 144
Gulf Stream, 23, 126, 150
Gulley, 81, 90
Gullfaks, 13
Gum coral, 19

H_2, 110, 128
H_2CO_3, 211
H_2S, 34, 119, 153
  smell, 44
Habitats, 7, 99
Hadean ocean, 129
Håkon Mosby, 24, 100
  mud volcano, 24
Haltenbanken, 22
Haltenpipe
  Reef Cluster, 22
  Reefs, 199
Hamar Lagdhad area, 170
Hancock Seamount, 171
Harstad Tidene, 83
Hatteras slope, 154
Hatton bank, 100, 101
Hawaii, 172
Hawaiian chain, 172
Haystacks, 44
Haystack-shaped mounds, 169
HCO^-, 211
Headless canyon, 1151
Head-wall, 90
Heat system, 231

Heavy silting, 93
Heidrun, 25
  field, 193
Helium flux, 176
Hermatypic, 10, 12
HERMES, 106
Heterotrophs, 36
Hexane, 58
Heywood Shoals, 103, 160
Hidden living structures, 214
High alkalinity, 119
High faunal diversity, 7
High methane flux areas
High-amplitude reflectors, 58
Higher hydrocarbons, 201
Higher trophic organisms, 28
High-resolution still photograph, 53
High-stand, 188
High-temperature, 129
HMMV, 24
Hola, 86
Holmestrand, 15
Holocene age, 122
Homogenous seafloor, 113
Hope, 212
Hornblende, 212
Hostile environment, 113
Hot vents, 231
Hotspot ecosystem research on Europe's
  deep-ocean margins, 106
Hovland Mounds, 184
HRC, 22, 191
Human
  flesh, 9
  -induced global warming, 210, 212
  -made structures, 93
Husmus
  Lophelia reefs, 62
  reefs, 62
Hydrates, 28, 153
  dissociation, 27
  pingoes, 27
  Ridge, 118
Hydraulic, 24, 191
  active conditions, 77
  activity, 92
  control, 133
  drag force, 123
  head, 126

passive, 204
theory, 4
Hydrocarbons, 144
  concentration, 56
  degraders, 179
  drilling, 209
  migration, 107
  plume, 201
  reservoir, 82
  -degrading microbes, 179, 201
Hydrocorals, 158
Hydrodynamical, 188
Hydrogen, 128
  sulphide, 44, 160
Hydrographic study, 100
Hydroids, 15
Hydrology, 77
Hydrothermal
  emission, 172
  input, 198
  sulphide deposit, 36
  system, 172
  vents, 49
Hydrozoa, 26
Hypercalcifying
  aragonite-producing organisms, 213
  calcite-producing organisms, 212
Hypothesis, 61, 203

Iberian, African, and Eurasian Plates, 129
Ice
  sheets, 188
  -caps, 13
  -flute marks, 194
  -gouged trough, 81
Iceberg plough-marks, 4
Iceland, 155
IFREMER, 29
IMR, 11, 194
Incoming current, 82, 151
Individual
  reefs, 57
  slabs, 99
Industrial revolution, 201
Industry, 145
  -imposed, 216
Inert, 204
Inferno Crater, 176
Inferred seep zones, 135

Inhomogeneities, 126
Inhospitable, 211
Inorganic, 59
  structure, 145
Insidious threat, 210
In situ formation, 145
Institute of Biology, 199
Instrumentation, 186
Instruments, 139
Integrated Ocean Drilling Program, 106
Interglacials, 188
Intergovernmental Panel for Climate
      Change, 214
Intermediate water, 144
Internal
  breaking waves, 183
  conduits, 176
  pore spaces, 90
Inter-ridge areas, 113
Interstitial fluids, 90
Intrusion, 166
Intrusive dikes, 166
Invertebrates, 7
IODP, 106
  Expedition 307, 114, 117
  Proposal 673-Pre2, 120
Ionian Sea, 158
IPCC, 214
Ireland, 23, 103
Irish Petroleum Department, 107
Iron, 129
  sulphide, 50
Iso-butane, 56
Isostatic uplift, 99
Italy, 158

Jago, 95
Jan Mayen, 24
Japan, 131
Jasper seamount, 171
Jellyfish, 9
Jesus Baraza mud volcano, 132
Joides Resolution, 114
Journal of Petroleum Geology, 111
Jurassic age, 56
Juvenile
  coral polyp, 10
  reefs, 120, 133

KA1, 71
KA2, 71
Kaolinite, 212
Karmt shoals, 160
Kattegat, 40
Kess-Kess Mounds, 163
Key to the present, 165
Kill and dissect, 120
Knolls, 107
Knot, 89
Korallbanken, 68
Kosterfjord reefs, 97
KP1, 71
Kristiania, 14
Kristiansund, 46
Kristin
  field, 25
  reefs
Krona, 14

Lador, 22
Lake Tanganyika, 174
Lake Van, 176
Laminations, 172
Landrover, 59
Landslides, 90
Langeled pipeline, 81
Large corals, 92
Large mound clusters, 12
Large-scale lineations, 68
Larson Seamounts, 172
Last Glacial Maximum, 13, 188
Late Cretaceous, 213
Late Devonian Kess-Kess mounds, 208
Late Pliocene, 158, 192
Laurentia, 13
Laurentian fan, 30
Lava flows, 120
Law enforcement, 209
Layered
  basalt, 128
  carbonate rock, 95
  sediments, 89
  upper sediments, 190
Leakage of
  fluids, 90
  gases, 91
Leaky hydrocarbon reservoir, 82

Lease Block A-384, 144
Lee side, 68
Leg 166, 153
Leg 210 of the ODP, 130
Legislation, 210
Lethal damage, 120
LGM, 13, 188
Libraries, 94
Life and death, 189
Lighthouse Reef Island, 148
Light hydrocarbons, 167
  hydrocarbon gases, 26
Lima excavata, 15
Limestone masses, 168
Linear reefs, 198
Lineations, 68
Liquefied muds, 31
Liquid, 145
  water, 214
Lithified sandstone, 99
Lithodes maja, 224
Lithoherms, 151
  theory, 164
Lithostratigraphy, 116
Little Bahama Bank, 150
Live
  corals, 13
  twigs, 131, 133
Lively mounds, 110
Living
  corals, 186
  fossil, 172
  tubeworms, 27
Lobster, 156, 228
Local
  depressions, 198
  fishermen, 92
  food source, 156
Localization, 167
Lofoten archipelago, 91
Logachev Mounds, 120
Loihi, 172
Long-lines, 82
Long-range sidescan sonar system, 123
Lophelia, 9
  pertusa, 3, 9, 15, 98, 209
  prolifera, 13
  remnants, 46
  skeleton, 55

worm, 11
-capped giant carbonate mounds, 104
colonies, 11
reef complexes, 57
Lost City, 129
Louisiana, 145
Lousy Bank, 100
Low relief, 125
Lower Congo Basin, 156
Lower portion, 89
Low-permeable material
Low-reflective, 125
Low-salinity water, 63
Low-stand, 188
Lucinid bivalves, 48, 51
Lucinoma borealis, 54
Lucinoma kazani, 54
Luhanga hydrothermal field, 174
Lush reef lush, 133
Lyngenfjord, 16
Låsø, 99

Macrofauna, 99
Mader Basin, 170
Madrepora, 101
  oculata, 3, 8
Magellan Mounds, 185
  Province, 189
Magellan reefs, 114
Magmas, 171
Magnesium, 128
  silicate, 212
Magnus Heinason, 100
Maintenance survey, 94
Major oceanographic change, 118
Malangen deep-water coral reef, 5
Malangs reef, 43, 88
Manifestation, 195
Manifesting species, 50
Manned submarine, 103, 142
Mantle, 128
  -derived gas, 176
Mapping campaign, 44
MAR, 155
Marble Canyon, 175
Mareano website, 86
Margin, 123
Maria S. Merian, 134
Mariana forearc, 129

Marine
  benthic fauna, 100
  biologists, 46
  Isotope Stage, 214
  Protected Area, 160
  sediments, 90
  worms, 11
Marinobacter, 201
Marion Dufresne, 134
Marl, 174
Marlim and Bubira fields, 142
Marsili Seamount, 172
Massive reef, 165
Master Surveyor, xiii, 16
Mat-forming communities, 173
Mathematics, 203
  of diversity, 203
Mauritania, 137
MBE, 74
Mbsf, 117, 133
MDAC, 25
Measurements, 78
Mechanical
  bearing capacity, 132
  disturbance, 216
Mechanism, 78
Mediterranean, 54
  Outflow Water, 119
  Sea, 141
Megaplume, 36
Mekenes mud volcano, 130
Membrane, 10
Mesophilic, 198
Messinian evaporites, 158, 203
Metabolism, 198
Methane, 21, 56, 58
  buffer, 110
  carbonate, 94
  -charged porewater, 170
  -derived authigenic carbonate rock, 25
  flux areas, 141
  to hexane, 58
  -oxidizing bacteria, 198
  to pentane, 138
  reef, 95
  sulfate transition, 117
  system of Earth, 214
Methanotrophic bacterial symbionts, 51
Methylocaldum, 75

Methylococcus, 75
Mg(OH)$_2$, 129
Mg, 128
Mg/Ca ratio, 213
Mg-rich fluids, 167
Michael Sars, 15
Microbes, 199
Microbial
  activity, 176
  analysis, 73
  bioherms, 175
  filter, 33
  populations, 36
Microbialites, 172, 176
Microbiological profiles, 115
Microfauna, 99
Micro-fossils, 168
Microorganisms, 39
Micro-seepage, 26
MICROSYSTEMS, 134
Mid-Atlantic Ridge, 129, 155
Middle Devonian, 169
Midgard field, 193
Mid-Norway, 13
Mid-ocean
  ridges, 214
  seamounts, 172
Mid-plate, 171
Mid-reef, 18
Migration
  pathway, 108
  pattern, 134
Migratory waves, 111
Mikkel, 193
Mineralized porewater, 171
Mineralogies, 168
Mingulay Reef Complex, 105
Miocene
  age, 115
  channels, 157
  sandstone, 116, 117
MIS 1, 214
MIS 11, 214
Mississippi Delta, 146
mM, 119
Mn, 36
  concentration, 36
Mn^{2+}, 36

Moats, 62, 198
  and other depressions, 198
Mobile silty sand, 111
Modeled scenarios, 210
Modeling studies, 210
Modern analogues, 163
Modern coral reef, 10
Molluscs, 11
Monferrato, 168
Monilifera, 30
Monterey
  Canyon, 151, 159
  Fan, 30
Moraine ridge, 23
Morainic sill, 98
Morocco, 120
Morphology, 127, 146
Morvin field, 25, 76
Moscow, 136
  State University, 136
Mother pockmark, 93
Mounds, 8
  base, 118
  formation, 187
  initiation, 187
  nucleation, 133
  roots, 115
  -building phases, 115
MOUNDFORCE, 134
Mound-like sonar targets, 125
Mouth, 8
MOW, 119
Mt Mazama, 35
Mucoid layer, 10
Mucus-lined tunnels
Mud Volcano, 76, 129, 133
Muleshoe Mound, 165
Multi-beam echosounder, 71
Munida sarsi, 228
Mushroom-shaped, 99
Mussel reef, 207
Mussels, 12
Myrtea, 54
Mytilidae, 51

N$_2$, 198
  fixation, 129
Nadir, 50
NADW, 100

Nakken (the "neck"), 92
Namibia, 216
Narrow inlets, 216
Natural
  concrete, 93
  leakage of fluids, 135
  oil and tar "pollution", 149
  silting process, 76
Navigation, 95
Navigational chart. 95
Near-field record, 47
Near-horizonatal strata, 120
Near-surface sediments, 61
Negative relief, 124
Nematocysts, 9
Neptunian dikes, 169
Nettle cells, 9
Network, 189
Newfoundland, 154
New Mexico, 165
New Zealand, 176
Newspaper, 83
NGU, 65, 194
NIOZ, 135
Nitrospira, 200
Nodule, 79
Non-deposition, 187
Non-destructive fishing, 82
Non-reef-building species, 44
Nordic waters, 31
Normal
  butane, 56
  fault, 156
  pentane, 56
  -sized pockmark, 77
Norman seafloor world, 6
Normand Tonjer, xiii
Normandville oil field, 166
North America, 13
North Atlantic Current, 23
North Atlantic Deep Water, 100
  coral reefs, 8
North Atlantic, 23, 184
  Ocean, 1
North Dakota Geological Survey, 167
North Porcupine Bank Mounds, 123
North Sea, 13
  oil platforms, 209
  fattyness, 202

North Sea Surveyor, xiii, 44
Northeast Denmark, 99
Northeast Europe, 13
Northern Norway, 85
Northwest Italy, 168
Norway, 9
Norway Redfish, 5, 219
Norwegian
  Academy of Science, 199
  coastline, 215
  Continental Shelf, 193
  coral reefs, 43
  fjord reefs, 43
  Geological Survey , 65
  Greenland Sea, 100
  Lophelia reefs, 176
  Sea, 40
  Sea Basin, 25, 100, 126
  Sula Ridge reefs, 115
  /Swedish border, 97
  Trench, 92
  waters, 6
Nova Scotia, 153
NR-1 US Navy nuclear submarine, 90
Nucleation, 133
Null hypothesis, 181
Nutrients, 10, 23
  enrichment, 93
  flux, 43
  source, 95
  -poor, 205
  -rich fluids, 191
  -rich water current, 23
Nyegga, 24

O. universa, 213
O_2, 34
  -depleted, 36
Oases in the desert, 82
Obligate hydrocarbon utilizers, 179, 201
Oblique view, 75
Obturata, 30
Occam's Razor, 188
Occluded, 56
Ocean
  acidification, 209
  surface temperature, 212
Oceaneering, 22

Oceanic, 171
  crust, 171
  plate, 171
Oceanographer, 15
Oceanographic change, 118
Octocorals, 101
ODP
  Leg 146, 117
  Leg 307, 165
Off-mound, 117
Off-axis seamounts, 172
OHCB, 179, 201
Oil
  and gasfield, 44
  and gas extraction industry, 201
  patch, 148
  rigs, 94
Oily liquids, 202
Oleispira, 202
Olenin mud volcano, 132
Oligobrachia sp, 32
Oligocene, 64, 65
  deltaic sandy fan deposit, 64
Olivine, 128
On-mound, 117
Ophiurids, 26
Optimum growth, 198
Orange fluffy mats, 34
Oregon, 35, 118
  Cascades, 35
Organic
  debris, 51
  films, 164
  -rich sediment mounds, 27
Organisms, 28, 73
Organosedimentary, 172
Orientation, 67
Origin, 215
  of life, 128
Ormen Lange, 25
Orphan Knoll, 154
Oslo, 14
Osterfjord reef, 43
Outpost, 100
Overburden thickness, 192
Over-pressured
  porewater, 90
  sediments, 136
  zones, 134

Oxidisers, 31
Oxygen
  minimum zone, 69
  -rich water, 53, 144
Øygarden fault zone, 91

Pacific Ocean, 52, 141, 158
Palaeocene
  age, 46, 55
  bedding planes, 193
  sandstone, 99
Paleoclimate, 175
  records, 115
Paleorecord, 211
Paleoseep, 168
Paleozoic–Mesozoic mounds, 118
Paradigm
  -induced expectations, 13
  shift, 7
Paradox, 13
Paragorgia arborea, 8, 76 P
Paragorgians, 19, 92
  octocoral, 89
Parasite, 136
Patch, 16
Patches of corals, 80
Patchy distribution, 168
Pavements, 99
Pavillion Lake, 175
PCR amplification, 179
PDB, 26
Peculiar morphology, 127
Pedee Belemnite standard, 26
Peepa, 172
Pen Duick escarpment, 120
Penecontemporaneous processes, 164
Pentane, 167
Perahora karst aquifer, 175
Peridotite, 128
Period, 93
Permeable, 139, 156
  silty sediment layer, 139
Permian extinction, 10
Permil PDB, 26
Permutations, 203
Perseverance Mound, 190
Perspective view, 58
Perth, 107
Pertinent observations, 203

Perviata, 30
Petrography, 26
Petroleum, 202
   -filled mounds, 167
pH, 129, 212
Phakellia sp., 217
Phormidium sp., 173
Phosphate, 129
Photic zone, 12
Photo-montage, 111
Photosynthesis, 141, 163
Photosynthetic production, 141
Phycis blennoides, 101
Phylotypes, 75
Phylum, 7, 29
Physical
   characteristics, 167
   experiments, 128
Physico-chemical factors, 181
Pie-shaped mud volcano, 136
Piip, 172
Pillars, 99
Pillow basalt, 232
Pingo, 28
Pink fish, 232
Pink *Lophelia pertusa*, 6
Pink shrimps, 84
Pinnacle, 155
Pipe-like structures, 99
Pipeline route, 20
Piping products, 73
Pisces III submersible, 16
Pitcairn, 170
Planctomycetes, 200
Planet's wildlife, 29
Plant cells, 10
Planula, 10
Plastic funnels, 21
Plate
   flanks, 171
   tectonics, 212
Platforms, 94, 209
Plethora of giant carbonate mounds, 127
Plexiglas, 232
Pliocene, 122
Pliocene–Pleistocene age, 115, 158
Pleistocene, 158
Ploughmarks, 4
Plumbing system, 168

Pluming, 202
pmoA
   clones, 75
   gene fragment, 75
Pockmarkable, 78
Pockmarks, 18, 53, 93
   -associated deep-water coral reefs, 143
   -associated reefs, 70
   crater, 69
   depression, 54
   -dwelling coral reefs, 69
   reefs, 69
Poecilosclerida, 75
Pogonophorans, 26
Poison into food, 163
Polar Circle, 37
Polar Queen, xiii
Policy-makers, 210
Pollachius sp., 5, 227
Pollock, 6
Pollutants
Pollution event, 201
Polygonal cells, 157
Polyp, 8
Poorly cemented, 99
Popcorn Ridge, 171
Porcupine Basin, 23
Porcupine Bight mounds, 119
Porcupine Bight, 139
Porcupine Seabight mounds, 184
Porcupine Seabight, 184
Pore-spaces, 90
Porewater, 43, 136, 171
Porosities, 90
Porous oceanic crust, 172
Porthole, 232
Portugal, 130
Portuguese Margin, 130
Positive topography, 27
Post-glacial sea-level rise, 55
Potential route, 80
ppb, 56
ppm, 119
Prebiotic organic synthesis, 128
Pre-Cambrian
   ocean, 176
   stromatolites, 176
Precipitation of methane-derived aragonite, 26

Pre-construction, 71
Pre-drilling, 209
Preliminary Report, 182
Prerequisite, 59
Presence of gas, 57
Pressure cylinder, 21
Pressurized porewater, 77
Prevailing current direction, 3
Prey, 9
Primary producers, 28, 204
Primnoa resedaeformis, 76
Privilege cruise, 134
Process in the seafloor, 3
Product, 73
Production phase, 71
Professor Logachev, 107
Project, 134
Prokaryote, 116, 118
Prolific
  megafauna, 93
  seepage, 79
  sponges, 80
Proof, 202
Propane, 56, 167
Proportion, 76
Protected against
  fishing, 98
  trawling, 158
Puddles, 20
Purple sea-anemones, 232
Puzzle, 22
Pycnoclinal, 126
Pycnogonids, 26
Pyroxenes, 212

Quarts, 212
Quaternary age, 65
Quaternary–Pliocene sequence, 192

Rabat mud volcano, 132
R&D, 22
R/V *Marion Dufresne*, 134
Radiative gas, 214
Radio carbon dating, 46
Radiolarians, 211
Rape, 207
Rare occasions, 79
Realization, 141

Reconnaissance survey, 47
Red spots, 89
Red Volcano, 172
Re-discovery, 22, 210
Redox potential, 128
Reduced compounds, 13
Reducing agent, 116
Reefs, 7, 204
  A, 221
  B, 217
  carbonte, 99
  of mussels, 232
Reflection seismograms, 114
Regulated, 147
Re-iteration, 204
Reliable nutrient provision, 192
Relict iceberg ploughmarks, 74
Remotely operated vehicles , 6
Remote-sensing, 139
  instruments, 139
Renard Ridge, 134
Research project, 110
Research vessels, 134
Reservoir, 82
Re-suspension, 186
Reykjanes Ridge, 155
Rhodes, 158
Rice Coral, 53
Ridge, 57, 172
  axes, 172
Risers, 70
Rising gas and liquid, 145
Riverine water, 175
Robustness, 113
Rock, 4
Rock overhang, 29
Rockall Bank, 16, 120
Rockall Trough, 120
Rooted on, 132
Rootlike structures, 34
Rosemary Bank, 100
Røst Reef, 88
  Complex, 90
ROV, 6
  -based documentation, 191
  -mounted, 45, 71
  -mounted multibeam echosounder, 71
Rubble, 123

Rugose corals, 169
Russia, 136

S^0, 128
Säcken, 98
Sahara Desert, 169, 176
  fossil reefs, 176
Saithe, 5, 227
Saline groundwater, 150
Salinity-defined, 3
Salt
  bodies, 149
  stocks, 145
Sampling, 100
Sand, 15
Sandstone, 99
Sandvaagen, 13
Sand volcanoes, 125
Sandwaves, 77
Santos Basin reefs, 142
São Paulo University
Sap, 90
Sapping, 90
Satellite pockmark, 93
Saturation point, 151
Sb, 176
Scandinavia, 13
Scandi Ocean, 22
Scarp, 90
Scenarios, 186
Schematic sketch, 166
Schiehallion, 100
Scientific community, 211
Scientific deep-sea drilling vessel, 114
Scientists, 85
Scleractinian
  coral, 11, 130
  skeletons, 213
Sclerolinum sp., 32
Scorpio, 17
Scotland, 15
Sea fans, 11
Sea pens, 89, 158
Sea spiders, 26
Sea surface, 120
Sea whips, 158
Sea, 120
Sea-anemones, 11, 224

Seabed
  fluid flow, 4
  mining, 209
Seafloor, 8, 17, 211
  "desert", 3
  build-ups, 103
  currents, 87
  plain, 39
  sediments, 20
Sea-level changes, 205
Seamounts, 154, 172
  shape, 159
Seascapes, 85
Seasonality, 172
Sea-urchins, 11, 226
Seawater, 10
  sampling, 60
Seaway Commander, xiii, 42
Sebastes sp., 5, 43, 76
Sebastes viviparus, 219
Secondary trophic level, 78
Sediment, 45
  bacteria, 75
  -bound hydrocarbons, 57
  column, 134
  -embedded rocks, 79
  samples, 45
Sedimentary rocks, 91
Sedimentation, 93
Sedimentological factors
Seep bioherm, 173
Seepage
  depression, 199
  -induced nutrient enrichment, 93
  of porewater, 43
  sites, 54
Seep-induced diagenesis, 167
Seeping fluids, 147
Seeps, 170
Seismic line, 68
Self-sealing seep, 134
Self-seeding population, 209
Semi-circular structures, 199
Septa, 8
Serpentinite-mud volcanism, 127
Serpentinization, 127
  -associated fluids, 127
Serpulid tubeworms, 20, 21
Seabed features database, 76

Sessile organisms, 172
Shaded digital terrain model, 60
Shallow-water
    corals, 7
    formations, 7
Shear-zones, 171
Shelf, 138
Shelf-break, 138
Shell Brent Spar Buoy, 93
Shetland Isles, 100
Shimmering, 232
    venting, 36
Shoulder of the continental shelf, 101
Shrimps, 11
Siboglinum ekmani, 30
Siboglinum fjordicum, 30
Siboglinum poseidoni, 30
Sicily, 158
Sidescan sonar, 46
Silicate rocks, 211
Silt, 78
Silting, 93
    episodes, 79
Sinter, 173
SiO_2, 211
Siphon, 53
Site survey, 22
Site
    U1316, 116
    U1317, 117
    U1318, 117
Site-specific mounds, 167
Skagerrak, 30
Skarnsundet, 15
Skeletal remains, 189
Skeletogenesis, 10
Skeleton, 8
    formation, 10
Skogn, 40
Skogsøy, 81
Slabs, 79
Sleipner, 22
Slick, 148
Slide disruption, 90
Sliding, 130
Slope
    break, 89
    gradient, 185
Soft corals, 92, 158

Soft mud, 132
Solemyidae, 51
Solenosmilia variabilis, 142
Solitary coral, 213
South Atlantic, 141, 170
South China Sea, 99
South of the 62nd degree parallel
Southern North Sea, 76
Southwest Ireland, 107
Southwest Sweden, 215
Spain, 130
Spanish Moroccan field, 130
Sparry, 168
Spawning, 219
Special animal community, 139
Species, 160
    richness, 204
Sponges, 6, 11, 72
    cells, 76
Spring
    conduits, 170
    -controlled bioherms, 173
    sapping, 90, 151
    water, 174
Squat lobster, 156
Stable nutrient source, 204
Stability, 132
Stable isotope probing, 200
Stable seabed sediments, 78
Stacked membranes, 75
Stadials, 130
Stalked crinoids, 26, 230
Start-up phase, 187
Statfjord, 94
Statistical properties, 123
Statoil, 17
    /BP, 56
    -funded, 22
    Hydro, 74
    -supported research, 54
Statpipe, 22
Stavanger, 46
Steeply dipping strata, 58
Stem, 94
Sticky clay, 16
Stjernsund reef, 43
Stolon corals, 158
Stomach, 219
Stömstad, 97

Stony corals, 10
Stony cup corals, 158
Storegga, 24
  slide, 24
  slope break, 89
Story, 207
Stoss side, 68
Strait of Gibraltar, 119
Strata, 55
Streams of petroleum, 202
Stromatolites, 172, 176
Strong currents, 3, 69
Structure, 93
Stylaster sp., 122
Sub-bottom profiler, 48, 229
Sub-circular, 123
Subcrop, 55
  topography, 190
Sub-cropping Palaeocene rocks, 58
Submarine
  canyon, 81
  drilling, 210
  pingoes, 28
  valley, 81
Sub-phyla, 30
Sub-seafloor
  biosphere, 39
  hydraulic activity, 92
Substratum, 190
  information, 89
  -related factors, 4, 181
Sub-surface
  bedding planes, 196
  hydrate dissociation, 27
Sub-zero ambient seawater temperatures, 27
Sula Deep, 193
Sula Ridge, 22
Sulfurivirga caldicuralii, 198
Sulfur-oxidizing bacteria, 198
Sulphate-to-methane transition zone, 133
Sulphide
  gradients, 33, 131
  oxidisers, 34
  -rich porewater, 53
  -rich vents, 50
Sunday Times, 105, 106
Sunlight, 10
Supertramp, 207

Supporting evidence, 114
Surf, 158
Surface temperatures, 10
Surface geochemical evidence, 114
Surfacing fault scarp, 144
Surprising conclusion, 122
Surrounding seabed, 208
Suspicion, 133
Svelvik, 15
Swath bathymetry, 85
Sweat, 90
Sweden, 13, 97
Swedish/Norwegian reefs, 97
Symbionts, 10, 79
Symbiosis, 30
Symmetric lung-shaped structures, 114
System noise, 49
Systematic sampling, 100

Tail, 125
Taketomi Island, 200
TANGANYDRO Group, 174
Tar lumps, 148
Tasyo mud volcano, 132
Taupo Volcanic Zone, 176
Tautra reefs, 40
  complex, 43
Taxa, 156
Taxonomic, 157
Team, 169
Tectonic
  cycle, 213
  forces, 170
Temperature change, 64
Tentacle, 8
Terminal ridge, 41
Terra Nova, 23
  article, 106
Terrain model, 60
Tertiary
  period, 46
  sedimentary rocks
Tethys, 129
Texas, 145
Thallassolituus, 201
Thamnoporides, 170
Thawing, 64
Theca, 10
Thérèse Mound, 110

Thermal
 field, 174
 property, 207
 spring settings, 173
Thermobarically stabilized gas hydrates, 62
Thermophilic, 198
Thicket, 146
Thiomicrospira spp, 198
Thioploca, 34
Thiothrix, 34
Threats, 207
Thyasiridae, 51
Tidal
 currents, 77, 184
 regime, 192
Timor Sea, 160
Tisler reef, 97
Tjärnö Marine Biological Laboratory, 97
Tjeldbergodden, 40
TOBI, 113
 Rockall Irish margin, 123
Tommeliten, 13
Topographical association, 136
Topographic highs, 23
Toppled block, 228
Torch-like geysers, 127
Tortuous path, 202
Towed
 ocean bottom instrument, 113
 video-rig, 89
Træna bank reefs, 66
Træna Deep, 65
 reefs, 65
 Reef Complex, 67
Training Through Research, 106
Transects, 73
Transform fault, 156
Transport accidents, 201
Trapping, 164
Travertine, 173
Trawl marks, 113
Trawler wires, 97
Trawling, 81, 94
Triassic, 10
Triggered mound formation, 187
Trilobites, 169
TRIM, 123
Tristan da Cunha, 170
Troll
 field, 22
 gasfield, 92
 pockmark, 94
Tromsø, 16
 University, 18
Trondheim, 13, 25
Trondheimsfjord, 15
Trophic organisms, 28
Tropical coral reef, 7
Tropics, 7
Troughs, 74
Trunk pipeline, 81
Truth, 120
TTR, 106
TTR-16 Cruise, 136
Tubeworm, 6, 168
Turbulence, 183
Turbulent eddies, 180
Turkey, 176
Tusk, 4
Twigs of living corals, 132
Type I methanotrophs, 75, 198
Types of sponges, 72
Tyrrhenian Basin, 172

U/Th dating, 158
UK, 80
 Block 22/4, 95
 sector of the North Sea, 102
Unambiguous interpretation, 194
Uncompacted, 168
Unconformity, 122
Underground
 fluids, 24
 H_2S–CH_4 hydrates, 119
 hydrology
Underlying rock, 94
Underwater
 photography, 18
 valley, 81
Undisturbed appearance, 169
UNESCO, 106, 136
Uneventful, 76
Unidentified sessile organisms, 26
Unintended extinction, 207
Unit pockmarks, 56
United Press International, 16
University of Alberta, 36
University of Bergen, 73

Unknown coral species, 221
Unnamed reefs, 75
Unprecedented rates, 211
Unusual fauna, 124
Up-dipping layers, 62
UPI, 16
Upper crust, 207
Upper Jurassic, 193
Upper mantle, 214
Upslope, 121
Upstream, 117, 151
Upward-migrating gas, 68
Upwelling of deep, nutrient-rich water, 81
US Navy, 90

Valley, 81
Variables, 204
Vegetation-free zones, 170
Verrucomicrobia, 200
Vertical scale, 85, 121
Vesicles, 171
Vesicomyidae, 50
Vessel, 16
Vesterålen, 36, 81
Vesterålsgrunnen, 86
Vestimentifera, 30
Video
   -grabbed image, 52, 222
   -rig, 89
   screens, 17
VISTA project, 199
Visual survey, 65
Vitenskapsakademiet, 199
Volcanic
   gases, 171
   island, 170
   material, 128
   processes, 171
   rock, 120
Volcanism, 159
Vøring Plateau, 24
Vortices, 3
Vrije University, 203
Vulcan Sub Basin, 107

Warm fluids, 36
Warming atmosphere, 213
Water, 181

column, 60
density forcing, 184
depth, 3
masses, 3
-related factors, 181
Waulsortian, 164
   -like mounds, 164
   mounds, 164, 166
Waves, 3
Weak acid, 211
Weekly report, 117
Weichselian, 193
   clays, 55
Wellbeing, 209
West Africa, 54
West Flower Garden Bank, 144, 145
West Vanguard blowout, 22
Western Australia, 107
Western Pacific, 54
Whale caracasses, 51
White "rind", 78
White bacterial mats, 27, 198
White portions, 78
White sponge, 84
Whitings, 153
Williston, 167
Wipe-out zone, 145
Wollastonite, 211
World Lake Database, 174
World War II, 180
World Wildlife Fund, 106
Worldwide, 141
Worldwide serpentinization, 214
Worms, 170
W-Tr, 124
WWF, 106
www.geomapapp.com, 15
www.ifm-geomar.de, 97
www.lophelia.org, 15
www.lophelia.org/images/jpeg/
   Aleutian_Garden.jpg, 15
www.mareano.no, 43
Wyville–Thomson ridge, 15

Xenophyophores, 124

Yampi Shelf, 160
YBP, 13

Year of the Planet Earth, ix
Years before present (YBP), 13
Yellow varieties, 98
Younger Dryas, 3, 43
Yttre Hvaler, 98
Yucatan peninsula, 147

Yuma mud volcano, 132

Zeepipe, 22
Zoarces viviparus, 32
Zone of least resistance, 192
Zooxanthellae, 10